# Lecture Notes in Physics

## Volume 899

T0210904

# The Lecture Notes in Physics

The series Lecture Notes in Physics (LNP), founded in 1969, reports new developments in physics research and teaching-quickly and informally, but with a high quality and the explicit aim to summarize and communicate current knowledge in an accessible way. Books published in this series are conceived as bridging material between advanced graduate textbooks and the forefront of research and to serve three purposes:

- to be a compact and modern up-to-date source of reference on a well-defined topic
- to serve as an accessible introduction to the field to postgraduate students and nonspecialist researchers from related areas
- to be a source of advanced teaching material for specialized seminars, courses and schools

Both monographs and multi-author volumes will be considered for publication. Edited volumes should, however, consist of a very limited number of contributions only. Proceedings will not be considered for LNP.

Volumes published in LNP are disseminated both in print and in electronic formats, the electronic archive being available at springerlink.com. The series content is indexed, abstracted and referenced by many abstracting and information services, bibliographic networks, subscription agencies, library networks, and consortia.

Proposals should be sent to a member of the Editorial Board, or directly to the managing editor at Springer:

Christian Caron
Springer Heidelberg
Physics Editorial Department I
Tiergartenstrasse 17
69121 Heidelberg/Germany
christian.caron@springer.com

More information about this series at
http://www.springer.com/series/5304

Philippe Blanchard • Jürg Fröhlich
Editors

# The Message of Quantum Science

## Attempts Towards a Synthesis

### With a Foreword by Serge Haroche

*Editors*
Philippe Blanchard
Fakultät für Physik
Universität Bielefeld
Bielefeld
Germany

Jürg Fröhlich
ETH Zürich
Institute for Theoretical Physics
Zürich
Switzerland

ISSN 0075-8450        ISSN 1616-6361   (electronic)
Lecture Notes in Physics
ISBN 978-3-662-46421-2        ISBN 978-3-662-46422-9   (eBook)
DOI 10.1007/978-3-662-46422-9

Library of Congress Control Number: 2015936047

Springer Heidelberg New York Dordrecht London

Printed on acid-free paper

Springer-Verlag GmbH Berlin Heidelberg is part of Springer Science+Business Media (www.springer.
com)

# Foreword

The paradox of quantum physics resides in the contrast between its extraordinary power and its strangeness. Every physicist will agree that it is the most successful theory ever invented. It has given us the keys to understand the microscopic world and to derive from this understanding the modern technologies that have revolutionized our lives. Indeed, there is hardly a single apparatus in use, nowadays, that does not in part or totally rely on quantum phenomena. Lasers, computers, atomic clocks, the GPS, magnetic resonance imaging, the cell phones to name only a few exploit in one way or another quantum concepts and they would have been unimaginable by a classical physicist. Yet, in spite of this huge power, quantum physics remains highly counterintuitive, leading to many conflicting interpretations some of which are discussed in this book.

The tension between these two aspects of quantum physics, its power and its strangeness, has constantly been present during the 100 years this theory has been with us. During the formative years of the theory (from 1900 to 1930 roughly), the bizarre quantum concepts have given rise to fierce debates between the founding fathers. Then when the successes of the calculations based on quantum ideas had become overwhelming (from the 1930s to the 1970s), the discussions about interpretation took a backseat, most physicists being content to use this powerful tool without too much afterthought, in order to understand the world and to master it. This was the *shut up and calculate!* period. Those physicists, including Einstein and de Broglie, who struggled to reconcile quantum concepts with their ideas about physical reality, were a minority. With their disciples, they lost contact with the mainstream of research of that time, and one might argue that they did not contribute much to the tremendous progresses of physics during that period.

Feynman, a leader of the successful physics school that used quantum concepts without challenging them, could dismiss their efforts by saying that the kind of paradoxes they were struggling with were just a contradiction between what Nature is and what they wanted it to be. In other words, Feynman agreed with the famous phrase Bohr is supposed to have told Einstein: *Stop telling God what to do with His dice.* At the same time, Feynman acknowledged, however, the strangeness of quantum physics by saying *nobody really understands quantum physics.* In this somewhat "tongue in cheek" way, this provocative sentence again echoes what Bohr is supposed to have said: *Anyone who is not shocked by quantum theory has not understood it.* This state of affairs is illustrated by an anecdote Steven Weinberg is telling in one of his books. In an elevator of the Physics Department, he once met a colleague and a former student whom he had lost track of. After the student left, he asked his colleague: "What happened with this guy?" and the colleague answered: *Oh, he is lost for physics, he got interested in the Interpretation Thing.*

I must say that during my early years in physics, I tended to be an adept of the *shut up and calculate* school. This is undoubtedly due to having learned quantum mechanics from Messiah's books and directly from the lectures of Claude Cohen-Tannoudji (who had not yet written his own textbook on the subject). He described the principles with great clarity and used them efficiently to lead us directly to calculations explaining important effects and beautiful historical experiments. And when I started my own research for my Ph.D. under his supervision and later on in my own lab with my students, I realized that quantum calculations allowed me to predict with high precision how atoms were behaving in the resonance cells I was experimenting with. I could not see the atoms directly then, but all observations pointed to the fact that they were certainly there and that they did exactly what quantum mechanics was predicting. Among these experiments, some had to do with observing cascades of successive photons emitted by atoms as they decayed step-by-step from an excited state. The pattern of emission of polarized light was predicted, beautifully and precisely by simple calculations based on the quantum theory of angular momentum, with the help of some Racah algebra.

At that time, Claude received the visit of a young and enthusiastic student working in another laboratory at Orsay who was interested in challenging the laws of quantum physics precisely by performing an experiment that would enable him to study the correlations of photons emitted by atoms in a fluorescence cascade. Alain Aspect, this is his name, was trying to implement in the lab an experiment suggested by a CERN physicist, then unknown to Claude and of course to me, namely John Bell. With this experiment, he was trying to improve on earlier work of the same kind performed by an American physicist, John Clauser. It is to the credit of Claude, whom we all considered the "pope" of orthodox quantum physics that not only did he not dissuade Alain Aspect, but he actually encouraged him to do the experiment. Claude told me about it and I remember being puzzled. I could very simply calculate what quantum physics was predicting in such a simple situation and that the result could be different was unthinkable for me. My reasoning was simple. If Bell's inequality is right and quantum mechanics wrong in this simple case, how come that its predictions have been vindicated in thousands of experiments (including mine) that were monitoring atomic cascades under similar conditions? We would have to build a new theory that should still agree with all the data accumulated so far and at the same time, explain why, under the particular conditions of Aspect's experiment, it was yielding another result.

Later on, I understood that Aspect (and Claude who had encouraged him) were right, even if in the end Alain's experiment vindicated quantum physics. In most experiments so far, the observations were dealing with big samples containing huge numbers of atoms, and were recording only average signals. Aspect's experiment was one of the first that revealed the correlations between the photons emitted by an atom in a single event (these correlations averaged, later, over many realizations of the same experiment in order to build the expression violating Bell's inequality in agreement with quantum theory). Instead of first averaging data and looking at the relationships between these averages, he was recording individual correlations *before* performing averages. It was important to find out whether the theory was right under these new conditions.

Here the concept of entanglement was central. Of course, we all knew that entanglement was a feature of quantum physics. After all, the ground state of the simplest of all atoms, Hydrogen is an entangled state of an electronic and nuclear spin, and this fact has been well known since the beginning of quantum physics. Entanglement, however, is not as spectacular when the entangled partners are only one Angström apart as when they are distant from each other by metres. Even if quantum theory did not put any limit to the distance at which entanglement should manifest itself, it was certainly worthwhile testing it. Future developments amply demonstrated that entanglement at a distance could lead to applications, unforeseen at that time, for quantum communication in particular and this largely explains the renewed interest in Bell's inequality tests decades after Aspect's early work.

For a long time however, Aspect's experiment remained an isolated tour de force. He went to work in other directions exploring first with Claude Cohen–Tannoudji, then with his own group, properties of atoms cooled and manipulated with laser light. Other physicists started cooling individual ions in traps and controlling with ever increasing precision their evolution. In my own group, with my colleagues Jean-Michel Raimond and Michel Brune, we focused on the study of photons trapped in high-Q superconducting cavities and interacting with Rydberg atoms, this domain being now known as Cavity Quantum Electrodynamics.

In all these experiments, single isolated quantum systems were monitored. Concepts that had been discussed in the context of thought experiments in the early days of quantum physics such as complementarity, quantum jumps and, of course, entanglement came back to the forefront of discussions among physicists since their manifestation became directly observable in real experiments. I remember that the existence of quantum jumps was challenged by some physicists, before such jumps became directly observable first in ion trap work, then in Cavity QED experiments. While, in the old times, experimentalists in atomic and molecular physics could explain their observations with the help of a density operator which dealt with ensemble averages evolving smoothly and without jumps, the new experiments required the description of single quantum trajectories, for which the density matrix approach was inadequate. Monte Carlo calculations, in which random quantum jumps were introduced in computer simulations of single quantum histories became the new tool replacing the Bloch equation approach of the density matrix formalism. With these tools, it became easy to compute high order correlations observed when measuring all kinds of observables in a long time sequence and to test the results of these predictions in increasingly complex experiments. So far, all of them have vindicated quantum mechanics.

One of the nagging questions remaining open has to do with the quantum-to-classical boundary, the so-called Schrödinger cat paradox. Is there a maximum size up to which quantum behaviour is directly observable? This question naturally arises in the field of quantum information, where we try to harness the strange laws of the quantum domain to communicate or calculate in new powerful ways. Quantum information science will make it necessary to manipulate quantum systems of macroscopic extension and made of large numbers of particles. We have of course to define the meaning of "size" for these systems. If we take it as meaning the distance between parts of a quantum object, we have learned from recent experiments, by the Gisin and Zeilinger groups notably, that entanglement can survive over many kilometres, without any indication of limitation so far. If we mean the number of particles in the system, we know that large molecules made of thousands of nuclei and electrons can give rise to interference effects and that fields made of hundreds of photons can exist in superposition states. Not to speak of superconducting circuits or degenerate quantum gases made of thousands to millions of particles, which can be prepared in state superpositions involving two or more components.

These superpositions are very fragile and are eventually destroyed by decoherence, a phenomenon linked to the coupling of the system to its environment. Some physicists believe that, beyond the mundane decoherence process, which involves entanglement with the environment well explained within the "orthodox" quantum theory, there may exist a yet undiscovered mechanism that makes the quantum laws invalid for large enough systems. I suspect that, behind this idea, there is the circumstance that some physicists still find it unacceptable that *God is playing dice*. The fact that the theory is at its heart probabilistic is bothering them and they would like to find a way to escape from it at least when large objects such as measuring devices are concerned. In fact, they are looking for a mechanism that would determine in which state the Schrödinger cat is after the box is opened. That decoherence has destroyed the coherence between the live and dead cat state is not enough for them. They want to explain how the fate of the cat is finally decided and they do not like the idea that it is left to pure chance.

In the end, the question of the validity of quantum physics for large objects will have to be decided by experiments. What I think about the likely outcome of such experiments or what those who are trying to perform them expect or hope is not really relevant. The answer will have to be given by Nature, and if quantum physics shows some kind of limitation, it will have to be modified (even if it is hard to imagine how). The development of quantum information theory and the concomitant multiplication of experiments manipulating quantum objects of all kinds—atoms, molecules, photons, quantum dots, superconducting circuits, mesoscopic cantilevers, etc.—make us much more conscious than in the 1960s or 1970s of the questions about quantum physics that can be answered by experiments. Developing methods to control or counteract decoherence—quantum error correction or quantum feedback—is not only necessary for implementing future quantum logic machines, but is also important to make us probe the true limits of the quantum world. If new ideas "à la John Bell" emerged to discriminate between various interpretations of quantum theory or to look for decoherence beyond the "environment induced model", we would certainly try to test them with the new tools we are developing. The right attitude for experimentalists towards quantum theory should thus be to *trust but verify*, rather than to *shut up and calculate*.

For now, we have to live with the present theory, which has been and still is so successful at explaining and predicting. This is not such a bad situation. Beyond that, do I find it bizarre that a physical quantity such as the polarization of a particle or the number of photons in a box has no meaning before it is measured? or that quantum systems undergo random jumps which cannot be predicted deterministically?, or else that it makes no sense to even talk about these jumps if there is not a detector to observe them? At some level, yes of course, I find all this weird, because this is contrary to my classical intuition, formed by the observation of macroscopic events, which occur even if you do not look at them.

I try to convince myself that there is even a Darwinian explanation to that apparent strangeness of the quantum world. Our brain has evolved over generations to adapt to the classical world, in which the underlying quantum phenomena are "veiled", according to the poetic statement by D'Espagnat. It is useful, for our survival, to have an intuition about the classical trajectory of a stone thrown at us, but not about how an atom crosses a double slit, in a superposition of trajectories. We have not even coined words in our everyday language to describe that weirdness. Thus, at some level, the world of atoms is indeed strange to us. We have, however, a beautiful language to describe it, the language of mathematics. Using simple math, we experimentalists in quantum physics have been generally able to predict what happens in our atomic beam machines or in our resonance cells when we design an experiment. This is a different form of intuition than the one of laymen, but it is still an intuition of sorts.

However, there are situations where this simple mathematical "intuition" eventually fails. Even with the help of the most powerful "classical" computers, we cannot solve the Schrödinger equation of a quantum system containing more than a few tens of two-level atoms and we are thus unable to predict in detail what happens in situations where complex massive entanglement is involved. Novel effects such as exotic quantum phases of matter in two or three dimensions may thus escape our understanding. We hope that, here, experiments will come to our help. We are developing methods using cold atoms in optical lattices or ions in traps, or connected superconducting circuits, to emulate these complex situations by reproducing, at a different scale, the precise conditions of the real situation involving tens to hundreds of particles. By having these artificial systems evolve and observing them in our labs, we hope to find out how real system behaves. These quantum simulators, predicted by Feynman in a prescient article 30 years ago, are the new tools we will try to use in order to keep probing the mysteries of the quantum world.

Serge Haroche

# Preface

A little less than 3 years ago, the editors of this book organized a program on recent developments and modern problems in quantum physics, entitled *The Message of Quantum Science—Attempts Towards a Synthesis*, which took place at the "Zentrum für interdisziplinäre Forschung" (ZiF) of the University of Bielefeld. Thus, between the middle of February and the middle of May 2012, a series of seminars and discussions and two workshops took place that attracted quite a number of distinguished theorists and experimentalists.

It was the very lively and stimulating atmosphere prevailing during our program and, in particular, during the two workshops that gave rise to the idea to put some of the insights gained in the course of our activities on record and attempt to publish a book containing essays by a certain number of participants. We are most grateful to *all* people who participated in and enriched our program and who helped to promote and deepen our understanding of quantum physics. Although not all of them have made contributions to this book, we would like to acknowledge that, without them, this book would not exist.

One might be tempted to say that Quantum Mechanics (QM) is so exceedingly well established and understood that attempting to publish a book like this one is a little like "carrying coals to Newcastle". Our distinguished colleague *Berthold-Georg Englert* has argued that *QM is spectacularly successful and reliable; there is no experimental fact, not a single one, that contradicts a quantum-theoretical prediction.*[1] Yet, it is an experience made very frequently that when grown-up physicists and, in particular, theorists start to discuss problems concerning the foundations and the interpretation of QM, it does not take long until a state of considerable confusion is reached, and their deliberations usually tend to become quite emotional. In his paper *Introductory Article: Quantum Theory*,[2] *Gianfausto Dell'Antonio* writes: *Quantum mechanics today is a refined and incredibly successful instrument ... but its internal consistency is still standing on a shaky ground.*

---

[1] Eur. Phys. J.D. (2013) 67 238.

[2] Encyclopedia of Mathematical Physics, Elsevier (2002).

Although we do not see any compelling reasons to doubt the internal consistency of QM, we *do* think that there are many issues concerning the foundations and the interpretation of QM that are still rather puzzling and not nearly as well understood as they ought to be. This opinion or conviction was among our reasons to organize a program on Quantum Science.

Other reasons why we were eager to convene specialists in quantum physics are related to—among others—the following exciting developments: Recent years have seen major experimental advances in the exploration of realms of the "quantum world" that had previously been inaccessible. These advances vastly augmented our capabilities to test fundamental features of Quantum Mechanics and quantum many-body systems and to manipulate quantum systems, such as individual atoms, atom gases and light. Some of this progress relies on major experimental and technological breakthroughs in exploiting the electromagnetic field and, in particular, its quantum properties and its interactions with matter—new lasers, laser cooling, optical lattices, magnetic traps, cavity QED, microscopy, etc.—and on advances in semi-conductor technology. The former has led, for example, to the experimental realization of Bose–Einstein condensates in dilute atom gases confined in magnetic traps and of other exotic quantum fluids, to the configuration of artificial crystals consisting of atoms located at the sites of optical lattices, and to numerous other exciting discoveries in the manipulation of quantum systems. The latter has given rise to new quantum Hall liquids (i.e., 2D electron gases exhibiting the quantum Hall effect) and novel possibilities of manipulating them and exploring their properties, e.g., measuring the fractional charges of quasi-particles, and to the discovery of novel states of matter in two and three dimensions called "topological insulators". One may also think of the discovery of graphene and its exotic quantum properties, such as the occurrence of "relativistic Dirac fermions" as quasi-particles. Some of these advances and discoveries were featured in lectures at our workshops, although it was not possible to do justice to all the exciting recent developments and breakthroughs.

Another direction in Physics that has seen tremendous progress, in recent years, is concerned with the study of the early universe and, in particular, with phenomena studied in cosmology and astro-particle physics that are suspected to belong to the realm of the "quantum world", such as structure formation in the early universe, dark matter and dark energy. This direction holds enormous promise for important future discoveries, including ones affecting the foundations of fundamental physics. Unfortunately, it could not be featured adequately, in our program—not least for lack of competence on the side of the organizers.

In order not to end up with too broad and voluminous a book, we had to decide to put the focus of this book on a more or less well-defined area in quantum physics. We have chosen to emphasize the foundations of Quantum Mechanics and the puzzling effects observed in the "quantum world". There are many new experiments in this general area, such as interference experiments with very large molecules passing through double-slits, ones that test the validity of the Kochen–Specker theorem, new tests of the violation of Bell's inequalities and of consequences of entanglement, new non-demolition measurements and tests of "wave-function collapse", experiments realizing quantum-teleportation, etc. One might also think of the progress in the study of open quantum systems, quantum transport and decoherence. Many of the effects encountered and studied in such experiments have real or tentative applications in the fields of quantum information science, quantum cryptography and quantum computation. Some of these applications have actually already been implemented in devices.

The experimental developments just alluded to have raised many challenging questions for theorists, some of which have been actively addressed and answered in recent years. All this has led to a new surge of interest in the foundations of Quantum Mechanics, which have puzzled physicists ever since the discovery of this theory, almost 90 years ago.

One main goal of our program was thus to gather experimentalists and theorists studying fundamental aspects of quantum physics and have them review and discuss the present state of affairs and draw our attention to some of the important open problems in their particular areas. We are deeply grateful to all the speakers in our seminars and at the workshops for the efforts they made to communicate their views to an interested audience, which, by and large, turned out to be very successful. Most contributions to this book have grown out of lectures presented to the participants of our program. We thank all the authors of the chapters appearing in this book for the care they took in writing their contributions and, in particular, *Serge Haroche* for having agreed to write an Introduction. We very much hope that the reader will find the material included in this book as stimulating and enlightening as the lectures were, and we wish him/her pleasant reading and much benefit in studying it.

To conclude, we have the pleasure to express our sincere gratitude to the direction and the staff of the ZiF for hosting our program and for generous support. We are very grateful to *Marina Hoffmann* for her invaluable help and assistance before and during our program at the ZiF. We thank *Hanne Litschewsky* for her dedicated help in collecting the manuscripts of the contributions to this book, editing them and preparing them for publication.

"In the name of all his friends we would like to dedicate this book to the memory of our dear colleague and friend Walter Schneider (1938–2014). It has been a great privilege for us to profit from his vast knowledge and his critical comments on innumerable occasions. His loyalty, generosity and fine humor have been exemplary and will be remembered. - Er wird uns fehlen!"

Bielefeld, Germany                                                                   Philippe Blanchard
Zürich, Switzerland                                                                        Jürg Fröhlich
August 2014

# Contents

# Chapter 1
# Theory of the Decoherence Effect in Finite and Infinite Open Quantum Systems Using the Algebraic Approach

Philippe Blanchard, Mario Hellmich, Piotr Ługiewicz, and Robert Olkiewicz

## 1.1 Preliminaries

Quantum mechanics is the greatest revision of our conception of the character of the physical world since Newton. Consequently, David Hilbert was very interested in quantum mechanics. He and John von Neumann discussed it frequently during von Neumann's residence in Göttingen. He published in 1932 his book *Mathematical Foundations of Quantum Mechanics*. In Hilbert's opinion it was the first exposition of quantum mechanics in a mathematically rigorous way. The pioneers of quantum mechanics, Heisenberg and Dirac, neither had use for rigorous mathematics nor much interest in it. Conceptually, quantum theory as developed by Bohr and Heisenberg is based on the positivism of Mach as it describes only observable quantities. It first emerged as a result of experimental data in the form of statistical observations of quantum noise, the basic concept of quantum probability.

The central concept in von Neumann's book is an abstract Hilbert space. The unexpected usefulness of Hilbert spaces arises from the fact that the equation of motion of quantum mechanics, Schrödinger's equation, is linear.

For the description of systems with infinitely many degrees of freedom the theory needs a generalization of the standard Hilbert space formulation. Von Neumann liked to spell out physical problems in an abstract and general language, and

P. Blanchard (✉)
Fakultät für Physik, Universität Bielefeld, Universitätsstr. 25, 33615 Bielefeld, Germany
e-mail: blanchard@physik.uni-bielefeld.de

M. Hellmich
Bundesamt für Strahlenschutz, Willy-Brandt-Str. 5, 38226 Salzgitter, Germany
e-mail: mhellmich@bfs.de

P. Ługiewicz • R. Olkiewicz
Institute of Theoretical Physics, University of Wrocław, pl. M. Borna 9, 50-204 Wrocław, Poland
e-mail: piotr.lugiewicz@ift.uni.wroc.pl; rolek@ift.uni.wroc.pl

© Springer-Verlag Berlin Heidelberg 2015
P. Blanchard, J. Fröhlich (eds.), *The Message of Quantum Science*, Lecture Notes in Physics 899, DOI 10.1007/978-3-662-46422-9_1

therefore formulated quantum mechanics as a theory of operators on Hilbert space (what is today known as the theory of C*-algebras and von Neumann algebras). He was influenced by Heisenberg's quantum matrix dynamics, a new and highly original approach to the mechanics of the atom, and obtained remarkable—though incomplete—results.

Quantum mechanics is incredibly successful: no phenomenon up to now has been found which contradicts it. The Copenhagen rules as formulated by Bohr can be used as a pragmatic recipe to arrive at experimentally testable conclusions from the Hilbert space formalism. Despite the puzzling nature of the measurement process in quantum mechanics, these rules work very well in practice, so they are justified "for all practical purposes" (FAPP), as it was put by John Bell.

Von Neumann's algebraic framework of quantum mechanics is general enough to accommodate both classical and quantum systems, and thus facilitates the description of situations in which quantum systems develop classical behavior in a FAPP fashion. For this reason, foundational issues of quantum physics are best discussed in this framework. Nevertheless, despite the possibility of formulating classical physics in the algebraic framework, the von Neumann epistemic principle claims that, at a fundamental level, there is only one kind physical laws, and these are the quantum principles—in this picture, classicality only emerges as their consequence.

In any experiment, two phases can be distinguished: the preparation of the system under study and the actual measurement. This situation can be idealized in the following way. Two systems, the observed system and the observing system, influence each other, and we observe for each preparation $\omega$ and each measurement $A$ one of several possible outcomes $\{a_i\}_{i \in I}, a_i \in \mathbb{R}$, the possible measurement results. In general, for given $\omega$ and $A$, the theory only determines a probability distribution $P_\omega^A(a_i)$, $i \in I$, for the individual outcomes, where $P_\omega^A(a_i) \geq 0$ and $\sum_{i \in I} P_\omega^A(a_i) = 1$. If for a given $\omega$ and $A$ there is always one unique (up to experimental error) outcome when the same experiment is repeated, then we have a *deterministic theory;* an example is classical mechanics. In contrast, classical statistical mechanics and quantum mechanics are examples for theories which are nondeterministic, or *probabilistic.*

Standard quantum mechanics is the only probabilistic theory where the probabilities are postulated ab initio and are not a consequence of hidden deterministic processes at a deeper level. Such processes are called *hidden variables* and reflect the ignorance of the observer. However, the majority of physicists today believe that the probabilities in quantum mechanics are not attributable to the ignorance of hypothetical hidden variables, but are of a fundamentally different nature. This is corroborated by Bell's inequalities, which hold in any theory with (local) hidden variables, and which were experimentally found to be violated.

If quantum theory is the fundamental principle of nature, the question arises how the laws of classical physics, which in particular govern the objects of our daily lives, follow from the more fundamental quantum laws. The most promising answer to this question seems to be the one offered by the program of environmental decoherence. Environmental decoherence contends that one has to take into account the fact that the objects of classical physics, and in particular macroscopic objects, are strongly interacting with their environment, and that precisely this interaction is the origin of classicality in the physical world. Thus classicality is a dynamically emergent phenomenon due to the unavoidable interaction of quantum systems with other quantum systems surrounding them.

Since decoherence is based on nothing else than the application of the standard formalism of quantum physics to the description of the interaction between a system and its environment, decoherence is neither an extraneous theory distinct from quantum physics nor something that we would freely choose to include or neglect. Decoherence is ubiquitous in nature and has to be taken into account to arrive at a realistic description of a physical system. Moreover, decoherence is not to be viewed as a disturbance of the system by its environment, on the contrary the system disturbs the environment: the quantum coherence immanent in the system is not lost but only delocalized in the environment. For recent reviews of the theory of decoherence see [25, 28, 31].

In our work the aim was to obtain a rigorous definition of decoherence in a general mathematical framework, which allows a classification of possible scenarios of decoherence, and which is sufficiently general to accommodate also systems with infinitely many degrees of freedom.

## 1.2 Algebraic Framework and Open Systems

Everybody agrees that concepts of classical and quantum physics are almost diametrically opposed. Therefore, in order to discuss for instance the emergence of classical behavior of quantum systems, we need a single mathematical framework which allows a coherent description of the quantum and classical worlds. Just as Newton invented calculus to describe classical mechanics, von Neumann invented a splendid theory of algebras of operators to describe quantum theory. The algebraic framework of quantum physics is an abstraction and generalization of von Neumann's formulation, which was pioneered by Segal, Haag, Kastler and Araki. It is a mathematical model for states, observables and their dynamics, covering all known physical applications and admitting a sufficiently rich structure to facilitate rigorous developments.

### 1.2.1 Algebraic Framework

Given a specific preparation $\omega$ (state) of a physical system and a specific measurement $A$ (observable), the role of the (kinematical part of the) theory is to predict a probability distribution $P_\omega^A(\cdot)$ on the set of all possible outcomes of the measurement. The set of all observables generate an operator algebra $\mathcal{N}$, which, for mathematical convenience, is taken to be a C*-algebra or, when represented on a Hilbert space, as a von Neumann algebra. The observables then correspond to the self-adjoint elements of $\mathcal{N}$, whereas the states are the positive and normalized linear functionals on $\mathcal{N}$. The probability distribution $P_\omega^A(\cdot)$ is obtained by spectral decomposition of a self-adjoint element $A \in \mathcal{N}$ in the same way as in the standard Hilbert space formulation of quantum mechanics. In this context, if $\mathcal{N}$ describes a classical system, it is commutative, whereas an algebra with a trivial center corresponds to a system with a pure quantum character.

### 1.2.2   *Time Evolution*

In modern physics the time evolution of physical systems is formulated using the Hamiltonian approach, in which the dynamical law is described by a Hamiltonian operator $H$. This assumption seems to be a very fundamental one and is valid on all energy scales encountered today (see, e.g., [23]). In particular, in the algebraic framework, the time evolution is given by a mapping on the algebra of observables $\mathcal{N}$ into itself, and is formulated in the so-called Heisenberg picture:

$$x(t) = e^{iHt} x e^{-iHt} \quad \text{for any } x \in \mathcal{N}, \ t \in \mathbb{R}, \tag{1.1}$$

where $x \in \mathcal{N}$ is an observable at time $t = 0$, and $x(t)$ the observable at time $t$. A key issue in theoretical physics is to build the Hamiltonian $H$ in an appropriate way, and the basic criterion for its acceptability is its suitability for modeling all phenomena of interest. To construct a Hamiltonian one first has to build its domain $D(H)$. This task is essentially related to the choice of the underlying Hilbert space $\mathcal{H}$ for the system. Let $\langle \cdot, \cdot \rangle_\mathcal{H}$ denote the scalar product of the Hilbert space $\mathcal{H}$. Then the Hamiltonian $H$ is given as a linear self-adjoint operator on $\mathcal{H}$ which is densely defined. When a Hamiltonian has been fixed, one can next decide which von Neumann subalgebra $\mathcal{N} \subset \mathcal{B}(\mathcal{H})$ can be used to represent the set of observables of the system, then the dynamics can be introduced by (1.1). The construction of Hamiltonians in physics is one of the most difficult tasks, and is in many situations an unsolved problem. Even the simplest models like the Hydrogen atom are far away from being mathematically trivial. Moreover, the inclusion of a all physical phenomena needed to describe the physical situation on hand makes the structure of Hamiltonians complex, even at a formal level without any attempts of mathematical rigor. A good example is the Hamiltonian of the electroweak interaction—a part of the standard model considered today as a one of the main achievements of theoretical physics [16]. Up to now there is still no mathematically rigorous theory describing all the physics inherent in the standard model, despite the many efforts directed towards this goal [11, 23].

   This fact determines our attitude in the sequel, where we emphasize the analysis of the general mathematical structure, avoiding discussions concerning particular constructions of physical Hamiltonians for concrete interactions.

There is some convenient mathematical abstraction of the idea associated with (1.1), which can be stated in a purely algebraic way. The mapping $\mathcal{N} \ni x \mapsto x(t) \in \mathcal{N}$, which is bijective for any fixed $t$, has some purely algebraic properties that can be summarized easily:

**a** $x(t)^* = (x^*)(t)$ for any $x \in \mathcal{N}$, here $*$ is the Hermitian conjugation in the operator algebra $\mathcal{B}(\mathcal{H})$,

**b** $(xy)(t) = x(t)y(t)$ for any $x, y \in \mathcal{N}$,

for any $t \in \mathbb{R}$. Any linear bijective mapping $\alpha : \mathcal{N} \longrightarrow \mathcal{N}$ obeying the algebraic rules **a** and **b** is called a *-*automorphism*. The dynamics of the quantum system is then described by a one-parameter group of *-automorphisms $\{\alpha_t\}_{t \in \mathbb{R}}$, called the *-*automorphic evolution,* which in above terms in given by $\alpha_t(x) = x(t)$. The group law reads $\alpha_{t+s} = \alpha_t \circ \alpha_s$ for all $t, s \in \mathbb{R}$, and $\circ$ denotes the ordinary composition of mappings. Some regularity property in the time variable $t$ are usually introduced. For instance, if the function $t \mapsto \langle \psi, \alpha_t(x)\varphi \rangle$ is continuous for all fixed vectors $\psi, \varphi \in \mathcal{H}$ and all $x \in \mathcal{N}$, we speak of *weak continuity* of the one-parameter group $\{\alpha_t\}_{t \in \mathbb{R}}$. Note that an arbitrary *-automorphic evolution is not of the form (1.1) for some Hamiltonian $H$, but can be more general. Finally, we mention that a *-automorphic evolution is *it completely positive,* which is a notion whose physical significance has been realized by Kraus, Lindblad, Gorini, Kossakowski and Sudarshan [1]. This concept of positivity can be explained in terms of physical requirements. Fix some instant $t$ and consider the mapping defined by $\mathcal{N} \ni x \mapsto \alpha_t(x) = x(t)$, where $x(t)$ is given by (1.1). Obviously, one obtains that $\alpha_t$ is a *positive map* on the algebra $\mathcal{N}$, i.e., $\alpha_t(x^2) \geq 0$ for any element $x \in \mathcal{N}$ such that $x^* = x$ (this follows, e.g., from property **a** and **b**). Consider now our system as a part of a bigger system by adding to our system another quantum system described by a $n$-dimensional Hilbert space $\mathbb{C}^n$, which has a trivial Hamiltonian $H_n = 0$. The algebra of observables for the joint system is the von Neumann algebra generated by all elements $x \otimes A \in \mathcal{B}(\mathcal{H} \otimes \mathbb{C}^n)$, with $x \in \mathcal{N}$ and $A \in \mathcal{B}(\mathbb{C}^n)$, the algebra of all $n \times n$-matrices with complex entries. We assume that the added system is far away from our original system, so that the interaction between the two systems is negligible. Let $\mathbb{1}_n$ be the unit matrix. The total Hamiltonian of the two noninteracting systems is then $H \otimes \mathbb{1}_n$, and the corresponding time evolution is $\alpha_t^n = \alpha_t \otimes \text{id}_n$, with $t \in \mathbb{R}$, and where $\text{id}_n$ is the trivial automorphism acting on the matrix algebra $\mathcal{B}(\mathbb{C}^n)$. Using the same argument as before we get $\alpha_t^n(x^2) \geq 0$ for any observable $x = x^*$ of the joint system. This positivity property holds true for all positive integers $n$, and therefore the mapping $\alpha_t$ is called a *completely positive* map. Every *-automorphism is completely positive, but we emphasize that not every positive linear map on a von Neumann algebra taking values in another von Neumann algebra is completely positive. Below we give some examples of such completely positive maps which are not arising from a *-automorphic evolution.

### 1.2.3 Open Systems

We already alluded to the fact that our understanding of the physical world is very limited and partial. A glimpse of this deficiency is present in the concept of an open system. We shall see in the next section that this concept is of major relevance for our understanding of appearance of classical physics in a world which, at a fundamental level, is governed by the quantum laws. Hence we elaborate a bit on this subject. An *open system* S is a physical system which is not (well) isolated from the influence of its surroundings. This is a typical situation; in fact, a perfectly isolated system is rather an exception since a perfect isolation can be achieved only approximately in practice. We include the interaction of the system S with its environment E, which can be thought of that part of the rest of the world which is in interaction with our system S, in the mathematical description of S. To be more specific we assume that the total Hamiltonian has the form

$$H_{\text{tot}} = H_{\text{S}} \otimes \mathbb{1}_{\text{E}} + \mathbb{1}_{\text{S}} \otimes H_{\text{E}} + H_{\text{int}},$$

where $H_{\text{S}}$ describes the time evolution of the perfectly isolated system, and similarly $H_{\text{E}}$ corresponds to the environment alone. The term $H_{\text{int}}$ describes the interaction of the system and its environment. The total system S + E can be considered as a perfectly isolated system, since it includes all interactions between all of its parts. It is this assumption that allows us to use Hamiltonian dynamics for the total system. Note that by assumption we do not perform any observation of the environment, all measurements are entirely confined to the system S (if we want to perform measurements on parts of E we would include those parts in S as well). In particular, using (1.1) one can write for the observables of the system S

$$x(t) = e^{iH_{\text{tot}}t}(x \otimes \mathbb{1}_{\text{E}})e^{-iH_{\text{tot}}t} \quad \text{for any } x \otimes \mathbb{1}_{\text{E}} \in \mathcal{N}, \ t \in \mathbb{R},$$

with $\mathcal{N} \equiv \mathcal{M} \hat{\otimes} \mathcal{M}_{\text{E}}$, where $\mathcal{M}$ is the algebra of observables of the system S and $\mathcal{M}_{\text{E}}$ is the algebra of observables for the environment E. Here $\mathcal{N}$ is the smallest von Neumann algebra containing $\mathcal{M} \otimes \mathbb{C}\mathbb{1}_{\text{E}}$ and $\mathbb{C}\mathbb{1}_{\text{S}} \otimes \mathcal{M}_{\text{E}}$.

In general, $x(t) \notin \mathcal{M} \otimes \mathbb{C}\mathbb{1}_{\text{E}}$ for $t \neq 0$. To mathematically describe the fact that the experimental capabilities are confined to measurements on S alone we need to find the resulting time evolution in $\mathcal{M}$. To do this we consider some fixed initial state $\omega_{\text{E}}$ of the environment and consider the linear mapping $\Pi^{\omega_{\text{E}}} : \mathcal{N} \longrightarrow \mathcal{M} \otimes \mathbb{C}\mathbb{1}_{\text{E}} \cong \mathcal{M}$, determined uniquely by the condition

$$\Pi^{\omega_{\text{E}}}(x \otimes X) = \omega_{\text{E}}(X)x \otimes \mathbb{1}_{\text{E}} \quad \text{for all } x \in \mathcal{M}, \text{ and } X \in \mathcal{M}_{\text{E}}.$$

Note that $\Pi^{\omega_{\text{E}}}$ is a norm-one projection since $\omega_{\text{E}}(\mathbb{1}_{\text{E}}) = 1$. The map $\Pi^{\omega_{\text{E}}}$ is called a *conditional expectation*. We shall often write $x$ as an abbreviation for $x \otimes \mathbb{1}_{\text{E}}$, which will cause no confusion. The state $\omega_{\text{E}}$ is usually called a *reference state*. The physical meaning is that the influence of the system S on the environment E is negligible, so that the state of E is not changed by the presence of S in a significant way. A typical example is an environment in thermodynamical equilibrium, whose temperature is unchanged after some exchange of energy due to the thermal contact with some small system. We extend this assumption to quantum states and neglect the influence on the environment due to its interaction with the system S (of course this assumption must be justified on a physical basis). As a matter of fact, this is not the only simplification that is usually made. The time evolution of the system S now follows the law

$$x(t) \equiv S_t(x) = \Pi^{\omega_E}\left[e^{iH_{tot}t}(x \otimes \mathbb{1}_E)e^{-iH_{tot}t}\right] \quad \text{for all } x \in \mathcal{M} \text{ and } t \in \mathbb{R}, \quad (1.2)$$

and this is not a *-automorphic evolution any longer. Moreover, the family of mappings $\{S_t\}_{t\geq 0}$ does not constitute a one-parameter group. The only properties inherited by the $S_t$ are complete positivity and contractivity, i.e., $\|S_t(x)\| \leq \|x\|$ for all $x \in \mathcal{M}$, where $\|\cdot\|$ is the norm on $\mathcal{M}$. The time evolution as given by (1.2) has in general a complicated form and is difficult to study. Therefore, some further simplifications are welcome. A simplification of great technical impact is achieved when the dynamics (1.2) is approximately memory free, i.e., the family $\{S_t\}_{t\geq 0}$ can be approximated by a semigroup $\{T_t\}_{t\geq 0}$. A semigroup satisfies the memoryless property $T_s \circ T_t = T_{t+s}$ for all $t, s \geq 0$. There are some physical regimes in which this kind of approximation can be justified rigorously. For a nice account of this approximation technique we refer to the book [1]. We remark that a Markovian approximation of the time evolution given by (1.2) is often unavoidable in order to arrive at concrete results. Nevertheless, below we give some examples where the dynamics $\{S_t\}_{t\geq 0}$ is given by a semigroup from the outset, without the need for any approximations.

### 1.2.4 Summary

To sum up our discussion: we are interested in an open quantum system S. The time evolution of such a system results from the interaction of S with its environment E, and we focus on cases where the time evolution can be well approximated by a Markovian dynamics, i.e., by a semigroup $\{T_t\}_{t\geq 0}$ of completely positive and contractive linear maps acting on the von Neumann algebra $\mathcal{M}$ representing the observables of the system S. For our applications to decoherence, a central issue is the long time behavior of such semigroups. In fact, we shall pay less attention to short time behavior of the semigroup here, which is intimately connected to regularity properties in the time parameter $t$ and the notion of the infinitesimal generator of the semigroup. We only mention weak* continuity as a natural notion of continuity in the von Neumann algebra setting. Let $\mathcal{M}^*$ be the dual space, consisting of all norm-continuous linear forms on $\mathcal{M}$. We consider the topology generated by all seminorms

$$P_{\{\xi_n,\zeta_n\}}(x) = \sum_{n=1}^{\infty} |\langle x\xi_n, \zeta_n\rangle_{\mathcal{H}}|$$

on $\mathcal{M}$, where $\xi_n, \zeta_n \in \mathcal{H}$ for $n = 1, 2, \ldots$ such that $\sum_{n=1}^{\infty} \|\xi_n\|_{\mathcal{H}} \cdot \|\zeta_n\|_{\mathcal{H}} < +\infty$. Here $\|\cdot\|_{\mathcal{H}} = \sqrt{\langle \cdot, \cdot \rangle_{\mathcal{H}}}$ denotes the Hilbert space norm of $\mathcal{H}$.

This topology is called the $\sigma$-*weak topology* on $\mathcal{M}$. Finally, we say that $\phi$ is normal if $\phi \in \mathcal{M}^*$ and if it is continuous in the $\sigma$-weak topology; the space of all normal linear functionals is denoted by $\mathcal{M}_*$. As a matter of fact the dual space of $\mathcal{M}_*$ is the algebra $\mathcal{M}$, and the $\sigma$-weak topology coalesces with the weak* topology when $\mathcal{M}$ is considered as the Banach space dual of $\mathcal{M}_*$. Sometimes we refer to $\mathcal{M}_*$ as the *predual space* of $\mathcal{M}$.

The Markov semigroup is weakly* continuous (or $\sigma$-weakly continuous) if the map $t \mapsto \phi(T_t(x))$ is continuous for all fixed $x \in \mathcal{M}$ and $\phi \in \mathcal{M}_*$. This notion of weak* continuity will be used in some constructions and in the examples below. We also shall use occasionally the *strong operator topology* on $\mathcal{M}$, which is given by the family of seminorms $x \mapsto \|x\varphi\|_{\mathcal{H}}$ with $\varphi \in \mathcal{H}$.

## 1.3 Decoherence

In the following we recall the notion of decoherence which we have proposed in the algebraic approach to quantum physics. It describes a physical process resulting from the interaction of an open system with its environment. Let $\mathcal{M}$ be the algebra of observables representing the kinematical degrees of freedom of the system. Due to its openness the time evolution of the system follows an irreversible dynamical law, given by a family of completely positive maps $\{T_t\}_{t \geq 0}$ on the algebra $\mathcal{M}$, i.e., $\mathcal{M} \ni x \mapsto T_t(x) \in \mathcal{M}$, where $t \geq 0$.

### 1.3.1 Definition of Decoherence

**Definition 1** We say that *decoherence takes place* if the following decomposition holds true: There exist two Banach subspaces $\mathfrak{M}$ and $\mathfrak{N}$ of $\mathcal{M}$ such that

$$\mathcal{M} = \mathfrak{M} \oplus \mathfrak{N},$$

where $\mathfrak{M}$ is a von Neumann subalgebra of $\mathcal{M}$. The restrictions $T_t \upharpoonright_{\mathfrak{M}}$ of the maps $T_t$ to $\mathfrak{M}$ are given by a reversible time evolution associated to some group of *-automorphisms $\{R_t\}_{t \in \mathbb{R}}$ of the subalgebra $\mathfrak{M}$, i.e., $R_t = T_t \upharpoonright_{\mathfrak{M}}$ for all $t \geq 0$. Moreover, we require that

$$\lim_{t \to \infty} \phi(T_t(x)) = 0 \quad \text{for all } \phi \in \mathcal{M}_* \text{ and } x \in \mathfrak{N}. \tag{1.3}$$

Let us make some comments concerning the physical content of this definition. Due to decoherence, any observable $x \in \mathfrak{N}$ starts to be out of the range of *any* measurement device after some lapse of time (which is finite in practice), i.e.,

$$|\phi(T_t(x))| < \varepsilon \quad \text{for any } t > t_d,$$

uniformly for any $x \in \mathfrak{N}$, where the bound $\varepsilon$ for the average values of the observables is small enough to be considered as *not* relevant in *all* measurements. The time $t_d$, which is found to be very short in all practical situations, is called the *decoherence time*. It marks a time scale after which the reversible dynamics $\{R_t\}_{t \in \mathbb{R}}$ on the subalgebra $\mathfrak{M}$ starts to be dominant in the system, and at the same time it effectively describes the system. This phenomenon is modeled by the requirement (1.3).

We call the algebra $\mathfrak{M}$ the algebra of *effective observables,* and the dynamics on $\mathfrak{M}$ given by $\{R_t\}_{t \in \mathbb{R}}$ the *effective dynamics* of the system [4, 17]. The pair $(\mathfrak{M}, \{R_t\}_{t \in \mathbb{R}})$ is called the *effective dynamical system.* We emphasize that due to the process of decoherence a part of the kinematical degrees of freedom represented by $\mathfrak{N}$ are suppressed as a consequence of the interaction of the system with its environment. Mathematically, this interaction is inherent in the family of maps $\{T_t\}_{t \geq 0}$. It is believed that the notion of a *strong decoherence* is more appropriate physically. For strong decoherence one assumes that the convergence in (1.3) is uniform for all observables from any set that is bounded in the operator norm $\|\cdot\|$.

Finally, note that it is sometimes convenient to replace condition (1.3) by the weaker requirement

$$\lim_{t \to \infty} \phi(T_t(x)) = 0 \quad \text{for all } \phi \in \mathcal{M}_* \text{ and } x \in \mathfrak{N}_0 \subset \mathfrak{N}, \qquad (1.4)$$

where $\mathfrak{N}_0$ is a subspace dense in $\mathfrak{N}$ with respect to the $\sigma$-weak topology (see below for an example where $\mathfrak{N}_0 = \mathfrak{N} \cap \mathfrak{C}$, with $\mathfrak{C}$ a C*-algebra).

### 1.3.2   Environment Induced Superselection Rules

If the family $\{T_t\}_{t \geq 0}$ of operators admits predual operators $T_{*t}$ acting on $\mathfrak{M}_*$ for any $t \geq 0$, then one can express decoherence in the language of dual objects, and in particular in terms of states, as it is often done in the physics literature.

Indeed, historically the notion of decoherence has origin in this dual picture. If one considers two states both represented by unit vectors $\varphi$ and $\psi$ from $\mathcal{H}$, then one can form their superposition, $\chi_\lambda = (\psi + \lambda\varphi)/\|\psi + \lambda\varphi\|_{\mathcal{H}}$, where $\lambda$ is some complex number. The vector $\chi_\lambda$, in agreement with the superposition principle of quantum mechanics, describes a new possible state of the system S. But in general there are some very serious obstacles to accept this point of view in its full generality. One of the basic example is provided by elementary particle physics, where no observation of any superposition of states with different electrical charges or a superpositions of states with integer and half-integer spin have ever observed in laboratories. More striking examples touching our everyday experience are obtained when one tries to understand the behavior of macroscopic systems using principles of quantum mechanics. The most famous example is the Gedanken experiment proposed by Schrödinger in 1935 [26]. A cat is enclosed in a nontransparent box together with a poisoning substance. A lethal portion of this substance is released when a Geiger counter inside the box registers the radioactive decay of one atom. The quantum mechanical description of this situation suggests that the cat is neither dead nor alive, but that its state is given by a superposition of "dead" and "alive" states of the cat until we open the box and check its condition. This grossly contradicts common sense. Obviously, if one opens the box then one finds the cat either dead or alive with some probability which is given by the laws of radioactive decay. This example touches not only the measurement problem in quantum theory, in particular the wave packet reduction postulate, but poses a fundamental question as well: Why does the superposition rule, valid in the micro-world of particles, ceases to be valid in the macro-world, which is made up from these microscopic particles?

The concept of superselection rules, originally introduced in the context of elementary particle physics [27], is a natural and simple solution: Superpositions of states not realizable in the physical world have to be excluded from the Hilbert space $\mathcal{H}$. This means, in particular, that the Hilbert space of physical states $\mathcal{H}$, has to be written as a direct sum of distinguished subspaces: $\mathcal{H} = \mathcal{H}_1 \oplus \mathcal{H}_2$, where the superposition rule is valid without any limitation exclusively within each *coherent subspace* $\mathcal{H}_1$ and $\mathcal{H}_2$, but not between elements of $\mathcal{H}_1$ and $\mathcal{H}_2$. In that case the algebra of observables $\mathcal{N}$ can no longer be equal to $\mathcal{B}(\mathcal{H})$, as was first realized by von Neumann. Indeed, assuming that the linear combinations $\chi_\lambda$ with $\psi \in \mathcal{H}_1$ and $\varphi \in \mathcal{H}_2$ have no physical meaning then the algebra of observables should not make any distinction between the statistical mixture $\frac{1}{1+|\lambda|^2} P_\psi + \frac{|\lambda|^2}{1+|\lambda|^2} P_\varphi$ and the state given by $P_{\chi_\lambda}$, where $P_{\chi_\lambda}$ is the orthogonal projection onto the one-dimensional subspace spanned by $\chi_\lambda$ (the projections $P_\psi$ and $P_\varphi$ are defined in a similar way). In other words,

$$\langle \chi_\lambda, x\chi_\lambda \rangle_\mathcal{H} = \frac{1}{1+|\lambda|^2} \langle \psi, x\psi \rangle_\mathcal{H} + \frac{|\lambda|^2}{1+|\lambda|^2} \langle \varphi, x\varphi \rangle_\mathcal{H} \quad \text{for all } x \in \mathcal{N}.$$

This means that the commutant $\mathcal{N}'$ contains orthogonal projectors $Q_1$ and $Q_2$ onto $\mathcal{H}_1$ and $\mathcal{H}_2$, respectively. Hence $\mathcal{N} \neq \mathcal{B}(\mathcal{H})$, since the commutant of $\mathcal{B}(\mathcal{H})$ is trivial, i.e., $\mathcal{B}(\mathcal{H})' = \mathbb{C}\mathbf{1}$. The superselection rules can thus be introduced by the so called superselection operators $Q$, which are given by all elements of the commutant $\mathcal{N}'$ of the algebra $\mathcal{N}$.

The addition of the postulate of superselection rules in quantum mechanics poses new questions for physicists: how can one justify them? This question, on a purely mathematical ground, partially overlaps with the question concerning the phenomenon of wave packet reduction in the measurement theory, or more generally, the question of a restriction of the superposition principle for macroscopic objects (i.e., the appearance of a classical world). Indeed, one wants to find a physical mechanism resulting in the following condition:

$$\langle \psi, x\varphi \rangle_{\mathcal{H}} = 0 \quad \text{for all } x \in \mathcal{N},$$

for all pairs states $\psi$ and $\varphi$ which cannot be superposed in the presence of superselection rules, or for all pairs of states $\psi$ and $\varphi$ which are macroscopically different, as in the case of measurement theory. The Hamiltonian dynamics has a fundamental character, but evidently it cannot account for the above goal. Since Hamiltonian evolution transforms each state $\chi \in \mathcal{H}$ into another state in $\mathcal{H}$, it cannot transform a pure state into a statistical mixture. In other words, the orthogonal projection $P_\chi$ will evolve into another orthogonal projection under any Hamiltonian evolution, but never into a mixture $\omega P_\psi + (1 - \omega) P_\varphi$, with $\omega \in (0, 1)$.

### 1.3.3  Zurek's Description of a Quantum Measurement

In this section we briefly present Zurek's idea of environment induced superselection rules, which was proposed in [29, 30] and further developed in [12, 21]. The starting point is the von Neumann scheme [26] of a measurement device A which measures an observable $S = \sum_s s|s\rangle\langle s|$ of a system S, which is initially prepared in the state $|\psi_0\rangle = \sum_s c_s |s\rangle$. Von Neumann's formulation of the measurement process can be divided into two phases. The first phase—often called the *premeasurement*—consists of forming quantum correlations between the measurement device A and the system S that is being measured, i.e.,

$$|\psi_0\rangle \otimes |A_0\rangle = \sum_s c_s |s\rangle \otimes |A_0\rangle \xrightarrow[H_{\text{int}}:S \leftrightarrow A]{} \sum_s c_s |s\rangle \otimes |A_s\rangle.$$

Note that initially there are no correlations between A and S. The second phase is often called the *wave packet reduction* and is connected to the transition from the superposition of states to a statistical mixture, which is also called the decoherence process. In symbols,

$$\sum_s c_s |s\rangle \otimes |A_s\rangle \implies \sum_s |c_s|^2 \cdot |s\rangle\langle s| \otimes |A_s\rangle\langle A_s|.$$

This transition cannot be realized by a Hamiltonian time evolution. Here the initial state $|\psi_0\rangle$ of S has been destroyed. Note that the coefficients $c_s$ have not been changed in the decoherence process, which means that certain information concerning the initial state $|\psi_0\rangle$ of the system S has been transported to the measuring device without any losses (we are considering an ideal measurement process).

Now Zurek argued that this scheme cannot be completely realistic since, in accordance with quantum theory, the state produced in the premeasurement can be represented by a superposition of eigenstates $|r\rangle$ of another observable $R = \sum_r r|r\rangle\langle r|$ as well, which need not commute with the original observable $S$ in general. As a particular consequence, this would mean that the measurement device itself would have contained the whole information about two non-commuting physical quantities at the same time. But as a consequence of their non-commutativity, they are subject to the Heisenberg uncertainty principle and hence cannot be determined simultaneously with arbitrarily high accuracy. Moreover, one could not say that the device A is related to a particular physical quantity in an unique way, contradicting our experience with real measurement devices in real laboratory situations. As a conclusion, if we believe that the device A obeys the quantum laws after the premeasurement process, we must contend that we still do not fully understand the wave packet reduction process. Zurek himself wrote:

> ...quantum mechanics alone, when applied to an isolated, composite object consisting of apparatus and a system, cannot in principle determine which observable has been measured.

The solution proposed by Zurek bridges the gap described above and removes the loopholes in our understanding of the premeasurement process, and at the same time it indicates the mechanism of the wave packet reduction process. In Zurek's proposal, the measurement device consists not only of the part A, but it is a composite system $A' = A + E$, where the part E is called by convention the *environment* of the device A. It seems to be natural that the measuring device consists of one part which has direct contact with the system S on which the measurement is performed. This part of the measurement device directly collects the information about the state $|\psi_0\rangle$ of S. It is identified in Zurek's scheme with the quantum subsystem A of the total system $A'$. The second part E decides which part of the information transferred by the interaction between S and A is actually displayed by the measurement apparatus $A'$, i.e., which observable it actually measures. Note that we consider the part E to interact only with A, but it does not influence the measured system S. Moreover, the part E is entirely ignored during the whole process of measurement. Mathematically, this corresponds to computing a partial average with respect to the ignored state of E. This averaging leads to a non-Schrödinger type dynamics in the state space of A, or equivalently, to an irreversible dynamics on the corresponding algebra of observables. This resulting dynamics is typically complicated, but at the same time it leaves room for the dynamical appearance of the phenomenon of wave packet reduction.

Summing up our discussion: On physical grounds we have argued that one has to take into account the essential openness of systems in order to understand the decoherence process. The main idea presented above is based on the pragmatic assumption that only partial information about the dynamics of the total system $A + E$ is available in practice, which is given by averaging over the degrees of freedom of E. Due to ignoring E the renouncement of the genuine *-automorphic dynamics is unavoidable.

## 1.4  Some General Results About Decoherence

Having clarified the basic mathematical notion of decoherence and having discussed its physical interpretation we are now ready to develop some mathematical consequences of the point of view we have chosen. We first state some theorems (without giving them in the most general form that is possible, instead we refer to the cited literature). After that, in the following section, we shall give some typical examples which give the flavor of the theory; however, they do not exhaust all known models to date (we have always chosen the simplest mathematical version and have avoided obvious generalizations). However, we keep our considerations mathematically rigorous. The price to pay for mathematical rigor is the loss of a complete connection to exact physical realizations as is met in the laboratory. We hope to be able to fill this gap in the future by a further development of the models. Nevertheless, the discussion below presents the main line of our idea in full detail.

It is clear that different scenarios for the algebra $\mathfrak{M}$ and the dynamics $\{R_t\}_{t\geq 0}$ are possible, including the case in which the dynamics $\{R_t\}_{t\geq 0}$ is trivial—this is a typical situation if the algebra $\mathfrak{M}$ is isomorphic to a discrete one, e.g., $\mathfrak{M} = \ell^\infty$, the space of all bounded sequences of complex numbers, which becomes a von Neumann algebra when multiplication is defined by $\{a_n\}\{b_n\} = \{a_n b_n\}$ and involution by $\{a_n\}^* = \{\bar{a}_n\}$. This situation is described by the Theorem 1 below (see also [20]).

Let $\mathcal{T}(\mathcal{H}) \subseteq \mathcal{B}(\mathcal{H})$ be the set of all linear operators of trace class, and let $\mathcal{T}_+(\mathcal{H}) \subseteq \mathcal{T}(\mathcal{H})$ the set of all non-negative operators. Moreover, $\mathcal{T}_{+,1}(\mathcal{H}) \subseteq \mathcal{T}_+(\mathcal{H})$ denotes the set of all non-negative elements with trace equal to one. This set is identified with the set of all states of the system. For the proof of the next theorem, recall that $\mathcal{T}(\mathcal{H})^* = \mathcal{B}(\mathcal{H})$. We assume that we are given a semigroup $\{T_t\}_{t\geq 0}$ of superoperators $T_t : \mathcal{B}(\mathcal{H}) \longrightarrow \mathcal{B}(\mathcal{H})$, for all $t \geq 0$, which describes the time evolution of an open system in the Markovian approximation—refer to our discussion in Sect. 1.2.3 for details. We assume that $\{T_t\}_{t\geq 0}$ satisfies the following properties:

a1   For each $t \geq 0$ we have $T_t(\mathbb{1}) \leq \mathbb{1}$.
a2   For each $t \geq 0$ there exists the map $T_{t*} : \mathcal{T}(\mathcal{H}) \longrightarrow \mathcal{T}(\mathcal{H})$, the predual operator of $T_t$.
a3   For each $t \geq 0$ the map $T_t$ is two-positive.
a4   For each $t \geq 0$ the map $T_t$ is contractive in the operator norm $\|\cdot\|_\infty$ on $\mathcal{B}(\mathcal{H})$, and $T_t \upharpoonright_{\mathcal{T}(\mathcal{H})}$ is contractive in the trace norm $\|\cdot\|_1$, defined on the trace class operators $\mathcal{T}(\mathcal{H})$.

We assumed that the initial algebra of observables $\mathcal{M}$ is given by the full operator algebra $\mathcal{B}(\mathcal{H})$, i.e., it is a factor of type I. This assumption is natural in the sense that we want to consider a system with a "maximal" quantum character, without any superselection rules, in which the whole Hilbert space $\mathcal{H}$ constitutes the set of physically realizable states. In this situation we want to investigate the appearance of possible superselection sectors due to decoherence. We assume two-positivity instead of the more physical assumption of complete positivity since it turns out that two-positivity is enough for our purposes. In particular, this positivity condition together with the conservation condition $T_t(\mathbb{1}) = \mathbb{1}$ guarantees that the density matrices $\rho \in \mathcal{T}_{+,1}(\mathcal{H})$ are transformed into density matrices $T_{t*}(\rho)$ by the mappings $T_{t*}$. This allows us to consider the dynamics in the state space in a natural way.

**Theorem 1** *Let $\{T_t\}_{t\geq 0}$ be a weak\* continuous semigroup on $\mathcal{B}(\mathcal{H})$ satisfying the above properties a1–a4. Then there exist linear subspaces $\mathfrak{M}$ and $\mathfrak{N}$ of $\mathcal{B}(\mathcal{H})$ having the properties described in Definition 1, together with the condition (1.4). If $\mathfrak{M}$ contains the unit $\mathbb{1}$, then it is additionally a von Neumann algebra which can be decomposed into a direct sum of factors $\mathfrak{M}_k$, $k = 0, 1, 2, \ldots$, i.e.,*

$$\mathfrak{M} = \bigoplus_k \mathfrak{M}_k, \tag{1.5}$$

*where each factor $\mathfrak{M}_k$ is of type I. Moreover, the effective evolution $\{R_t\}_{t\in\mathbb{R}}$ is given by a unitary evolution*

$$R_t(x) = e^{itH} x e^{-itH} \quad \text{for all } x \in \mathfrak{M},$$

*where H is a self-adjoint operator.*

*Sketch of Proof* Let $HS(\mathcal{H}) \subseteq \mathcal{B}(\mathcal{H})$ denote the set of Hilbert–Schmidt operators, which can be considered as a Hilbert space with inner product $\langle x, y \rangle_2 = \text{tr}\, x^* y$, and define a subspace of $HS(\mathcal{H})$ by

$$\mathfrak{K} = \{x \in HS(\mathcal{H}) : \|T_{t*}(x)\|_2 = \|T_t(x)\|_2 = \|x\|_2 \text{ for all } t \geq 0\}, \tag{1.6}$$

where $\|\cdot\|_2$ denotes the Hilbert–Schmidt norm, which is defined by $\|x\|_2 = \sqrt{\text{tr}(x^*x)}$. We assume that $\mathfrak{K}$ is nontrivial, i.e., $\mathfrak{K} \neq \{0\}$. As a matter of fact, the map $T_t$, when restricted to $HS(\mathcal{H})$, can be considered as the Hilbert space adjoint of $T_{t*}$ acting on $HS(\mathcal{H})$, i.e. the extension of $T_{t*}$, which is denoted in (1.6) again by $T_{t*}$ and is given by the restriction of the dual operator $T_t \upharpoonright_{\mathcal{T}(\mathcal{H})^*}$ to $HS(\mathcal{H})$. The space $\mathfrak{K}$ is invariant under the action of both $T_{t*}$ and its adjoint operator. The same holds true for the orthogonal complement $HS(\mathcal{H}) \ominus \mathfrak{K}$. The actions of $T_{t*}$ and $T_t$ are isometric on $\mathfrak{K}$ and suppressing on $HS(\mathcal{H}) \ominus \mathfrak{K}$, i.e., $\lim_{t\to\infty} \text{tr}(\varphi T_t(x)) = 0$ for all $x \in HS(\mathcal{H}) \ominus \mathfrak{K}$ and all $\varphi \in HS(\mathcal{H})$. Let $P$ be the orthogonal projection onto the subspace $\mathfrak{K}$. One can show that $\text{tr}\, P(\varphi) \leq \text{tr}\, \varphi$ for all $\varphi \in \mathcal{T}_+(\mathcal{H})$. In particular, $P$ transforms $\mathcal{T}(\mathcal{H})$ into $\mathcal{T}(\mathcal{H})$. Hence, one can consider the restriction $P \upharpoonright_{\mathcal{T}(\mathcal{H})}$, which induces a decomposition of the space $\mathcal{T}(\mathcal{H})$ into the isometric (iso) and suppressing (s) parts

$$\mathcal{T}(\mathcal{H}) = \mathcal{T}(\mathcal{H})_{\text{iso}} \oplus \mathcal{T}(\mathcal{H})_{\text{s}},$$

where $T(\mathcal{H})_{\mathrm{iso}} = P(T(\mathcal{H})) = \mathfrak{K} \cap T(\mathcal{H})$, and $T(\mathcal{H})_{\mathrm{s}} = (\mathrm{id} - P)(T(\mathcal{H}))$. Finally, one defines the algebra $\mathfrak{M}$ as the dual space of $T(\mathcal{H})_{\mathrm{iso}}$. Let $E$ be equal to $(P \upharpoonright_{T(\mathcal{H})})^*$, i.e., the dual operator of $P$ when considered as an operator in $T(\mathcal{H})$. Then $\mathfrak{M} = E(\mathcal{B}(\mathcal{H}))$ and $\mathfrak{N} = (\mathrm{id} - E)\mathcal{B}(\mathcal{H})$. For any compact operator $x$ from $\mathfrak{N}$ one has the suppression property, $\lim_{t \to \infty} \mathrm{tr}(\varphi T_t(x)) = 0$ for all $\varphi \in T(\mathcal{H})$. This can be easily seen if one notes that $\mathrm{HS}(\mathcal{H})$ is a dense subset of the set of all compact operators in the operator norm topology of $\mathcal{B}(\mathcal{H})$.

The decomposition of the algebra $\mathfrak{M}$ in (1.5) is described in detail in [20]. Having established this decomposition, we note that the dynamics should preserve each subalgebra $\mathfrak{M}_k$ and hence cannot generate "transitions" between different sectors $\mathfrak{M}_k$. Hence we need to analyze the restriction $T^{(k)} \equiv T_t \upharpoonright_{\mathfrak{M}_k}$, which is a *-automorphic evolution. A closer analysis shows that $\mathfrak{K}$ is generated by orthogonal projections (denote the collection of all of them by $\mathcal{P}(\mathfrak{K})$), which are finite-dimensional since $\mathfrak{K}$ is a subset of the space of Hilbert–Schmidt operators. Moreover, $T_{t*}$ and $T_t$, as operators on $\mathfrak{K}$, transform orthogonal projectors into orthogonal projectors of the same finite dimension, and they are evidently bijective mappings when restricted to $\mathcal{P}(\mathfrak{K})$, as $T_{t*}(T_t(e)) = e$ for all $e \in \mathcal{P}(\mathfrak{K})$. They satisfy the multiplication rule, i.e., $T_t(ee') = T_t(e)T_t(e')$ for all $e, e' \in \mathcal{P}(\mathfrak{K})$. The operator $T_t$ is normal, and $\mathcal{P}(\mathfrak{K})$ generates $\mathfrak{M}$, hence each $T_t$ is a *-automorphism of $\mathfrak{M}$. The same conclusion holds for each $T_t^{(k)}$. Now, as $\mathfrak{M}$ is a factor of type I, it is spatially isomorphic to $\mathcal{B}(\mathcal{H}_k)$, where $\mathcal{H}_k$ is some Hilbert space. Let $\gamma : \mathfrak{M}_k \longrightarrow \mathcal{B}(\mathcal{H}_k)$ denote this isomorphism. Finally, each *-automorphism of $\mathcal{B}(\mathcal{H}_k)$ is inner, so we conclude that $\gamma \circ T_t^{(k)}(\cdot) \circ \gamma^{-1} = V_t^{(k)}(\cdot)V_t^{(k)*}$ for some unitary operator $V_t$ in $\mathcal{H}_k$. The family $\{V_t^{(k)}\}_{t \in \mathbb{R}}$, with $V_{-t}^{(k)} = V_t^{(k)*}$ for $t \geq 0$, is a strongly continuous one-parameter group of unitary operators in $\mathcal{H}_k$.                    □

Note that when $\{T_{t*}\}_{t \geq 0}$ is relatively compact in the strong operator topology one obtains strong decoherence. We shall not dwell on all the different cases covered by this theorem as it is more adapted for the discussion of superselection rules (which are not our central issue here) and the measurement process if $\mathcal{M} \simeq \ell^\infty$, giving a set of discrete outputs.[1]

To address situations with nontrivial continuous effective dynamics $\{R_t\}_{t \in \mathbb{R}}$ on an abelian algebra $\mathfrak{M}$ we need to go beyond the realm of type I von Neumann algebras. Theorem 2 below covers the cases of type II von Neumann algebras, but it can be stated in a more general form which applies to type III von Neumann algebras as well (we invite interested readers to consult [18]). Let us first introduce some notation. Let $\iota$ a be semifinite faithful and normal trace on the factor $\mathcal{M}$. One defines the set $M_\tau \subseteq \mathcal{M}$ by

$$M_\tau = \mathrm{lin}\{x \in \mathcal{M}_+ : \tau(x) < +\infty\}.$$

---

[1] We do not consider random dynamics on $\ell^\infty$ given by some continuous Markov chain.

We make similar assumptions as in the case of Theorem 1:

A1    For each $t \geq 0$ we have $T_t(\mathbb{1}) \leq \mathbb{1}$.
A2    For each $t \geq 0$ there exists $T_{t*} : \mathcal{M}_* \longrightarrow \mathcal{M}_*$, i.e., the predual operator of $T_t$.
A3    For each $t \geq 0$ the mapping $T_t$ is two-positive.
A4    For each $t \geq 0$ the mapping $T_t$ is contractive in the operator norm $\|\cdot\|_\infty$ on $\mathcal{M}$, and $T_t \upharpoonright_{M_\tau}$ is contractive in the trace norm $\|\cdot\|_1$ defined on the trace class operators $M_\tau$.

**Theorem 2** *Let $\{T_t\}_{t \geq 0}$ be a weak\* continuous semigroup on $\mathcal{M}$ satisfying the above properties A1–A4. Then there exist $T_t$-invariant linear subspaces $\mathfrak{M}$ and $\mathfrak{N}$ such that the decomposition $\mathcal{M} = \mathfrak{M} \oplus \mathfrak{N}$ holds. Both $\mathfrak{M}$ and $\mathfrak{N}$ are weak\* closed and invariant under the involution in $\mathcal{M}$. Moreover, $\mathfrak{M}$ is a subalgebra of $\mathcal{M}$. If $\mathbb{1} \in \mathfrak{M}$ then $\mathfrak{M}$ is a von Neumann algebra. Moreover, $T_t \upharpoonright_{\mathfrak{M}}$ gives rise to the \*-automorphic evolution $\{R_t\}_{t \in \mathbb{R}}$. We also have*

$$\lim_{t \to \infty} \varphi(T_t(x)) = 0 \quad \text{for all } \varphi \in \mathcal{M}_* \text{ and } x \in \mathfrak{N} \cap \mathfrak{C}, \tag{1.7}$$

*where $\mathfrak{C}$ is a C\*-algebra which is weak\* dense in $\mathcal{M}$. If $\{T_{t*}\}_{t \geq 0}$ is relatively compact in the strong operator topology then we get strong decoherence. Finally, if the trace $\tau$ is finite on $\mathcal{M}$ then the condition (1.7) is satisfied with $\mathfrak{C} = \mathcal{M}$.*

To further simplify our discussion we formulate our next Theorem 3 only for the case of a finite von Neumann algebra of type II$_1$. We consider a normalized trace on $\mathcal{M}$, i.e., $\tau(\mathbb{1}) = 1$. First note that in this case any von Neumann subalgebra $\mathfrak{M}$ of $\mathcal{M}$ can be considered as a result of some decoherence process induced by a semigroup $\{T_t\}_{t \geq 0}$ of operators satisfying the above conditions A1–A4. We are particularly interested in the case in which $\mathfrak{M}$ is a maximal abelian subalgebra in $\mathcal{M}$. In this case one can describe the effective dynamics $\{R_t\}_{t \in \mathbb{R}}$ in a canonical way. Indeed, by the Riesz representation theorem $\mathfrak{M} \simeq L^\infty([0, 1], \beta, \mathrm{d}x) \equiv L^\infty$, where $\beta$ is the Borel $\sigma$-algebra of $[0, 1]$ and $\mathrm{d}x$ is the normalized Lebesgue measure. To simplify the notation we identify both algebras $\mathfrak{M}$ and $L^\infty$. According to Theorem 2, $R_t$ is a \*-automorphism of $L^\infty$ preserving the trace $\tau$, i.e., one gets

$$\int_0^1 (R_t(f))(x)\, \mathrm{d}x = \int_0^1 f(x)\, \mathrm{d}x \quad \text{for all } f \in L^\infty.$$

Using classical theorems about Lebesgue spaces the existence of bijective mappings $\Phi_t$ can be established, which have the property that both $\Phi_t$ and $\Phi_t^{-1}$ are measurable and invariant under the Lebesgue measure, and which are related to $R_t$ by

$$R_t(\chi_A) = \chi_{\Phi_t^{-1}(A)} \quad \text{for all } A \in \beta,$$

where $\chi_A$ denotes the characteristic function of the Borel set $A$. Using the group law of $\{R_t\}_{t \in \mathbb{R}}$ and the above relation one gets $\Phi_t \circ \Phi_r = \Phi_{t+r}$ for all $t, r \in \mathbb{R}$. Hence we have obtained a measurable flow $\{\Phi_t\}_{t \in \mathbb{R}}$. We can now formulate the next result.

**Theorem 3** *Let $\mathcal{M}$ be a factor of type $\mathrm{II}_1$ with a normalized trace $\tau$. If the effective algebra $\mathfrak{M}$ is a maximal abelian subalgebra of $\mathcal{M}$ then the effective dynamics $\{R_t\}_{t \in \mathbb{R}}$ can be viewed as a classical dynamics on the phase space $[0, 1]$ which is given by a measurable flow $\Phi_t : [0, 1] \longrightarrow [0, 1]$ for $t \in \mathbb{R}$, where $R_t(x) = \Phi_{t,*}(x)$ for all $x \in \mathfrak{M}$, with $(\Phi_{t,*}(x))(\omega) = x(\Phi_t(\omega))$ a. e. with respect to Lebesgue measure on $[0, 1]$.*

## 1.5   Examples

Let us now present some examples. We focus our attention on cases with nontrivial effective dynamics as they seem to be most interesting. First, we consider an example with a factor of type $\mathrm{II}_1$. Then we move on to examples of infinite von Neumann algebras.

### 1.5.1   Newtonian Motion on the Circle

One of the simplest C*-algebras describing infinite systems is the Glimm algebra, which we now introduce. Consider the set of all complex matrices with $2^n$ rows and columns ($n = 0, 1, 2, \ldots$). We shall denote this set by $\mathcal{A}_n$. One then defines the involution $*_n$ on $\mathcal{A}_n$ as the Hermitian conjugation of matrices. The norm $\|x\|_n$ of an element $x \in \mathcal{A}_n$ is defined as the square root of the biggest eigenvalue of the matrix $x^{*_n}x$. Notice that $\mathcal{A}_n$ is the algebra of observables for the spin degrees of freedom of a set of $n$ spin-$\frac{1}{2}$ particles (the Planck constant $\hbar$ is set to 1). For $n < m$ each C*-algebra $\mathcal{A}_n$ is *-isometrically included in the C*-algebra $\mathcal{A}_m$, i.e., $\|x\|_n = \|x\|_m$ and $x^{*_n} = x^{*_m}$, where the element $x \in \mathcal{A}_n$ is identified with its image in $\mathcal{A}_m$. This inclusion can be easily constructed. For the sake of simplicity let us take $m = n+1$. Then the matrices of the algebra $\mathcal{A}_{n+1}$, which are built from diagonal blocks of size $2 \times 2$ of the form $\begin{pmatrix} \lambda & 0 \\ 0 & \lambda \end{pmatrix}$, can be identified with elements of the algebra $\mathcal{A}_n$ in a natural way. Using this inclusion $\mathcal{A}_n \hookrightarrow \mathcal{A}_{n+1}$ it is intuitively clear that one gets a normed *-algebra $\bigcup_{n=0}^{\infty} \mathcal{A}_n$. This algebra can be next completed with respect to the norm $\|\cdot\|$, where $\|\cdot\| \upharpoonright_{\mathcal{A}_n} = \|\cdot\|_n$. The resulting C*-algebra is called the *Glimm algebra*. This construction can be stated in a more precise fashion if one uses the concept of inductive limit [14], which we do not introduce here. The Glimm algebra describes the spin degrees of freedom of an infinite system of spin-$\frac{1}{2}$ particles.

Consider the factor generated by the Glimm algebra $\mathcal{A}$ in its trace representation via the GNS construction, i.e., $\mathcal{M} = \pi_{\omega_0}(\mathcal{A})''$. We shall consider the dynamics on $\mathcal{M}$ given by

$$\frac{d}{dt}x = \delta(x) + L_0(x) = i[H, x] + L_0(x),$$

where $\delta$ generates a weak* continuous one-parameter group of *-automorphisms of the algebra $\mathcal{M}$, and $L_0$ is the dissipative part of the dynamics, which is responsible for the suppression of certain observables due to decoherence. These two ingredients are chosen as follows:

*Step 1:*   Let $D_n$ be the algebra of diagonal $2^n \times 2^n$-matrices and let $\mathcal{U}_n$ be the set of all unitary matrices $U$ of size $2^n \times 2^n$ which preserve the diagonal algebra $D_n$. Let $U(\frac{1}{2^n})$ be the element of the group $\mathcal{U}_n$ which is defined as

$$U\left(\frac{1}{2^n}\right)^* \mathrm{diag}(d_{11}, d_{22}, \ldots)U\left(\frac{1}{2^n}\right) = \mathrm{diag}(d_{2^n 2^n}, d_{11}, d_{22}, \ldots),$$

where $\mathrm{diag}(d_{11}, d_{22}, \ldots)$ denotes the diagonal matrix with entries $d_{11}, d_{22}, \ldots$ on the diagonal. Since there is a natural homomorphic inclusion of the groups $\mathcal{U}_n \subseteq \mathcal{U}_{n+1}$, one can consider the set $\bigcup_{n=0}^{+\infty} \mathcal{U}_n$, which is an abelian group isomorphic to the group of dyadic numbers $\mathcal{D}$ of the interval $[0, 1]$ with addition modulo 1,

$$\mathcal{D} = \left\{ \frac{k}{2^n} : k = 0, 1, 2 \ldots, 2^n - 1, \text{ and } n = 1, 2, \ldots \right\}.$$

This isomorphism is given as follows: to each dyadic number $d = \frac{k}{2^n}$ we attach a unitary operator $U(d) = U(\frac{1}{2^n})^k$. Now we can give the representation of the group $\mathcal{D}$ of dyadic numbers as a subgroup of the group of inner *-automorphisms of the algebra $\mathcal{M}$, i.e.

$$\mathcal{D} \ni d \mapsto \alpha(d) \in \mathrm{Aut}\,\mathcal{M},$$

where $\alpha(d)x \equiv U(d)^* x U(d)$ for any $x \in \mathcal{M}$. The following result has been shown in [17]: There is a weak* continuous group homomorphism $\alpha$ : $\mathbb{R} \ni t \mapsto \mathrm{Aut}\,\mathcal{M}$ such that $\alpha(m) = \mathrm{id}_\mathcal{M}$ for any integer $m \in \mathbb{Z}$, and for all dyadic numbers $d \in \mathcal{D}$ we have $\alpha(d) = U(d)^* \cdot U(d)$. Moreover, any *-automorphism $\alpha(t)$ is spatial, i.e., $\alpha(t) = U(t)^* \cdot U(t)$ for some strongly continuous group $\{U(t)\}_{t \in \mathbb{R}}$ of unitary operators in the Hilbert space $\mathcal{H}_{\omega_0}$, the space on which the representation $\pi_{\omega_0}$ acts. In particular, we can represent the generator $\delta$ of the group $\{\alpha_t\}_{t \in \mathbb{R}}$ in the form $\delta(\cdot) = i[H, \cdot]$, where $H$ is a self-adjoint operator in the Hilbert space $\mathcal{H}_{\omega_0}$, which, in turn, generates the group $\{U(t)\}_{t \in \mathbb{R}}$.

*Step 2:*    Let $\mathfrak{D} \subseteq \mathcal{A}$ be the Banach algebra generated by the infinite matrices of the form $\sigma_k^3 = \mathbb{1} \otimes \cdots \otimes \mathbb{1} \otimes \sigma_3 \otimes \mathbb{1} \otimes \cdots$, with $\mathbb{1} = \begin{pmatrix} 1 & 0 \\ 0 & 1 \end{pmatrix}$ and with the Pauli matrix $\sigma_3 = \begin{pmatrix} 1 & 0 \\ 0 & -1 \end{pmatrix}$, which appears in the $k$-th position of the above tensor product. The algebra $\mathfrak{D}$ is a maximal abelian subalgebra algebra (m. a. s. a.). Moreover, $\mathfrak{D}$ is *-isomorphic to the algebra $C(\mathcal{C})$ of all continuous complex valued functions on the Cantor set $\mathcal{C}$.

*Step 3:*    We consider a system similar to the one considered by Bell [3] in his discussion of the wave packet reduction.[2] So we have a single particle which plays the role of the environment for our spin system. The particle moves past the spins, interacts with them and thereby induces changes in the spin system. In the laboratory one observes only the macroscopical behavior of the measurement device, forgetting about the small particle whose presence can be registered only indirectly by a measuring device. In our example the measurement device is represented by the spin system, and the particle is a part of its environment. We "forget" about the particle by looking only at the degrees of freedom of the spin system. The state space of the particle is the Hilbert space $\mathcal{H}_E = L^2(\mathbb{R}, dm)$, where $dm$ is again the Lebesgue measure on the real line $\mathbb{R}$. Its kinematical degrees of freedom are described by the algebra of all bounded operators on $\mathcal{H}_E$, i.e., $\mathfrak{M}_E = \mathcal{B}(\mathcal{H}_E)$. The algebra of the joint system is given by the von Neumann algebra $\mathfrak{M} \otimes \mathfrak{M}_E$ which acts on the Hilbert space $\mathcal{H}_{\omega_0} \otimes \mathcal{H}_E$. The interaction Hamiltonian is taken to be

$$H_{\text{int}} = \pi_{\omega_0} \left( \sum_{k=1}^{+\infty} \frac{1}{2^k} \sigma_k^3 \right) \otimes \hat{p}, \qquad (1.8)$$

where $\hat{p}$ is the momentum operator of the particle in $\mathcal{H}_E$. The reduced dynamics is obtained by

$$T_t^0(x) = \Pi^{\omega_E}(e^{itH_{\text{int}}} x \otimes \mathbb{1}_E e^{-itH_{\text{int}}}),$$

where $\Pi^{\omega_E} : \mathfrak{M} \otimes \mathfrak{M}_E \longrightarrow \mathfrak{M}$ is the conditional expectation with respect to the reference state $\omega_E = |\psi\rangle\langle\psi|$, where the vector $|\psi\rangle$ is given by

$$\psi(x) = \frac{1}{\sqrt{2\pi}} \int_{\mathbb{R}} \frac{e^{ipx}}{\sqrt{\pi(1+p)}} \, dm(p), \qquad (1.9)$$

and where $\Pi^{\omega_E}(x \otimes A) = \omega_E(A)(x \otimes \mathbb{1}_E) \equiv \omega_E(A)x$.

---

[2]Bell worked within the Hamiltonian formulation and without averaging over the degrees of freedom of the environment.

The following fact has been proved in [17]: The family of maps $\{T_t^0\}_{t\geq 0}$ is a semigroup on $\mathcal{M}$ which satisfies the above properties A1–A4. Moreover, the generator $L_0$ of this semigroup is a bounded operator on $\mathcal{M}$. It follows that we have a well defined generator $\delta + L_0$, and we denote the corresponding semigroup by $\{T_t\}_{t\geq 0}$. Then the following theorem [17] can be established.

**Theorem 4** *The von Neumann subalgebra $\mathfrak{M}$ of effective observables is *-iso-morphic to the algebra $\pi_{\omega_0}(\mathfrak{D}) = L^\infty(\mathcal{C}, d\mu)$. Moreover, the group $\{R_t\}_{t\in\mathbb{R}}$ of *-automorphisms is isomorphic to the group of *-automorphisms of the alge-bra $L^\infty(S^1, dm)$ generated by the smooth flow $\Phi_t : S^1 \longrightarrow S^1$ ($t \in \mathbb{R}$), which describes the uniform motion of a particle along a unit circle $S^1$. More precisely, for $x \in L^\infty(S^1, dm)$ one gets $x \mapsto x_t$ with $x_t(\alpha) = x(\alpha \hat{+} 2\pi t)$ for $\alpha \in [0, 1)$, where $\hat{+}$ denotes addition modulo $2\pi$.*

### 1.5.2  Newtonian Motion in Euclidean Space

In this example we show that a similar construction can be carried through for the case of the classical system with a non-compact configuration space. Let $G = D \times D \times D$ be Cartesian product of the group $D$ of dyadic numbers on the real line $\mathbb{R}$. Let $\mathcal{H} = \bigoplus_G L^2(\mathbb{R}^3, dm)$, where $dm$ is the Lebesgue measure on $\mathbb{R}^3$, which we take as the Hilbert space describing the system dynamics, which will be defined below. The group $G$ acts on the algebra $L^\infty \equiv L^\infty(\mathbb{R}^3, dm)$ by $\alpha_g(f)(r) = f(r + g)$ for any $r \in \mathbb{R}^3$. This action is free and ergodic [14]. The kinematical degrees of freedom are described by the von Neumann algebra $\mathcal{M} = L^\infty \otimes_\alpha G$, where $\otimes_\alpha$ denotes the crossed product [14]. The application of the crossed product to physical problems was pioneered by Landsman in [15]. He used it to obtain a quantization procedure of classical systems. We will use the crossed product in the opposite direction since our result can be regarded as a "dequantization", i.e., we start from a system with quantum character and then arrive at a classical one.

The algebra $\mathcal{M}$ is a factor of type $II_\infty$. We define a faithful normal semifinite trace $\tau : \mathfrak{M}_+ \longrightarrow [0, +\infty]$ by $\tau(x) = \int x(0, 0) \, dm$. Let $\pi_\alpha$ be the canonical normal *-isomorphism of the algebra $L^\infty$ into $\mathcal{M}$. Then $\pi_\alpha(L^\infty)$ is a commutative subalgebra of the von Neumann algebra $\mathcal{M}$. The environment is described by the Hilbert space $\mathcal{H}_E = L^2(\mathbb{R}^3, dm)$ and by the algebra $\mathcal{B}(\mathcal{H}_E)$ of all bounded operators on $\mathcal{H}_E$. The algebra of the total system is then given by the von Neumann algebra $\mathcal{M} \otimes \mathcal{B}(\mathcal{H}_E)$, acting on $\mathcal{H} \otimes \mathcal{H}_E$. In the following theorem the total Hamiltonian of the system and its environment is constructed. For a proof refer to [18].

**Theorem 5** *The total Hamiltonian of the total system is given by the following essentially self-adjoint operator*

$$H_{SE} = -v \cdot \hat{P} \otimes \mathbb{1}_E + \frac{1}{2}\mathbb{1} \otimes \hat{p} \cdot \hat{p} + \sum_{k=1}^{3} c_k \pi_\alpha(\hat{x}_k) \otimes \hat{p}_k,$$

where $c_k > 0$ with $k = 1, 2, 3$, $v = (v_1, v_2, v_3) \in \mathbb{R}^3$, $\hat{x} = (\hat{x}_1, \hat{x}_2, \hat{x}_3)$ is the position operator and $\hat{p} = (\hat{p}_1, \hat{p}_2, \hat{p}_3)$ is the momentum operator in $L^2(\mathbb{R}^3, dm)$. Moreover, $\hat{P} = (\hat{P}_1, \hat{P}_2, \hat{P}_3)$, $\hat{P}_k = \bigoplus_G \hat{p}_k$ and $\pi_\alpha(\hat{x}_k) = \int \lambda \pi_\alpha(dE_k(\lambda))$, where $dE_k$ is the spectral measure of the operator $\hat{x}_k$. The domain of $H$ is given by

$$D(H) = \left\{ \tilde{\xi} \in \mathcal{H} \otimes \mathcal{H}_E : \tilde{\xi}(g) \in \mathcal{S}(\mathbb{R}^3 \times \mathbb{R}^3), \; \tilde{\xi}(g) = 0 \text{ for a. a. } g \in G \right\},$$

where $\mathcal{S}$ is the space of Schwartz test functions vanishing at infinity together with all their derivatives faster than any polynomial.

We choose the state of the environment as $\omega_E = |\psi\rangle\langle\psi|$, where the vector $\psi \in L^2(\mathbb{R}^3, dm)$ is defined by $\psi(r) = \prod_k \psi_0(r_k)$, and where $\psi_0$ is determined by the Fourier transform of (1.9). The reduced dynamics of the system is given by

$$T_t(x) = \Pi^{\omega_E}(e^{iH_{SE}} x \otimes 1_E e^{-iH_{SE}}) \tag{1.10}$$

for all $x \in \mathcal{M}$, which, in fact, is a semigroup [18]. Then the following theorem [18] can be proved.

**Theorem 6** *The algebra of effective observables induced by the semigroup* (1.10) *has the form* $\mathfrak{M} = \pi_\alpha(L^\infty)$, *and for all* $\eta \in \mathcal{M}_*$ *and* $x \in \mathfrak{N}$ *one gets* $\eta(T_t(x)) \to 0$ *when* $t \to \infty$. *Moreover,* $(\mathfrak{M}, R_t) \simeq (L^\infty, \Phi_t)$. *Consequently, the effective dynamics* $\{R_t\}_{t \in \mathbb{R}}$ *leads to the flow* $\Phi$ *on* $\mathbb{R}^3$ *which is given by*

$$\Phi_t(r) = r + tv, \quad \text{for all } r \in \mathbb{R}^3.$$

*This flow describes a uniform motion on the Euclidean space* $\mathbb{R}^3$.

With this example we finish the review of basic models leading to an effective dynamical system which has a structure as in classical mechanics.

### 1.5.3 Spin System Coupled to a Phonon Bath

The previous two examples show that in the algebraic framework it is possible to arrive at a consistent description of classical aspects by decoherence in a very natural way, including Newtonian dynamics, starting from the rules of quantum mechanics.

The next model shows that it is possible to obtain an effective dynamical system $(\mathfrak{M}, R_t)$ having again a pure quantum character [6]. To this end we consider the same spin algebra $\mathcal{M} = \pi_{\omega_0}(\mathcal{A})''$ as above, where $\mathcal{A}$ is the Glimm algebra. This spin system is now coupled to the free phonon bath of a one dimensional harmonic crystal at nonzero temperature $T$ (we write $\beta = \frac{1}{kT}$ for the inverse temperature, with $k$ the Boltzmann constant [10]). The Hilbert space of a spin-zero phonon is the space $\mathcal{H}_f = L^2(\mathbb{R}, dm)$. The Hilbert space describing the whole crystal is given by $\mathcal{F} \otimes \mathcal{F}$, where $\mathcal{F}$ is the boson Fock space over the single particle space $\mathcal{H}_f$ [22].

The phonon quantum field $\phi$ is given by the formula $\phi(f) = \frac{1}{\sqrt{2}}(a(f) + a^*(f))$, with $f \in \mathcal{H}_f$. Here $a(f)$ and $a^*(f)$ are the annihilation and creation operators in the Araki–Woods representation [2], which are given by

$$a(f) = a_F\big((1 + \rho)^{\frac{1}{2}} f\big) \otimes \mathbb{1} + \mathbb{1} \otimes a_F^*\big(\rho^{\frac{1}{2}} \bar{f}\big)$$
$$a^*(f) = a_F^*\big((1 + \rho)^{\frac{1}{2}} f\big) \otimes \mathbb{1} + \mathbb{1} \otimes a_F\big(\rho^{\frac{1}{2}} \bar{f}\big)$$

and where $a_F$, $a_F^*$ are annihilation and creation operators in Fock space [22] (at zero temperature). Let $\rho$ be the Planck distribution

$$\rho(k) = \frac{1}{e^{\beta \omega(k)} - 1}$$

with dispersion relation $\omega(k) = |k|$. The Hamiltonian describing the system of noninteracting phonons is given by $H_E = H_0 \otimes \mathbb{1} + \mathbb{1} \otimes H_0$ with $H_0 = \int \omega(k) a_F^*(k) a_F(k) \, dk$. The representation of the CCR algebra (cf. also Sect. 1.5.4) corresponds to the state $\omega_E$ given by

$$\omega_E(a^*(f) a(g)) = \int \rho(k) \bar{g}(k) f(k) \, dk.$$

The state $\omega_E$ is the reference state of the environment which we will use below to determine the dynamics of the spin system. The dynamics of the total system is now given by the Hamiltonian

$$H = \pi_{\omega_0}(H_S^0) \otimes \mathbb{1}_E + \mathbb{1}_S \otimes H_E + \lambda \sum_{l=1}^{\infty} \pi(\sigma_l^1) \otimes \phi(f_l),$$

where $H_S^0 = \sum_{k=1}^{\infty} h_l \sigma_l^3$. The matrix $\sigma_l^1$ is the Pauli matrix $\begin{pmatrix} 0 & 1 \\ 1 & 0 \end{pmatrix}$ corresponding to the $l$-th spin of the lattice. For the sake of simplicity we consider the coupling of the spin system with only a single mode of the phonon field, i.e., $f_n = a_n g$, where $a_n$ is a sequence such that $a_n \geq \sum_{l=n+1}^{\infty} a_l$ for $n = 1, 2, \ldots$, for instance we can take $a_n \sim \frac{1}{2^n}$. Under some technical conditions concerning the field mode $g$ (the interested reader is referred to [5, 6]) one can derive the following dynamics of the spin system, making use of a singular coupling limit [1]: The reduced dynamics on $\mathcal{M}$ is approximated by a semigroup $\{T_t\}_{t \geq 0}$ with generator

$$L(x) = i[\pi_{\omega_0}(H_S^0) + b\pi_{\omega_0}(A)^2, x] + \lambda a \pi_{\omega_0}(A) x \pi_{\omega_0}(A) - \frac{\lambda a}{2}\{\pi_{\omega_0}(A)^2, x\},$$

i.e. $T_t = e^{tL}$, where $A = \sum_{l=1}^{\infty} a_l \sigma_l^1$, and where the constants $a > 0$ and $b \in \mathbb{R}$ depend on the environment parameters. The bracket $\{\cdot, \cdot\}$ above denotes the anti-commutator. Below we present some long-time analysis of the semigroup generated by this operator. Note that the standard approach [9] based on the analysis of some prelimit expressions for the generator restricted to local algebras generated by $\mathcal{A}_n$ cannot be used here as the local algebras are not preserved by the generator (infinite tails involving all local algebras are present). Some global analysis is therefore welcome at the beginning.

First, we observe that the algebra of effective observables $\mathfrak{M}$ can be represented in a purely algebraic way by

$$\mathfrak{M} = \bigcap_{l=0}^{\infty} \ker(L_{\mathrm{D}} \circ \delta_{H_{\mathrm{S}}}^l), \qquad (1.11)$$

where $\delta_{H_{\mathrm{S}}}(\cdot) = \mathrm{i}[H_{\mathrm{S}}, \cdot]$, with the effective Hamiltonian of the system $H_{\mathrm{S}} = \pi_{\omega_0}(H_{\mathrm{S}}^0) + b\pi_{\omega_0}(A)^2$. Here $L_{\mathrm{D}}$ is the dissipative part $L_{\mathrm{D}} = L - \delta_{H_{\mathrm{S}}}$. Keeping in mind this remark we can establish the following theorem.

**Theorem 7** *Let the sequences* $\{h_l\}_{l=1,2,\ldots}$ *and* $\{a_l\}_{l=1,2,\ldots}$ *with values in the interval* $(0, +\infty)$ *be chosen such that the following conditions are satisfied:*

$$h_m \geq \sum_{l=m+1}^{\infty} h_l \quad and \quad a_m \geq \sum_{l=m+1}^{\infty} a_l \quad for\ any\ m = 1, 2, \ldots.$$

*Then* $\mathfrak{M} = \mathbb{C}\mathbb{1}$, *i.e., the system is ergodic.*

We sketch the proof of this theorem in the Appendix.

Thus we have discussed in detail the ergodic property of the spin system. More generally, the coupling of the spin system with the harmonic crystal can be chosen in a specific way to generate a nontrivial effective algebra $\mathfrak{M}$ with a nontrivial effective dynamics $\{R_t\}_{t \in \mathbb{R}}$. This can be done quite intuitively, e.g., just by excluding some (not necessarily finite) set of spins from the interaction. This procedure leads to a quite interesting variety of examples. But here we restrict ourselves only to the simplest case, i.e., we put $a_1 = 0$ and thereby exclude just one spin from the interaction. In this case the following theorem can be proven.[3]

**Theorem 8** *Suppose that the assumptions of Theorem 5 are satisfied, except that we now put* $a_1 = 0$. *Then* $\mathfrak{M} \cong \mathcal{A}_1$. *Moreover, the effective dynamics* $\{R_t\}_{t \in \mathbb{R}}$ *is given by*

$$R_t(x) = U_t^* x U_t \quad for\ every\ x \in \mathcal{A}_1,$$

*where* $U_t = \mathrm{e}^{-\mathrm{i}th_1\sigma_3}$, *i.e., the unitary Schrödinger evolution.*

---

[3] We recall that the algebra $\mathcal{A}_1$ is the algebra of all complex $2 \times 2$-matrices.

To save space we do not provide the proof here, instead we refer the interested reader to [6].

### 1.5.4  Dissipative Quantum Dynamical Systems

The construction of semigroups for the evolution of physically interesting open systems is one of the most outstanding problems to be solved. Therefore, this section is devoted to some constructions of semigroups [8] which can be used to obtain decoherence-induced effective dynamical systems [7]. Here we provide some introduction to the subject by studying a simple toy model. Consider the standard Schrödinger representation of the canonical commutation relations (CCR) in $L^2(\mathbb{R}, dm)$, where $dm$ is the Lebesgue measure on the real line $\mathbb{R}$. They can be described by the position and momentum operators $\hat{q}$ and $\hat{p}$ acting on the Hilbert space $L^2(\mathbb{R}, dm)$. The domain of $\hat{q}$ and $\hat{p}$ can be chosen in the standard way, but we consider them on a core which is most convenient for our purpose, namely the Schwartz space $\mathcal{S}(\mathbb{R})$ of smooth complex valued functions decreasing at infinity together with all their derivatives faster than any polynomial. Let $\psi'$ denote the derivative of $\psi$ with respect to $q$. The position and momentum operators are defined by $(\hat{q}\psi)(q) = q\psi(q)$ and $(\hat{p}\psi)(q) = -i\psi'(q)$, respectively, for all $q \in \mathbb{R}$ and $\psi \in \mathcal{S}(\mathbb{R})$. Then $[\hat{q}, \hat{p}] \equiv \hat{q}\hat{p} - \hat{p}\hat{q} = i\mathbb{1}$ on $\mathcal{S}(\mathbb{R})$ (in appropriate units with $\hbar = 1$). We consider the symplectic form $\sigma$ on $\mathbb{R}^2$ given by $\sigma(\xi, \zeta) = \xi_+\zeta_- - \xi_-\zeta_+$, where we write $\xi = (\xi_+, \xi_-) \in \mathbb{R}^2$. Let $\Lambda_+\xi = (\xi_+, 0)$ and $\Lambda_-\xi = (0, \xi_-)$. In the symplectic space $(\mathbb{R}^2, \sigma)$ one can introduce a complex structure $J : \mathbb{R}^2 \longrightarrow \mathbb{R}^2$ by $J(\xi_+, \xi_-) = (-\xi_-, \xi_+)$. Any element $\xi$ can be expressed as $(\xi_+, \xi_-) = \xi_+(1, 0) + \xi_- J(1, 0)$. By noting that $J^2 = -\mathrm{id}_{\mathbb{R}^2}$ the space $\mathbb{R}^2$ can be regarded as the complex plane $\mathbb{C}$ using the identification $(\xi_+, \xi_-) \equiv \xi_+ + i\xi_-$. The symplectic form can then be written as $\sigma(\xi, \zeta) = \mathrm{Im}(\bar{\xi}\zeta)$, the bar denoting complex conjugation. Having the complex structure available one can introduce the Euclidean bilinear form $s(\xi, \zeta) = \sigma(\xi, J\zeta)$ for all $\xi, \zeta \in \mathbb{R}^2$. Recall that a linear map $S : \mathbb{R}^2 \longrightarrow \mathbb{R}^2$ is called *symplectic* if $\sigma(S\xi, S\zeta) = \sigma(\xi, \zeta)$ for all $\xi, \zeta \in \mathbb{R}^2$.

We consider the field operator $\varphi$ in the "zero-dimensional" space, i.e., a linear operator-valued function $\mathbb{R}^2 \ni \xi \mapsto \varphi(\xi)$. It is defined by the relations $\varphi(1, 0) = \hat{q}$ and $\varphi(0, 1) = \hat{p}$. The commutation relation for the field operator reads

$$[\varphi(\xi), \varphi(\zeta)] = i\sigma(\xi, \zeta) \quad \text{for all } \xi, \zeta \in \mathbb{R}^2,$$

which in turn can be represented by the unitary Weyl operators $W(\xi) = e^{i\varphi(\xi)}$, since the operators $\varphi(\xi)$ are essentially self-adjoint on $\mathcal{S}(\mathbb{R})$. Using the Weyl operators the commutation relation can be expressed as

$$W(\xi)W(\zeta) = e^{-i\sigma(\xi,\zeta)}W(\zeta)W(\xi) \quad \text{for all } \xi, \zeta \in \mathbb{R}^2. \tag{1.12}$$

Finally, the Weyl algebra is defined as the C*-subalgebra of $\mathcal{B}(L^2(\mathbb{R}^2, dm))$ that is generated by all Weyl operators $W(\xi)$, $\xi \in \mathbb{R}^2$. We shall denote it by $\mathcal{W}$.

The simplest way to introduce a *-automorphic dynamics on $\mathcal{W}$ is by using a symplectic mapping $S : \mathbb{R}^2 \longrightarrow \mathbb{R}^2$ and the characters $\chi$ of the group $\mathbb{R}^2$ [19]. Let us introduce the notion of a *dynamical character* $\{\chi_t\}_{t \in \mathbb{R}}$ with respect to a generator $\delta$ of some one-parameter group of symplectic transformations: Each $\chi_t$ is a character of the group $\mathbb{R}^2$, and additionally we require them to satisfy the relation

$$\chi_t(\xi)\chi_\tau(e^{t\delta}\xi) = \chi_{t+\tau}(\xi) \quad \text{for all } t, \tau \in \mathbb{R} \text{ and } \xi \in \mathbb{R}^2. \tag{1.13}$$

Then a family of *-automorphisms $\alpha_t : \mathcal{W} \longrightarrow \mathcal{W}$ can be uniquely defined by the formula

$$\alpha_t(W(\xi)) = \chi_t(x)W(e^{t\delta}\xi) \quad \text{for all } \xi \in \mathbb{R}^2.$$

It is easy to show that the family $\{\alpha_t\}_{t \in \mathbb{R}}$ is a one parameter group,

$$\begin{aligned}
\alpha_t \circ \alpha_\tau(W(\xi)) &= \chi_\tau(\xi)\alpha_t(W(e^{\tau\delta}\xi)) \\
&= \chi_\tau(\xi)\chi_t(e^{\tau\delta}\xi)W(e^{t\delta}(e^{\tau\delta}\xi)) \\
&= \chi_{t+\tau}(\xi)W(e^{(t+\tau)\delta}\xi) \\
&= \alpha_{t+\tau}(W(\xi)).
\end{aligned}$$

Instead of analyzing the most general case [7, 8, 19] we consider a particular example where everything can be checked directly without much technical complication. Consider a dynamical character of the form

$$\chi_t(\xi) = e^{-itE\xi}e^{-it^2E\xi_+/2} \quad \text{for all } t \in \mathbb{R} \text{ and } \xi \in \mathbb{R}^2.$$

Here $E$ is a real parameter whose physical meaning will be seen shortly. Next we choose a symplectic evolution as $\delta_+ = J\Lambda_+$ and hence $e^{t\delta_+}\xi = \xi + tJ\Lambda_+\xi$ (we shall also consider $\delta_- = J\Lambda_-$ and the corresponding one-parameter group). A direct inspection shows that the family $\{\chi_t\}_{t \in \mathbb{R}}$ indeed constitutes a dynamical character with respect to $\delta_+$. Note that the evolution is not unitary in the complex Hilbert space $\mathbb{C}$. Let us put

$$\alpha_t(W(\xi)) = e^{-itE\xi}e^{-it^2F\xi_+/2}W(\xi + tJ\Lambda_+\xi) \quad \text{for all } t \in \mathbb{R} \text{ and } \xi \in \mathbb{R}^2. \tag{1.14}$$

As a matter of fact, this dynamics can not only be extended to linear combinations of Weyl operators, but to the whole Weyl algebra $\mathcal{W}$. Moreover, it can be extended to the whole algebra $\mathcal{B}(L^2(\mathbb{R}^2, dm))$ as well. To see this consider the Hamilton operator $H = \frac{1}{2}\varphi^2(0, 1) + E\varphi(1, 0)$, which is essentially self-adjoint on the set $\mathcal{S}(\mathbb{R})$. Its closure has purely continuous spectrum. The physical interpretation is clear: the Hamiltonian describes a charged quantum particle (with unit charge) moving in an external electrostatic field of strength $E$. The time evolution of the field $\varphi$ is easily found to be

$$e^{itH}\varphi(\xi)e^{-itH} = \varphi(\xi) + t(\xi_+\varphi(0,1) - \xi_- E) - \frac{1}{2}\xi_+ Et^2 \quad \text{for all } t \in \mathbb{R}.$$

Hence the evolution of the Weyl operators is spatial, and it leads to a natural extension of the dynamics introduced at the beginning,

$$\alpha_t(x) = e^{itH}xe^{-itH} \quad \text{for all } t \in \mathbb{R} \text{ and } x \in \mathcal{B}(L^2(\mathbb{R}^2, dm)).$$

We use the same symbol $\alpha_t$ for this extended group of automorphisms. So far the time evolution is Hamiltonian. Now we consider a perturbation of this evolution by introducing a modification of the dynamical character $\chi_t$,

$$\Theta_t(\xi) = \chi_t(\xi)\exp\left(-\int_0^t s(\xi, e^{-\tau\delta-}Qe^{\tau\delta+}\xi)\,d\tau\right) \quad \text{for all } t \geq 0 \text{ and } \xi \in \mathbb{R}^2.$$

$$(1.15)$$

Here $Q$ is an arbitrary non-negative matrix on $\mathbb{R}^2$. The family $\{\Theta_t\}_{t\in\mathbb{R}}$ satisfies an equation similar to (1.13),

$$\Theta_t(\xi)\Theta_\tau(e^{t\delta+}\xi) = \Theta_{t+\tau}(\xi) \quad \text{for all } t, \tau \geq 0 \text{ and } \xi \in \mathbb{R}^2,$$

hence the expression

$$T_t(W(\xi)) = \Theta_t(\xi)W(e^{t\delta+}\xi)$$

defines a semigroup. Again it can be extended to the algebra $\mathcal{B}(L^2(\mathbb{R}^2, dm))$. To see this note that the function in (1.15) which perturbs $\chi_t$ is the Fourier transform of some family of probability measures, i.e.,

$$\exp\left(-\int_0^t s(\xi, e^{-\tau\delta-}Qe^{\tau\delta+}\xi)\,d\tau\right) = \int_{\mathbb{R}^2} e^{is(\zeta,\xi)}\,d\mu_t(\zeta),$$

so the canonical commutation relations in Weyl form (1.12) give

$$\int_{\mathbb{R}^2} \alpha_t(W(J\zeta)W(\xi)W(-J\zeta))\,d\mu_t(\zeta) = \Theta_t(\xi)W(e^{t\delta+}\xi).$$

The extension (using the same notation $T_t$) is then given by an appropriate Bochner integral

$$T_t(x) = \int_{\mathbb{R}^2} e^{itH} W(J\zeta) x W(-J\zeta) e^{-itH} \, d\mu_t(\zeta)$$

for all $t \geq 0$ and $x \in \mathcal{B}(L^2(\mathbb{R}, dm))$. To illustrate the dissipative behavior of this dynamical system let us assume for simplicity that

$$Q = \begin{pmatrix} a & -1 \\ 1 & 0 \end{pmatrix} \quad \text{with} \quad a > 0.$$

Then we obtain

$$\Theta_t(\xi) = e^{-itE\xi_-} e^{-it^2 E\xi_+/2} e^{-at\xi_+^2} \quad \text{for all } t \geq 0 \text{ and } \xi \in \mathbb{R}^2.$$

Note that $\lim_{t \to +\infty} \Theta_t(\xi) = 0$ for any $\xi$ with the property that $\xi_+ \neq 0$. If $\xi_+ = 0$ then we see that $\Theta_t(\xi) = e^{-itE\xi_-}$, and moreover $e^{t\delta+}\xi = \xi$. Hence the Weyl operators $W(\xi)$ with $\xi_+ \neq 0$ are suppressed and vanish as time goes to infinity. On the other hand, Weyl operators $W(\xi)$ with $\xi_+ = 0$ evolve in time in a unitary way only by a phase factor, i.e., $W(\xi) \mapsto e^{-itE\xi_-} W(\xi)$ if $\xi_+ = 0$. This behavior can be extended to more general elements of the Weyl algebra $\mathcal{W}$. To this end let us introduce some auxiliary notation. Let $V_-$ be the one-dimensional subspace of $\mathbb{R}^2$ generated by the vector $(0, 1)$. Then we have a decomposition

$$\mathcal{W} = \mathcal{W}(V_-) \oplus \mathcal{W}(\mathbb{R}^2 \backslash V_-)$$

of the Weyl algebra, where $\mathcal{W}(\Omega)$ is the Banach subspace of $\mathcal{W}$ generated by the linear hull of the Weyl operators $W(\xi)$ with $\xi \in \Omega$ for some subset $\Omega \subseteq \mathbb{R}^2$. As $V_-$ is a linear subspace of $\mathbb{R}^2$ the corresponding Banach space $\mathcal{W}(V_-)$ is a C*-subalgebra of $\mathcal{W}$. Moreover, the algebra $\mathcal{W}(V_-)$ is commutative. We conclude that

$$\lim_{t \to +\infty} T_t(x) = 0 \quad \text{for any } x \in \mathcal{W}(\mathbb{R}^2 \backslash V_-)$$

relative to the norm topology of the Weyl algebra. On the other hand,

$$T_t(x) = e^{itH} x e^{-itH} \quad \text{for all } t \geq 0 \text{ and } x \in \mathcal{W}(V_-).$$

This semigroup can be extended in a natural way to a one-parameter group of *-automorphisms $\{R_t\}_{t \in \mathbb{R}}$. This group is spatial, but not inner as the Hamiltonian $H$ is not affiliated with the von Neumann algebra $\mathcal{W}(V_-)''$. Note that this effective dynamics $\{R_t\}_{t \in \mathbb{R}}$ on $\mathcal{W}(V_-)$ has got a classical counterpart. This can be seen if one uses the momentum representation, i.e., the Fourier image of the position representation. The algebra $\mathcal{W}(V_-)$ can be identified with the algebra $\mathcal{AC}(\mathbb{R})$ of all almost periodic continuous functions on $\mathbb{R}$. The action of[4] $R_t$ can be expressed as $(R_t x)(p) = x(p - tE)$ for all $p \in \mathbb{R}$, $t \in \mathbb{R}$ and $x \in \mathcal{AC}(\mathbb{R})$. Again, the effective dynamics is given by a classical evolution on $\mathbb{R}$, i.e., $\Phi_t(p) = p - tE$ for all $p \in \mathbb{R}$ and $t \in \mathbb{R}$. The parameter $-E$ corresponds to the rate of change of the "observed" momentum $p$, i.e., it corresponds to the force, in full agreement with the interpretation of the parameter $-E$ as the field strength of an electrostatic field, which indeed is equal to the electrostatic force acting on the (positive) unit charge.

It is interesting to look at the predual semigroup $\{T_{t*}\}_{t \geq 0}$. It describes the wave packet reduction in the momentum representation in a similar way as in [13], i.e., one obtains the evolution of the density matrices in the momentum representation

$$T_{t*}(\rho(p,q)) = e^{-at(p-q)^2} \rho(\Phi_t(p), \Phi_t(q)).$$

The novel aspect in comparison to [13] is the presence of the nontrivial classical evolution $\Phi_t$ in addition to the wave packet reduction. This model, as well a model given below, can be generalized to von Neumann algebras of type III [8]. In particular, one can consider the Araki–Woods nonzero temperature representation [2] as a starting point for this generalized construction and analysis.

If one chooses a different $Q$, e.g.,

$$Q = \begin{pmatrix} 0 & -1 \\ 1 & a \end{pmatrix} \quad \text{with} \quad a > 0,$$

then the dynamics $\{T_t\}_{t \geq 0}$ is ergodic, i.e.,

$$\lim_{t \to +\infty} T_t(x) = 0 \quad \text{for any } x \in \mathcal{W}(\mathbb{R}^2 \setminus \{0\})$$

relative to the norm topology of the Weyl algebra. Moreover, $T_t(\mathbb{1}) = \mathbb{1}$ for all $t \geq 0$.

Using the unitary equivalence of the operators $\hat{q}$ and $\hat{p}$, one can repeat the analysis discussed above for a Hamiltonian of the type $H = \varphi(0, 1) + \frac{1}{2}\varphi^2(1, 0)$ on the domain $\mathcal{S}(\mathbb{R})$. This Hamiltonian, in some sense, is similar to the one Bell used in his discussion of the wave packet reduction [3]. Again, one either arrives at a uniform motion on $\mathbb{R}$, namely, $\Phi_t(q) = q - t$ for all $q \in \mathbb{R}$ and $t \in \mathbb{R}$, or one obtains ergodic behavior by an appropriate choice of the matrix $Q$.

---

[4]We mean the image of $R_t$ under the identification of $\mathcal{W}(V_-)$ and $\mathcal{AC}(\mathbb{R})$.

## Appendix

*Sketch of Proof of Theorem 7* The proof is organized by successively proving the following three properties.

**P1**   $x \in L^\infty(\pi_{\omega_0}(A))$ if and only if $[\pi_{\omega_0}(A), [\pi_{\omega_0}(A), x]] = 0$.
**P2**   $L^\infty(\pi_{\omega_0}(A)) \cap L^\infty(\pi_{\omega_0}(H_S^0)) = \mathbb{C}\mathbb{1}$.
**P3**   $\mathfrak{M} = \mathbb{C}\mathbb{1}$ if and only if $\ker L_D \cap L^\infty(H_S)' = \mathbb{C}\mathbb{1}$.

Given these properties the proof is easily carried through. If $x \in \ker L_D \cap L^\infty(H_S)'$ then $[\pi_{\omega_0}(A), [\pi_{\omega_0}(A), x]] = 0$ and $[H_S, x] = 0$. Consequently, by **P1** we get $x \in L^\infty(\pi_{\omega_0}(A))$, and hence $[\pi_{\omega_0}(H_S^0), x] = [H_S, x] = 0$, i.e., $x \in L^\infty(\pi_{\omega_0}(A)) \cap L^\infty(\pi_{\omega_0}(H_S^0)) = \mathbb{C}\mathbb{1}$ according to **P2**. In this way we obtained $\ker L_D \cap L^\infty(H_S)' = \mathbb{C}\mathbb{1}$, i.e., $\mathfrak{M} = \mathbb{C}\mathbb{1}$ according to **P3**. To complete the proof we therefore need to establish properties **P1**, **P2** and **P3**.

*Proof of P1:*   We start by proving the direction "$\Leftarrow$". Let us define the derivation $\delta_x(\cdot) = i[\cdot, x]$. If $[\pi_{\omega_0}(A), [\pi_{\omega_0}(A), x]] = 0$ then $[\pi_{\omega_0}(A), x] \in L^\infty(\pi_{\omega_0}(A))$ as $L^\infty(\pi_{\omega_0}(A))$ is m. a. s. a. (see the proof of **P2** below). Let $P$ be any polynomial, then

$$\delta_x(P(\pi_{\omega_0}(A))) = i[P(\pi_{\omega_0}(A)), x] = i[\pi_{\omega_0}(A), xP'(\pi_{\omega_0}(A))] \in L^\infty(\pi_{\omega_0}(A)).$$

This means that $\delta_x(L^\infty(\pi_{\omega_0}(A))) \subseteq L^\infty(\pi_{\omega_0}(A))$ since $\delta_x$ is continuous in the weak operator topology. But $L^\infty(\pi_{\omega_0}(A))$ is abelian, thus $\delta_x(E) = 0$ for any projector $E$ from the domain of the derivation $\delta_x$. On the other hand, $\delta_x$ is defined on the whole algebra and in particular $L^\infty(\pi(A))$ is contained in its domain. Hence

$$\delta_x \upharpoonright_{L^\infty(\pi_{\omega_0}(A))} = 0.$$

In particular, $[\pi_{\omega_0}(A), x] = -i\delta_x(\pi_{\omega_0}(A)) = 0$. But $L^\infty(\pi_{\omega_0}(A))$ is a m. a. s. a., hence $x \in L^\infty(\pi_{\omega_0}(A))$. The proof of the converse is obvious.
*Proof of P2:*   Let $C_3 \subseteq A$ be C*-algebra generated by the set

$$\{\sigma_{i_1} \otimes \cdots \otimes \sigma_{i_n} \otimes \mathbb{1} \otimes \cdots, \ i_k = 0, 3 \text{ for } k = 1, \ldots, n, \ n = 1, 2, \ldots\}.$$

Then $\pi_{\omega_0}(C_3)''$ is a m. a. s. a. If we substitute $\sigma_3$ by $\sigma_1$ we can define $C_1$ in a similar fashion and get another m. a. s. a. $\pi_{\omega_0}(C_1)''$. Evidently, $\pi_{\omega_0}(C_3)'' \cap \pi_{\omega_0}(C_1)'' = \mathbb{C}\mathbb{1}$. Now the choice of the sequences $\{h_l\}_{l=1,2,\ldots}$ and $\{a_l\}_{l=1,2,\ldots}$ in the statement of the Theorem ensures that $L^\infty(\pi_{\omega_0}(H_S^0)) = \pi_{\omega_0}(C_3)''$, and $L^\infty(\pi_{\omega_0}(A)) = \pi_{\omega_0}(C_1)''$ as is proven below. Let us introduce some notation. As the Cantor set $C$ is homeomorphic to $\{0, 1\}^{\times \infty}$ we shall use the representation of elements $\omega \in C$ by

$$\omega \equiv \{i_1, i_2, \ldots\} \quad \text{with } i_1, i_2, \ldots \in \{0, 1\}.$$

We say that two elements $\omega_0$ and $\omega_1$ constitute a pair of adjoint points if

$$\omega_0 = \{i_1, i_2, \ldots, i_m, 0, 1, 1, 1, \ldots\} \quad \text{and}$$

$$\omega_1 = \{i_1, i_2, \ldots, i_m, 1, 0, 0, 0, \ldots\}$$

for some non-negative integer $m$. Let $C_0$ be the set of pairs of adjoint points. Moreover, let $\Phi : \pi_{\omega_0}(C_3) \longrightarrow C(\mathcal{C})$ denote the Gelfand–Naimark isomorphism, where $C(\mathcal{C})$ is the set of all continuous functions defined on Cantor set. Let us consider the unique extension of $\Phi$ to the normal isomorphism $\Psi : \pi_{\omega_0}(C_3)'' \longrightarrow L^\infty(\mathcal{C}, \mu)$, where $\mu = \prod \mu_0$ with the measure $\mu_0$ on $\{0, 1\}$ defined by $\mu_0(\{0\}) = \mu_0(\{1\}) = \frac{1}{2}$.

Finally, $C_0(\mathcal{C})$ is the set of all continuous functions $f$ such that $f(\omega_0) = f(\omega_1)$ for some pair of adjoint points $\omega_0$ and $\omega_1$. Let $h = \Phi(\pi_{\omega_0}(H_S^0))$, then

$$h(\{i_1, i_2, \ldots\}) = \sum_{l=1}^{\infty} (-1)^{i_l} h_l.$$

We can evaluate the difference at the pair of adjoint points to obtain

$$h(\omega_0) - h(\omega_1) = 2\left(h_{m+1} - \sum_{l=m+2}^{\infty} h_l\right) \geq 0,$$

so the function $h$ takes different values to different points, except perhaps a pair of adjoint points of $C_0$. Repeating the standard argument we get $C_0(\mathcal{C}) \subseteq \Phi(C^*(\mathbb{1}, \pi_{\omega_0}(H_S^0)))$, where $C^*(D)$ denotes the smallest C*-algebra generated by the set $D$.

Let $P_{i_1 i_2 \cdots i_m} = \pi_{\omega_0}(P_{i_1} \otimes P_{i_2} \otimes \cdots \otimes P_{i_m} \otimes \mathbb{1} \otimes \cdots)$ be one of the generating projectors of $\pi(C_3)$. Then

$$P_{i_1 i_2 \cdots i_m} \in L^\infty(\pi_{\omega_0}(H_S^0)) \tag{1.16}$$

for any $i_1, i_2, \ldots, i_m \in \{0, 1\}$, and $m = 1, 2, \ldots$. Consequently, $\pi_{\omega_0}(C_3) \subseteq L^\infty(\pi_{\omega_0}(H_S^0))$ and hence $\pi_{\omega_0}(C_3)'' \subseteq L^\infty(\pi_{\omega_0}(H_S^0))$. We give the proof of (1.16) only in the special case of the projector $P_0$ ($m = 1, i_1 = 0$) to avoid the complex notation of the general case. Let $\{f_n\}$ be the sequence in $C_0(\mathcal{C})$ given by

$$f(\{i_1, i_2, \ldots\}) = \begin{cases} 1 & : i_1 = 0 \\ \dfrac{1}{2} + \displaystyle\sum_{l=1}^{\infty} (-1)^{i_n+l} \dfrac{1}{2^{l+1}} & : i_1 = 1, \ \displaystyle\sum_{l=2}^{n} i_l = 0 \\ 0 & : i_1 = 1, \ \displaystyle\sum_{l=2}^{n} i_l \neq 0 \end{cases},$$

where $0 \leq f_{n+1} \leq f_n \leq 1$ for $n = 1, 2, \ldots$. As $f_n \in C_0(\mathcal{C})$, there exists $F_n \in C^*(\mathbb{1}, \pi_{\omega_0}(H_S^0))$ such that $f_n = \Phi(F_n)$ and moreover $0 \leq F_{n+1} \leq F_n \leq \mathbb{1}$ for all $n = 1, 2, \ldots$. Hence $F_n \to F$ in the strong operator topology. The map $\Psi$ is normal, hence $1 - \Psi(F) = \sup(1 - f_n) = 1 - \Psi(P_0)$, where we have used $\mu(\{1, 0, 0, \ldots\}) = 0$. Consequently, $P_0 = F \in L^\infty(\pi_{\omega_0}(H_S^0))$. This ends the proof of **P2**.

*Proof of P3:*   We start by proving the direction "$\Leftarrow$". Using (1.11) we deduce that $\delta_{H_S}(\mathfrak{M}) \subseteq \mathfrak{M}$, so we can introduce another derivation defined by

$$\delta_1 \equiv \delta_{H_S} \upharpoonright \mathfrak{M} .$$

The derivation $\delta_1$ is inner, i.e., $\delta_1(\cdot) = \mathrm{i}[H_1, \cdot]$ for some Hermitian operator $H_1 \in \mathfrak{M}$ (see, e.g., [24]). In particular, for each spectral projector $P \in L^\infty(H_1)$ we get $P \in \ker L_D$. On the other hand, $[H_S, P] = -\mathrm{i}\delta_1(P) = [H_1, P] = 0$, hence $P \in L^\infty(H_S)'$. Summarizing, we have $P \in \ker L_D \cap L^\infty(H_S)' = \mathbb{C}\mathbb{1}$. This means that $H_1 = \lambda\mathbb{1}$, and consequently $[H_S, x] = -\mathrm{i}\delta_1(x) = [\lambda\mathbb{1}, x] = 0$ for any $x \in \mathfrak{M}$. We conclude that $\mathfrak{M} \subseteq \ker L_D \cap L^\infty(H_S)' = \mathbb{C}\mathbb{1}$.

For the direction "$\Rightarrow$" notice that for a projector $P \in L^\infty(H_S)'$ we have $[H_S, P] = 0$ and hence $L_D \circ \delta_{H_S}^m = 0$ for any $m = 1, 2, \ldots$. This means that $P \in \mathfrak{M}$ if and only if $P \in \ker L_D$. Consequently, if $P \in \ker L_D \cap L^\infty(H_S)'$ then $P \in \mathfrak{M} = \mathbb{C}\mathbb{1}$. This in turn means that $P = \mathbb{1}$ or $P = 0$. Hence $\ker L_D \cap L^\infty(H_S)' = \mathbb{C}\mathbb{1}$.

$\square$

# References

1. Alicki, R., Lendi, K.: Quantum Dynamical Semigroups and Applications. Lecture Notes in Physics, vol. 717, 2nd edn. Springer, Berlin (2007)
2. Araki, H., Woods, E.J.: Representations of the canonical commutation relations describing a nonrelativistic infinite free Bose gas. J. Math. Phys. **4**, 637–662 (1963)
3. Bell, J.S.: On wave packet reduction in the Coleman–Hepp model. Helv. Phys. Acta **48**, 93–98 (1975)
4. Blanchard, Ph., Olkiewicz, R.: Decoherence induced transition from quantum to classical dynamics. Rev. Math. Phys. **15**, 217–243 (2003)
5. Blanchard, Ph., Olkiewicz, R.: Decoherence induced continuous pointer states. Phys. Rev. Lett. **90**, 010403 (2003)
6. Blanchard, Ph., Ługiewicz, P., Olkiewicz, R.: From quantum to quantum via decoherence. Phys. Lett. A **314**, 29–36 (2003)
7. Blanchard, Ph., Hellmich, M., Ługiewicz, P., Olkiewicz, R. Quantum dynamical semigroups for finite and infinite Bose systems. J. Math. Phys. **48**, 012106 (2007)
8. Blanchard, Ph., Hellmich, M., Ługiewicz, P., Olkiewicz, R.: Continuity and generators of dynamical semigroups for infinite Bose systems. J. Funct. Anal. **256**, 1453–1475 (2009); Corrigendum: J. Funct. Anal. **259**, 2455–2456 (2010)
9. Bratteli, O., Robinson, D.W.: Operator Algebras and Quantum Statistical Mechanics, vol. 2, 2nd edn. Springer, New York (1997)
10. Huang, K.: Statistical Mechanics, 2nd edn. Wiley, New York (1987)

11. Jaffe, A.: Constructive quantum field theory. In: Mathematical Physics 2000, p. 111. Imperial College Press, London (2000)
12. Joos, E.: Decoherence through interaction with the environment. In: Giulini, D., et al. (eds.) Decoherence and the Appearance of a Classical World in Quantum Theory, 2nd edn. Springer, Berlin (2003)
13. Joos, E., Zeh, H.D.: The emergence of classical properties through interaction with the environment. Z. Phys. B: Condens. Matter 59, 223–243 (1985)
14. Kadison, R.V., Ringrose, J.R.: Fundamentals of the Theory of Operator Algebras: Advanced Theory, 2nd print. American Mathematical Society, Providence (2002)
15. Landsman, N.P.: Quantization and superselection sectors I. Transformation group C*-algebras. Rev. Math. Phys. 2, 45–72 (1990)
16. Leader, E., Predazzi, E.: An Introduction to Gauge Theories and the New Physics. Cambridge University Press, Cambridge (1982)
17. Ługiewicz, P., Olkiewicz, R.: Decoherence in infinite quantum systems. J. Phys. A: Math. Gen. 35, 6695–6712 (2002)
18. Ługiewicz, P., Olkiewicz, R.: Classical properties of infinite quantum open systems. Commun. Math. Phys. 239, 241–259 (2003)
19. Manuceau, J.: C*-algèbre de relations de commutation. Ann. Inst. Henri Poincare 2, 139–161 (1968)
20. Olkiewicz, R.: Environment-induced superselection rules in Markovian regime. Commun. Math. Phys. 208, 245–265 (1999)
21. Paz, J.P., Zurek, W.H.: Environment-induced decoherence, classicality and consistency of quantum histories. Phys. Rev. D 48, 2728–2738 (1993)
22. Reed, M., Simon, B.: Methods of Modern Mathematical Physics: Fourier Analysis, Self-Adjointness, vol. 2. Academic, New York (1975)
23. Sahlmann, H.: Loop quantum gravity—a short review. arXiv:1001.4188v3 [gr-qc]. http://arxiv.org/abs/1001.4188 (2011)
24. Sakai, S.: Operator Algebras in Dynamical Systems. Cambridge University Press, Cambridge (1991)
25. Schlosshauer, M.: Decoherence and the Quantum-to-Classical Transition. Springer, Berlin (2007)
26. Schrödinger, E.: Die gegenwärtige Situation in der Quantenmechanik. Die Naturwissenschaften 23, 807–812, 823–828, 844–849 (1935)
27. Wick, G.C., Wightmann, A.S., Wigner, E.P.: The intrinsic parity of elementary particles. Phys. Rev. 88, 101–105 (1952)
28. Zeh, H.D.: Physik ohne Realität: Tiefsinn oder Wahnsinn? Springer, Berlin (2012)
29. Zurek, W.: Pointer basis of quantum apparatus: into what mixture does the wave packet collapse? Phys. Rev. D 24, 1516 (1981)
30. Zurek, W.: Environment induced superselection rules. Phys. Rev. D 26, 1862 (1982)
31. Zurek, W.H.: Decoherence and the transition from quantum to classical revisited. In: Quantum Decoherence, Poincaré Seminar 2005. Progress in Mathematical Physics, vol. 48. Birkhäuser, Basel (2007)

# Chapter 2
# Quantum Systems and Resolvent Algebras

Detlev Buchholz and Hendrik Grundling

## 2.1 Introduction

The conceptual backbone for the modeling of the kinematics of quantum systems is the Heisenberg commutation relations which have found their mathematical expression in various guises. There is an extensive literature analyzing their properties, starting with the seminal paper of Born, Jordan and Heisenberg on the physical foundations and reaching a first mathematical satisfactory formulation in the works of von Neumann and of Weyl.

These canonical systems of operators may all be presented in the following general form: there is a real (finite or infinite dimensional) vector space $X$ equipped with a non-degenerate symplectic form $\sigma : X \times X \to \mathbb{R}$ and a linear map $\phi$ from $X$ onto the generators of a polynomial *-algebra $\mathcal{P}(X, \sigma)$ of operators satisfying the canonical commutation relations

$$[\phi(f), \phi(g)] = i\sigma(f, g)\, \mathbf{1}, \quad \phi(f)^* = \phi(f).$$

In the case that $X$ is finite dimensional, one can reinterpret this relation in terms of the familiar quantum mechanical position and momentum operators, and if $X$ consists of Schwartz functions on some manifold one may consider $\phi$ to be a bosonic quantum field. As is well-known, the operators $\phi(f)$ cannot all be bounded. Moreover, the algebra $\mathcal{P}(X, \sigma)$ does not admit much interesting dynamics acting on it by automorphisms; in fact there are in general only transformations induced

---

D. Buchholz (✉)
Institut für Theoretische Physik, Universität Göttingen, 37077 Göttingen, Germany
e-mail: buchholz@theorie.physik.uni-goettingen.de

H. Grundling
Department of Mathematics, University of New South Wales, Sydney, NSW 2052, Australia

© Springer-Verlag Berlin Heidelberg 2015
P. Blanchard, J. Fröhlich (eds.), *The Message of Quantum Science*, Lecture Notes in Physics 899, DOI 10.1007/978-3-662-46422-9_2

by polynomial Hamiltonians which leave it invariant [7]. Thus $\mathcal{P}(X, \sigma)$ is not a convenient kinematical algebra in either respect.

The inconveniences of unbounded operators can be evaded by expressing the basic commutation relations in terms of bounded functions of the generators $\phi(f)$. In the approach introduced by Weyl, this is done by considering the C*-algebra generated by the set of unitaries $W(f) \doteq \exp(i\phi(f))$, $f \in X$ (the Weyl operators) satisfying the Weyl relations

$$W(f)W(g) = e^{-i\sigma(f,g)/2} W(f+g), \quad W(f)^* = W(-f).$$

This is the familiar Weyl (or CCR) algebra $\mathcal{W}(X, \sigma)$. Yet this algebra still suffers from the fact that its automorphism group does not contain physically significant dynamics [9]. This deficiency can be traced back to the fact that the Weyl algebra is simple, whereas any unital C*-algebra admitting an expedient variety of dynamics must have ideals [4, Sec. 10], cf. also the conclusions.

For finite systems this problem can be solved by proceeding to the twisted group algebra [10] derived from the unitaries $W(f)$, $f \in X$. By the Stone–von Neumann theorem this algebra is isomorphic to $\mathcal{K}(\mathcal{H})$, the compact operators on a separable Hilbert space, for any finite dimensional $X$. This step solves the problem of dynamics for finite systems, but it cannot be applied as such to infinite systems since there $X$ is not locally compact. Moreover, one pays the price that the original operators, having continuous spectrum, are not affiliated with $\mathcal{K}(\mathcal{H})$. So one forgets the specific properties of the underlying quantum system.

This unsatisfactory situation motivated the formulation of an alternative version of the C*-algebra of canonical commutation relations, given in [4]. Here one considers the C*-algebra generated by the resolvents of the basic canonical operators which are formally given by $R(\lambda, f) \doteq (i\lambda \mathbf{1} - \phi(f))^{-1}$ for $\lambda \in \mathbb{R} \backslash \{0\}$, $f \in X$. All algebraic properties of the operators $\phi(f)$ can be expressed in terms of polynomial relations amongst these resolvents. Hence, in analogy to the Weyl algebra generated by the exponentials, one can abstractly define a unital C*-algebra $\mathcal{R}(X, \sigma)$ generated by the resolvents, called the resolvent algebra.

In accordance with the requirement of admitting sufficient dynamics the resolvent algebras have ideals. Their ideal structure was recently clarified in [1], where it was shown that it depends sensitively on the size of the underlying quantum system. More precisely, the specific nesting of the primitive ideals encodes information about the dimension of the underlying space $X$. This dimension, if it is finite, is an algebraic invariant which labels the isomorphism classes of the resolvent algebras. Moreover, the primitive ideals are in one-to-one correspondence to the spectrum (dual) of the respective algebra, akin to the case of commutative algebras. The resolvent algebras are postliminal (type I) if the dimension of $X$ is finite and they are still nuclear if $X$ is infinite dimensional. Thus these algebras not only encode specific information about the underlying systems but also have comfortable mathematical properties.

The resolvent algebras already have proved to be useful in several applications to quantum physics such as the representation theory of abelian Lie algebras of derivations [5], the study of constraint systems and of the BRST method in a C*-algebraic setting [4, 6], the treatment of supersymmetric models on non-compact spacetimes and the rigorous construction of corresponding JLOK-cocycles [3]. Their virtues also came to light in the formulation and analysis of the dynamics of finite and infinite quantum systems [2, 4].

In the present article we give a survey of the basic properties of the resolvent algebras and an outline of recent progress in the construction of dynamics, shedding light on the role of the ideals. The subsequent section contains the formal definition of the resolvent algebras and some comments on their relation to the standard Weyl formulation of the canonical commutation relations. Section 2.3 provides a synopsis of representations of the resolvent algebras and some structural implications and Sect. 2.4 contains the discussion of observables and of dynamics. The article concludes with a brief summary and outlook.

## 2.2  Definitions and Basic Facts

Let $(X, \sigma)$ be a real symplectic space; in order to avoid pathologies we make the standing assumption that $(X, \sigma)$ admits a unitary structure [11]. The pre-resolvent algebra $\mathcal{R}_0(X, \sigma)$ is the universal *-algebra generated by the elements of the set $\{R(\lambda, f) : \lambda \in \mathbb{R} \setminus \{0\}, \ f \in X\}$ satisfying the relations

$$R(\lambda, f) - R(\mu, f) = i(\mu - \lambda)R(\lambda, f)R(\mu, f) \tag{2.1}$$

$$R(\lambda, f)^* = R(-\lambda, f) \tag{2.2}$$

$$[R(\lambda, f), R(\mu, g)] = i\sigma(f, g) R(\lambda, f) R(\mu, g)^2 R(\lambda, f) \tag{2.3}$$

$$\nu R(\nu\lambda, \nu f) = R(\lambda, f) \tag{2.4}$$

$$R(\lambda, f)R(\mu, g) = R(\lambda + \mu, f + g)\big(R(\lambda, f) + R(\mu, g)$$
$$+ i\sigma(f, g)R(\lambda, f)^2 R(\mu, g)\big) \tag{2.5}$$

$$R(\lambda, 0) = -\tfrac{i}{\lambda}\mathbf{1} \tag{2.6}$$

where $\lambda, \mu, \nu \in \mathbb{R} \setminus \{0\}$ and $f, g \in X$, and for (2.5) we require $\lambda + \mu \neq 0$. That is, start with the free unital *-algebra generated by $\{R(\lambda, f) : \lambda \in \mathbb{R} \setminus \{0\}, \ f \in X\}$ and factor out by the ideal generated by the relations (2.1) to (2.6) to obtain the *-algebra $\mathcal{R}_0(X, \sigma)$.

*Remarks* (a) Relations (2.1), (2.2) encode the algebraic properties of the resolvent of some self-adjoint operator, (2.3) amounts to the canonical commutation relations and relations (2.4) to (2.6) correspond to the linearity of the initial map $\phi$ on $X$.

(b) The *-algebra $\mathcal{R}_0(X, \sigma)$ is nontrivial, because it has nontrivial representations. For instance, in a Fock representation $(\pi, \mathcal{H})$ one has self-adjoint operators $\phi_\pi(f)$, $f \in X$ satisfying the canonical commutation relations over $(X, \sigma)$ on a sufficiently big domain in the Hilbert space $\mathcal{H}$ so that one can define $\pi(R(\lambda, f)) \doteq (i\lambda \mathbf{1} - \phi_\pi(f))^{-1}$ to obtain a representation $\pi$ of $\mathcal{R}_0(X, \sigma)$.

It has been shown in [4, Prop. 3.3] that the following definition is meaningful.

**Definition 2.1** Let $(X, \sigma)$ be a symplectic space. The supremum of operator norms with regard to all cyclic *-representations $(\pi, \mathcal{H})$ of $\mathcal{R}_0(X, \sigma)$

$$\|R\| \doteq \sup_{(\pi,\mathcal{H})} \|\pi(R)\|_{\mathcal{H}}, \quad R \in \mathcal{R}_0(X, \sigma)$$

exists and defines a C*-seminorm on $\mathcal{R}_0(X, \sigma)$. The resolvent algebra $\mathcal{R}(X, \sigma)$ is defined as the C*-completion of the quotient algebra $\mathcal{R}_0(X, \sigma)/\ker \|\cdot\|$, where here and in the following the symbol ker denotes the kernel of the respective map.

Of particular interest are representations of the resolvent algebras, such as the Fock representations, where the abstract resolvents characterized by conditions (2.1), (2.2) (sometimes called pseudo-resolvents) are represented by genuine resolvents of self-adjoint operators.

**Definition 2.2** A representation $(\pi, \mathcal{H})$ of $\mathcal{R}(X, \sigma)$ is said to be regular if for each $f \in X$ there exists a densely defined self-adjoint operator $\phi_\pi(f)$ such that one has $\pi(R(\lambda, f)) = (i\lambda \mathbf{1} - \phi_\pi(f))^{-1}$, $\lambda \in \mathbb{R}\backslash\{0\}$. (This is equivalent to the condition that all operators $\pi(R(\lambda, f))$ have trivial kernel.)

The following result characterizing regular representations, cf. [4, Thm. 4.10 and Prop. 4.5], is of importance, both in the structural analysis of the resolvent algebras and in their applications. It implies in particular that the resolvent algebras have faithful irreducible representations (e.g. the Fock representations), so their centers are trivial.

**Proposition 2.3** *Let $(\pi, \mathcal{H})$ be a representation of $\mathcal{R}(X, \sigma)$.*

*(a) If $(\pi, \mathcal{H})$ is regular it is also faithful, i.e. $\|\pi(R)\|_{\mathcal{H}} = \|R\|$ for $R \in \mathcal{R}(X, \sigma)$.*
*(b) If $(\pi, \mathcal{H})$ is faithful and the weak closure of $\pi(\mathcal{R}(X, \sigma))$ is a factor, then $(\pi, \mathcal{H})$ is regular.*

The regular representations of the resolvent algebras are in one-to-one correspondence with the regular representations of the Weyl-algebras, cf. [4, Cor. 4.4]. (Recall that a representation $(\pi, \mathcal{H})$ of $\mathcal{W}(X, \sigma)$ is regular if the maps $\nu \in \mathbb{R} \mapsto \pi(W(\nu f))$ are strong operator continuous for all $f \in X$.) In fact one has the following result.

**Proposition 2.4** *Let $(X, \sigma)$ be a symplectic space and*

*(a) let $(\pi, \mathcal{H})$ be a regular representation of the resolvent algebra $\mathcal{R}(X, \sigma)$ with associated self-adjoint operators $\phi_\pi(f)$ defined above. The exponentials*

$W_\pi(f) \doteq \exp(i\phi_\pi(f))$, $f \in X$ satisfy the Weyl relations and thus define a regular representation of the Weyl algebra $\mathcal{W}(X,\sigma)$ on $\mathcal{H}$;

(b) let $(\pi, \mathcal{H})$ be a regular representation of the Weyl algebra $\mathcal{W}(X,\sigma)$ and let $\phi_\pi(f)$ be the self-adjoint generators of the Weyl operators. The resolvents $R_\pi(\lambda, f) = (i\lambda\mathbf{1} - \phi_\pi(f))^{-1}$ with $\lambda \in \mathbb{R}\backslash\{0\}$, $f \in X$ satisfy relations (2.1) to (2.6) and thus define a regular representation of the resolvent algebra $\mathcal{R}(X,\sigma)$ on $\mathcal{H}$.

Whilst this proposition establishes the existence of a bijection between the regular representations of $\mathcal{R}(X,\sigma)$ and those of $\mathcal{W}(X,\sigma)$, there is no such map between the non-regular representations of the two algebras. In order to substantiate this point consider for fixed nonzero $f \in X$ the two commutative subalgebras $C^*\{R(1, sf) : s \in \mathbb{R}\} \subset \mathcal{R}(X,\sigma)$ and $C^*\{W(sf) : s \in \mathbb{R}\} \subset \mathcal{W}(X,\sigma)$. These algebras are isomorphic respectively to the continuous functions on the one point compactification of $\mathbb{R}$, and the continuous functions on the Bohr compactification of $\mathbb{R}$. Now the point measures on the compactifications having support in the complement of $\mathbb{R}$ produce non-regular states (after extending to the full C*-algebras by Hahn–Banach) and there are many more of these for the Bohr compactification than for the one point compactification of $\mathbb{R}$. Proceeding to the GNS-representations it is apparent that the Weyl algebra has substantially more non-regular representations than the resolvent algebra.

## 2.3 Ideals and Dimension

Further insight into the algebraic properties of the resolvent algebras is obtained by a study of its irreducible representations. In case of finite dimensional symplectic spaces these representations have been completely classified [4, Prop. 4.7], cf. also [1].

**Theorem 3.1** Let $(X, \sigma)$ be a finite dimensional symplectic space and let $(\pi, \mathcal{H})$ be an irreducible representation of $\mathcal{R}(X,\sigma)$. Depending on the representation, the space $X$ decomposes as follows, cf. Fig. 2.1.

(a) There is a unique subspace $X_R \subset X$ such that there are self-adjoint operators $\phi_\pi(f_R)$ satisfying $\pi(R(\lambda, f_R)) = (i\lambda\mathbf{1} - \phi_\pi(f_R))^{-1}$ for $\lambda \in \mathbb{R}\backslash\{0\}$, $f_R \in X_R$.

(b) Let $X_T \doteq \{f \in X_R : \sigma(f,g) = 0 \text{ for all } g \in X_R\}$ Then $\psi_\pi$ restricts on $X_T$ to a linear functional $\varphi : X_T \to \mathbb{R}$ such that $\pi(R(\lambda, f_T)) = (i\lambda - \varphi(f_T))^{-1}\mathbf{1}$ for $f_l \in X_T$, $\lambda \in \mathbb{R}\backslash\{0\}$.

(c) For $f_S \in X_S \doteq X\backslash X_R$ and $\lambda \in \mathbb{R}\backslash\{0\}$ one has $\pi(R(\lambda, f_S)) = 0$.

Conversely, given subspaces $X_T \subset X_R \subset X$ and a linear functional $\varphi : X_T \to \mathbb{R}$ there exists a corresponding irreducible representation $(\pi, \mathcal{H})$ of $\mathcal{R}(X,\sigma)$, unique up to equivalence, with the preceding three properties.

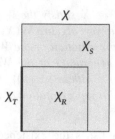

**Fig. 2.1** Decomposition of $X$ fixed by an irreducible representation

This result may be regarded as an extension of the Stone–von Neumann uniqueness theorem for regular representations of the CCR algebra. It shows that the only obstruction to regularity is the possibility that some of the underlying canonical operators are infinite and the corresponding resolvents vanish. This happens in particular if there are some canonically conjugate operators having sharp (non-fluctuating) values in a representation, as is the case for constraint systems [4, Prop. 8.1]. But, in contrast to the Weyl algebras, the non-regular representations of the resolvent algebras only depend on the values of these canonical operators. So the abundance of different singular representations of the Weyl algebras shrink to a manageable family on the resolvent algebras.

The preceding theorem is the key to the structural analysis of the resolvent algebras for symplectic spaces of arbitrary finite dimension. We recall in this context that the primitive ideals of a C*-algebra are the (possibly zero) kernels of irreducible representations and that the spectrum of the algebra is the set of unitary equivalence classes of irreducible representations. The following result has been established in [1].

**Theorem 3.2** *Let $(X, \sigma)$ be a finite dimensional symplectic space.*

(a) *The mapping $\hat{\pi} \mapsto \ker \hat{\pi}$ from the elements $\hat{\pi}$ of the spectrum (dual) of the resolvent algebra $\mathcal{R}(X, \sigma)$ to its primitive ideals $\ker \hat{\pi}$ is a bijection.*
(b) *Let $L \doteq \sup \{l \in \mathbb{N} : \ker \hat{\pi}_1 \subset \ker \hat{\pi}_2 \cdots \subset \ker \hat{\pi}_l\}$ be the maximal length of proper inclusions of primitive ideals of $\mathcal{R}(X, \sigma)$. Then $L = \dim(X)/2 + 1$.*

*Remarks* Property (a) is a remarkable feature of the resolvent algebras, shared with the abelian C*-algebras. It rarely holds for non-commutative algebras and also fails if $X$ is infinite dimensional. The quantity $L$ defined in (b) is an algebraic invariant, so this result shows that the dimension $\dim(X)$ of the underlying systems is algebraically encoded in the resolvent algebras. As a matter of fact, $L$ is a complete algebraic invariant of resolvent algebras in the finite dimensional case.

As indicated above, there is an algebraic difference between the resolvent algebras for finite dimensional $X$ and those where $X$ has infinite dimension. A further difference is seen through the minimal (nonzero) ideals [1].

**Proposition 3.3** *Let $(X, \sigma)$ be a symplectic space of arbitrary dimension and let $\mathcal{I} \subset \mathcal{R}(X, \sigma)$ be the intersection of all nonzero ideals of $\mathcal{R}(X, \sigma)$.*

(a) *If* $\dim(X) < \infty$ *then* $\mathcal{I}$ *is isomorphic to the* $C^*$-*algebra* $\mathcal{K}(\mathcal{H})$ *of compact operators. Moreover, in any irreducible regular representation* $(\pi, \mathcal{H})$ *one has* $\pi(\mathcal{I}) = \mathcal{K}(\mathcal{H})$.

(b) *If* $\dim(X) = \infty$ *then* $\mathcal{I} = \{0\}$. *In fact, there exists no nonzero minimal ideal of* $\mathcal{R}(X, \sigma)$ *in this case.*

If $(X, \sigma)$ is infinite dimensional the resolvent algebra $\mathcal{R}(X, \sigma)$ is the $C^*$-inductive limit of the net of its subalgebras $\mathcal{R}(Y, \sigma)$ where $Y \subset X$ ranges over all finite dimensional non-degenerate subspaces of $X$, cf. [4, Thm. 4.9]. This fact in combination with the first part of the preceding result is a key ingredient in the construction of dynamics, see below. It also enters in the proof of the following statement [1].

**Proposition 3.4** *Let* $(X, \sigma)$ *be a symplectic space of arbitrary dimension.*

(a) $\mathcal{R}(X, \sigma)$ *is a nuclear* $C^*$-*algebra,*
(b) $\mathcal{R}(X, \sigma)$ *is a postliminal (type I)* $C^*$-*algebra if and only if* $\dim(X) < \infty$.

Recall that a $C^*$-algebra is said to be postliminal (type I) if all of its irreducible representations contain the compact operators and that postliminal $C^*$-algebras as well as their $C^*$-inductive limits are nuclear, i.e. their tensor product with any other $C^*$-algebra is unique. It should be noted, however, that the resolvent algebras are not separable [4, Thm. 5.3]. With this remark we conclude our outline of pertinent algebraic properties of the resolvent algebras.

## 2.4 Observables and Dynamics

The main virtue of the resolvent algebras consists of the fact that it includes many observables of physical interest and admits non-trivial dynamics. In order to illustrate this important feature we discuss in detail a familiar example of a finite quantum system and comment on infinite systems at the end of this section.

Let $(X, \sigma)$ be a finite dimensional symplectic space, i.e. $\dim(X) = 2N$ for some $N \in \mathbb{N}$. Since regular representations of the resolvent algebras are faithful, cf. Proposition 2.3, it suffices to consider any regular irreducible representation $(\pi_0, \mathcal{H}_0)$ of $\mathcal{R}(X, \sigma)$ (which is unique up to equivalence). Choosing some symplectic basis $f_k, g_k \in X$ and putting $P_k \doteq \phi_{\pi_0}(f_k)$, $Q_k \doteq \phi_{\pi_0}(g_k)$, $k = 1, \dots N$ we identify the self-adjoint operators fixed by the corresponding resolvents with the momentum and position operators of $N$ particles in one spatial dimension.

The (self-adjoint) quadratic Hamiltonian

$$H_0 \doteq \sum_{k=1}^{N} \left( \frac{1}{2m_k} P_k^2 + \frac{m_k \omega_k^2}{2} Q_k^2 \right)$$

describes the free, respectively oscillatory motion of these particles, where $m_k$ are the particle masses and $\omega_k \geq 0$ the frequencies of oscillation, $k = 1, \ldots N$. The interaction of the particles is described by the operator

$$V \doteq \sum_{1 \leq k < l \leq N} V_{kl}(Q_k - Q_l)$$

where we assume for simplicity that the potentials $V_{kl}$ are real and continuous, vanish at infinity, but are arbitrary otherwise. Since $V$ is bounded, the Hamiltonian $H \doteq H_0 + V$ is self-adjoint on the domain of $H_0$ and its resolvents are well defined.

**Proposition 4.1** *Let $H$ be the Hamiltonian defined above. Then*

$$(i\mu\mathbf{1} - H)^{-1} \in \pi_0(\mathcal{R}(X, \sigma)), \quad \mu \in \mathbb{R} \backslash \{0\}.$$

*Remark* Since $\pi_0$ is faithful its inverse $\pi_0^{-1} : \pi_0(\mathcal{R}(X, \sigma)) \to \mathcal{R}(X, \sigma)$ exists, so this result shows that $H$ is affiliated with the resolvent algebra. Note that this is neither true for the Weyl algebra $\mathcal{W}(X, \sigma)$ nor for the corresponding twisted group algebra $\mathcal{K}(\mathcal{H})$ if one of the frequencies $\omega_k$ vanishes. Thus $\mathcal{R}(X, \sigma)$ contains many more observables of physical interest than these conventional algebras.

*Proof* Let $X_k \subset X$ be the two-dimensional subspaces spanned by the symplectic pairs $(f_k, g_k)$, let $\sigma_k \doteq \sigma \upharpoonright X_k \times X_k$ and let $(\pi_k, \mathcal{H}_k)$ be regular irreducible representations of $\mathcal{R}(X_k, \sigma_k)$, $k = 1, \ldots N$. Then $\pi_0 \doteq \pi_1 \otimes \cdots \otimes \pi_N$ defines an irreducible representation of the C*-tensor product $\mathcal{R}(X_1, \sigma_1) \otimes \cdots \otimes \mathcal{R}(X_N, \sigma_N)$ on the Hilbert space $\mathcal{H}_0 \doteq \mathcal{H}_1 \otimes \cdots \otimes \mathcal{H}_N$. It extends by regularity to the Weyl algebra $\mathcal{W}(X, \sigma) \simeq \mathcal{W}(X_1, \sigma_1) \otimes \cdots \otimes \mathcal{W}(X_N, \sigma_N)$ and hence to a regular representation of $\mathcal{R}(X, \sigma)$, cf. Proposition 2.4.

One has $H_{0k} \doteq (i\mu\mathbf{1} - \frac{1}{2m_k}P_k^2 - \frac{m_k\omega_k^2}{2}Q_k^2)^{-1} \in \pi_k(\mathcal{R}(X_k, \sigma_k))$, $k = 1, \ldots N$, disregarding tensor factors of $\mathbf{1}$. If $\omega_k > 0$ this follows from the fact that the resolvent of the harmonic oscillator Hamiltonian is a compact operator and hence belongs to the compact ideal of $\pi_k(\mathcal{R}(X_k, \sigma_k))$, cf. Proposition 3.3. If $\omega_k = 0$ one resorts to the fact that the abelian C*-algebra generated by the resolvents $(i\lambda\mathbf{1} - P_k)^{-1}$, $\lambda \in \mathbb{R} \backslash \{0\}$ coincides with $C_0(P_k)$, the algebra of all continuous functions of $P_k$ vanishing at infinity. Hence $C_0(P_k) \subset \pi_k(\mathcal{R}(X_k, \sigma_k))$. Since $(i\mu\mathbf{1} - \frac{1}{2m_k}P_k^2)^{-1} \in C_0(P_k)$ the preceding statement holds also for $\omega_k = 0$.

As is well known $C_0(\mathbb{R}_+^N) = \overset{N}{\overbrace{C_0(\mathbb{R}_+) \otimes \cdots \otimes C_0(\mathbb{R}_+)}}$ and it is also clear that $u_1, \ldots, u_N \mapsto (i\mu - u_1 \cdots - u_N)^{-1}$ is an element of $C_0(\mathbb{R}_+^N)$. Since the resolvents of the positive self-adjoint operators $H_{0k}$ generate the abelian C*-algebras $C_0(H_{0k})$, $k = 1, \ldots, N$, it follows from continuous functional calculus that $(i\mu\mathbf{1} - H_0)^{-1} = (i\mu\mathbf{1} - H_{01} \cdots - H_{0N})^{-1} \in C_0(H_{01}) \otimes \cdots \otimes C_0(H_{0N}) \subset \pi_0(\mathcal{R}(X, \sigma))$.

Similarly, for the interaction potentials one uses the fact that the abelian C*-algebras generated by the resolvents $(i\lambda\mathbf{1} - (Q_k - Q_l))^{-1}$, $\lambda \in \mathbb{R} \backslash \{0\}$ coincide with $C_0(Q_k - Q_l)$. So as $V_{kl} \in C_0(\mathbb{R})$, one also has that

$$V = \sum_{1 \leq k < l \leq N} V_{kl}(Q_k - Q_l) \in \pi_0(\mathcal{R}(X, \sigma)).$$

In summary one gets $(1 - (i\mu\mathbf{1} - H_0)^{-1}V) \in \pi_0(\mathcal{R}(X, \sigma))$. Its inverse exists if $|\mu| > \|V\|$ and $(i\mu\mathbf{1} - H)^{-1} = (1 - (i\mu\mathbf{1} - H_0)^{-1}V)^{-1}(i\mu\mathbf{1} - H_0)^{-1} \in \pi_0(\mathcal{R}(X, \sigma))$ for such $\mu$. The statement for arbitrary $\mu \in \mathbb{R} \setminus \{0\}$ then follows from the resolvent equation for $H$, completing the proof.

As a matter of fact, the preceding proposition holds for a much larger class of interaction potentials, including discontinuous ones. It does not hold, however, for certain physically inappropriate Hamiltonians such as that of the anti-harmonic oscillator [4, Prop. 6.3]. The characterization of all Hamiltonians which are affiliated with resolvent algebras is an interesting open problem.

We turn now to the analysis of the dynamics induced by the Hamiltonians given above. The exponentials of the quadratic Hamiltonians $H_0$ induce symplectic transformations, so one has $(\mathrm{Ad}\, e^{itH_0})(\pi_0(\mathcal{R}(X, \sigma))) = \pi_0(\mathcal{R}(X, \sigma))$ for $t \in \mathbb{R}$. For the proof that the resolvent algebra is also stable under the adjoint action of the interacting dynamics the crucial step consists of showing that the cocycles $\Gamma(t) = e^{itH}e^{-itH_0}$ are elements of $\pi_0(\mathcal{R}(X, \sigma))$. Putting $V(t) = (\mathrm{Ad}\, e^{itH_0})(V)$ one can present the cocycles in the familiar form of a Dyson series

$$\Gamma(t) = 1 + \sum_{n=1}^{\infty} i^n \int_0^t dt_1 \int_0^{t_1} dt_2 \ldots \int_0^{t_{n-1}} dt_n\, V(t_n) \cdots V(t_1)$$

and this series converges absolutely in norm since the operators $V(t)$ are uniformly bounded. Moreover, the functions $t \mapsto V(t)$ have values in the algebra $\pi_0(\mathcal{R}(X, \sigma))$; but since they are only continuous in the strong operator topology it is not clear from the outset that their integrals, defined in this topology, are still contained in this algebra. Here again the specific structure of the resolvent algebra matters. It allows to establish the desired result.

**Proposition 4.2** *Let $H$ be the Hamiltonian defined above. Then*

$$(\mathrm{Ad}\, e^{itH})(\pi_0(\mathcal{R}(X, \sigma))) = \pi_0(\mathcal{R}(X, \sigma)), \quad t \in \mathbb{R}.$$

*Remark* Since $\pi_0$ is faithful it follows from this result that $\alpha_t \doteq \pi_0^{-1}(\mathrm{Ad}\, e^{itH})\pi_0$, $t \in \mathbb{R}$ defines a one-parameter group of automorphisms of $\mathcal{R}(X, \sigma)$. It should be noted, however, that its action is not continuous in the strong (pointwise norm) topology of $\mathcal{R}(X, \sigma)$.

*Proof* Let $k, l \in 1, \ldots, N$ be different numbers, let $(f_k, g_k)$ and $(f_l, g_l)$ be symplectic pairs as in the previous proof and let $X_{kl} \subset X$ be the space spanned by $h_{kl}(t) \doteq ((\cos \omega_k t)\, g_k - (\cos \omega_l t)\, g_l + (\sin \omega_k t)/m_k\omega_k\, f_k - (\sin \omega_l t)/m_l\omega_l\, f_l)$, $t \in \mathbb{R}$, where we stipulate $(\sin \omega t)/\omega = t$ if $\omega = 0$. This space is non-degenerate and, depending on the masses and frequencies, either two or four dimensional. We put $\sigma_{kl} \doteq \sigma \upharpoonright X_{kl} \times X_{kl}$. Let $V_{kl}(t) \doteq (\mathrm{Ad}\, e^{itH_0})(V_{kl}(Q_k - Q_l))$, where $V_{kl}(Q_k - Q_l)$ is any one of the two-body potentials contributing to $V$. Then, for any $t \in \mathbb{R}$,

$$V_{kl}(t) = V_{kl}((\cos \omega_k t)Q_k - (\cos \omega_l t)Q_l + (\sin \omega_k t)/m_k \omega_k P_k - (\sin \omega_l t)/m_l \omega_l P_l)$$

$$\in \pi_0(\mathcal{R}(X_{kl}, \sigma_{kl})).$$

Now the function $s_1, \ldots s_d \mapsto V_{kl}(s_1) \cdots V_{kl}(s_d)$ is continuous in the strong operator topology and, for almost all $s_1, \ldots s_d$, an element of the compact ideal of $\pi_0(\mathcal{R}(X_{kl}, \sigma_{kl}))$, provided $d \geq \dim(X_{kl})$. The latter assertion follows from the fact that $V_{kl}(s)$ is, for given $s$, an element of the abelian C*-algebra generated by the resolvents $\pi_0(R(\lambda, h_{kl}(s)))$, $\lambda \in \mathbb{R}\backslash\{0\}$ and that the compact ideal coincides with the principal ideal of $\pi_0(\mathcal{R}(X_{kl}, \sigma_{kl}))$ generated by $\pi_0(R(\lambda_1, h_1) \cdots R(\lambda_d, h_d))$ for any choice of $\lambda_1, \ldots \lambda_d \in \mathbb{R}\backslash\{0\}$ and of elements $h_1, \ldots h_d \in X_{kl}$ which span $X_{kl}$ [2]. It is then clear that $\left(\int_0^t ds\, V_{kl}(s)\right)^d = \int_0^t ds_1 \cdots \int_0^t ds_d\, V_{kl}(s_1) \cdots V_{kl}(s_d)$ is contained in the compact ideal of $\pi_0(\mathcal{R}(X_{kl}, \sigma_{kl}))$. But this is then also true for the operator $\int_0^t ds\, V_{kl}(s)$ since it is self-adjoint. As $k, l$ were arbitrary this implies $\int_0^t dt_1 V(t_1) \in \pi_0(\mathcal{R}(X, \sigma))$.

That all other terms in the Dyson series are elements of $\pi_0(\mathcal{R}(X, \sigma))$ is seen by induction. Let $I_n(t) \doteq \int_0^t dt_1 \int_0^{t_1} dt_2 \ldots \int_0^{t_{n-1}} dt_n\, V(t_n) \cdots V(t_1) \in \pi_0(\mathcal{R}(X, \sigma))$, $t \in \mathbb{R}$; then $I_{n+1}(t) = \int_0^t dt_1 I_n(t_1)V(t_1)$, where the integrals are defined in the strong operator topology. Now $t \mapsto I_n(t)$ is continuous in norm, hence $I_{n+1}(t)$ can be approximated according to

$$I_{n+1}(t) = \lim_{J \to \infty} \sum_{j=1}^{J} I_n(jt/J) \int_{(j-1)t/J}^{jt/J} dt_1 V(t_1),$$

where the limit exists in the norm topology. Since each term in this sum is an element of $\pi_0(\mathcal{R}(X, \sigma))$ according to the induction hypothesis it follows that $I_{n+1}(t) \in \pi_0(\mathcal{R}(X, \sigma))$. Because of the convergence of the Dyson series this implies $\Gamma(t) \in \pi_0(\mathcal{R}(X, \sigma))$, $t \in \mathbb{R}$, completing the proof of the statement.

Having illustrated the virtues of the resolvent algebras for finite systems we discuss now the situation for infinite systems. There the results are far from being complete, though promising. For the sake of concreteness we consider an infinite dimensional symplectic space $(X, \sigma)$ with a countable symplectic basis $f_k, g_k \in X$, $k \in \mathbb{Z}$. Similarly to the case of finite systems one can analyze the observables and dynamics associated with $\mathcal{R}(X, \sigma)$ in any convenient faithful representation $(\pi_0, \mathcal{H}_0)$, such as the Fock representation.

As before, we identify the self-adjoint operators fixed by the resolvents with the momentum and position operators of particles, $P_k \doteq \phi_{\pi_0}(f_k)$, $Q_k \doteq \phi_{\pi_0}(g_k)$, $k \in \mathbb{Z}$. In view of Haag's Theorem [8] it does not come as a surprise that global observables, such as Hamiltonians having a unique ground state or the particle number operator are no longer affiliated with the resolvent algebras of such infinite systems. In fact, one has the following general result [2].

**Lemma 4.3** *Let* $(X, \sigma)$ *be an infinite dimensional symplectic space, let* $(\pi_0, \mathcal{H}_0)$ *be a faithful irreducible representation of* $\mathcal{R}(X, \sigma)$ *and let* $N$ *be a (possibly unbounded) self-adjoint operator on* $\mathcal{H}_0$ *with an isolated eigenvalue of finite multiplicity. Then* $(i\mu \mathbf{1} - N)^{-1} \notin \pi_0(\mathcal{R}(X, \sigma))$ *for* $\mu \in \mathbb{R} \backslash \{0\}$, *i.e.* $N$ *is not affiliated with* $\mathcal{R}(X, \sigma)$.

Observables corresponding to finite subsystems of the infinite system are still affiliated with $\mathcal{R}(X, \sigma)$. Relevant examples are the partial Hamiltonians of the form given above,

$$H_\Lambda \doteq \sum_{k \in \Lambda} \left( \frac{1}{2m_k} P_k^2 + \frac{m_k \omega_k^2}{2} Q_k^2 \right) + \sum_{k,l \in \Lambda} V_{kl}(Q_k - Q_l),$$

where $\Lambda \subset \mathbb{Z}$ is any finite set. By exactly the same arguments as in the proof of Proposition 4.1 one can show that any such $H_\Lambda$ is affiliated with $\mathcal{R}(X, \sigma)$. Clearly, these Hamiltonians may have isolated eigenvalues, but these have infinite multiplicity. By the preceding arguments one can also show that the resolvent algebra is stable under the time evolution induced by the partial Hamiltonians. Moreover, for suitable potentials the evolution converges to some global dynamics in the limit $\Lambda \nearrow \mathbb{Z}$. The precise results are as follows.

**Proposition 4.4** *Let* $H_\Lambda$, $\Lambda \subset \mathbb{Z}$ *be the partial Hamiltonians introduced above, where* $V_{kl}$ *are continuous functions tending to* 0 *at infinity,* $k, l \in \mathbb{Z}$.

*(a) Then* $(\text{Ad } e^{itH_\Lambda})(\pi_0(\mathcal{R}(X, \sigma))) = \pi_0(\mathcal{R}(X, \sigma))$, $t \in \mathbb{R}$.
*(b) Let* $C, D$ *be positive constants such that* $\|V_{kl}\| \leq C$ *and* $V_{kl} = 0$ *for* $|k - l| \geq D$, $k, l \in \mathbb{Z}$. *Then* $\lim_{\Lambda \nearrow \mathbb{Z}} (\text{Ad } e^{itH_\Lambda})$, $t \in \mathbb{R}$ *exists pointwise on* $\pi_0(\mathcal{R}(X, \sigma))$ *in the norm topology.*

A proof of this statement is given in [2]. It generalizes the results on a class of models describing particles which are confined to the points of a one-dimensional lattice by a harmonic pinning potential and interact with their nearest neighbors [4]. In the present more general form it also has applications to other models of physical interest. These results provide evidence to the effect that the resolvent algebras are an expedient framework also for the discussion of the dynamics of infinite systems. Yet a full assessment of their power for the treatment of such systems requires further analysis.

## 2.5 Conclusions

In the present survey we have outlined some recent structural results and instructive applications of the theory of resolvent algebras. These algebras are built from the resolvents of the canonical operators in quantum theory and their algebraic relations encode the basic kinematical features of quantum systems just as well as the Weyl algebras. But, as we have shown, the novel approach cures several shortcomings of this traditional algebraic setting.

The resolvent algebras comply with the condition that kinematical algebras of quantum systems must have ideals if they are to carry various dynamics of physical interest. This requirement can easily be inferred from the preceding arguments in case of a single particle: there the cocycles $\Gamma(t) = e^{itH}e^{-itH_0}$ appearing in the interaction picture have the property that the differences $(\Gamma(t) - 1)$ are compact operators for generic interaction potentials. Hence $(e^{itH}We^{-itH} - e^{itH_0}We^{-itH_0})$ is a compact operator for any choice of bounded operator $W$. It is then clear that any unital C*-algebra which is stable under the action of these dynamics must contain compact operators and consequently have ideals.

The resolvent algebras, respectively their subalgebras corresponding to finite subsystems, contain these ideals from the outset. As we have demonstrated by several physically significant examples, the ideals play a substantial role in the construction of dynamics of finite and infinite quantum systems. For they accommodate the terms in the Dyson expansion of the cocycles resulting from the interaction picture and thereby entail the stability of the resolvent algebras under the action of the perturbed dynamics. In order to cover a wider class of models it would, however, be desirable to invent some more direct argument, avoiding this expansion and the ensuing questions of convergence.

The ideals of the resolvent algebras also play a prominent role in their classification. The nesting of primitive ideals encodes precise information about the size of the underlying quantum system, i.e. its dimension. It is a complete algebraic invariant in the finite dimensional case. There is also a sharp algebraic distinction between finite and infinite quantum systems in terms of their minimal ideals. In either case the resolvent algebras have comfortable algebraic properties: they are nuclear, thereby allowing to form unambiguously tensor products with other algebras which plays a role in the discussion of coupled systems.

In company with the resolvents of the canonical operators all their continuous functions vanishing at infinity are contained in the resolvent algebras. This feature ensures, as we have shown, that many operators of physical interest are affiliated with the resolvent algebras. It also implies that these algebras contain multiplicative mollifiers for unbounded operators which appear in the algebraic treatment of supersymmetric models [3] or of constraint systems [4, 6]. Thus the resolvent algebras provide in many respects a natural and convenient mathematical setting for the discussion of finite and infinite quantum systems.

# References

1. Buchholz, D.: The resolvent algebra: ideals and dimension. J. Funct. Anal. **266**, 3286–3302 (2014)
2. Buchholz, D., Grundling, H.: The resolvent algebra: dynamics of finite and infinite quantum systems (in preparation)

3. Buchholz, D., Grundling, H.: Algebraic supersymmetry: a case study. Commun. Math. Phys. **272**, 699–750 (2007)
4. Buchholz, D., Grundling, H.: The resolvent algebra: a new approach to canonical quantum systems. J. Funct. Anal. **254**, 2725–2779 (2008)
5. Buchholz, D., Grundling, H.: Lie algebras of derivations and resolvent algebras. Commun. Math. Phys. **320**, 455–467 (2013)
6. Costello, P.: The mathematical structure of the quantum BRST constraint method. e-print arXiv:0905.3570 (preprint appeared in 2009)
7. Dixmier, J.: Sur les algébres de Weyl. Bull. Soc. Math. France **96**, 209–242 (1968)
8. Emch, G.G.: Algebraic Methods in Statistical Mechanics and Quantum Field Theory. Wiley, New York (1972)
9. Fannes, M., Verbeure, A.: On the time evolution automorphisms of the CCR-algebra for quantum mechanics. Commun. Math. Phys. **35**, 257–264 (1974)
10. Kastler, D.: The C*-algebras of a free Boson-field. Commun. Math. Phys. **1**, 14–48 (1965)
11. Robinson, P.L.: Symplectic pathology. Q. J. Math. **44**, 101–107 (1993)

# Chapter 3
# What the Philosophical Interpretation of Quantum Theory Can Accomplish

**Martin Carrier**

## 3.1 Introduction: What Philosophy Can Do for Science

Some physicists are anxious to understand quantum theory without being misled by philosophers [11, pp. 111–112]. However, approaching a subject matter with a philosophical attitude tends to make us aware of problems, issues and all sorts of contentious claims that might remain hidden and unconscious otherwise. Philosophy can be good at making issues explicit, while it usually fails to resolve these issues. Plato realized that being amazed or astonished or unsettled, *taumazein*, is the first step toward the gain of knowledge. Speaking more specifically, philosophy of science approaches the sciences in a reflective attitude. It aims at clarification. Philosophy is unable to supply data or to decide between rival theories. It rather aims at sorting out concepts, elucidating relations and illuminating the broader impact a scientific theory might have on understanding the pertinent phenomena. In particular, conceptual, epistemological, and ontological issues constitute primary challenges to philosophical reflection. I will attend to quantum theory in what follows, but as a preliminary, let me briefly address the similar case of space-time theory.

Philosophical reflection on space and time represents an earlier version of this reflecting attitude that also characterizes quantum philosophy. Thus, the former can serve as a role model of the latter in methodological respect. As regards *conceptual analysis*, a major challenge has been to reconstruct clearly in which sense general relativity theory has abandoned, or still retains, an absolute understanding of spatiotemporal properties. Such a clarification can be achieved by introducing suitable counterconcepts. An absolute space-time property, in contrast

M. Carrier (✉)
Bielefeld University, Bielefeld, Germany
e-mail: martin.carrier@uni-bielefeld.de

© Springer-Verlag Berlin Heidelberg 2015
P. Blanchard, J. Fröhlich (eds.), *The Message of Quantum Science*, Lecture Notes
in Physics 899, DOI 10.1007/978-3-662-46422-9_3

to a relational one, can stand alone and be couched without reference to what happens in space-time. An absolute space-time property, in contradistinction to a relative one, is independent of the frame of reference chosen. An absolute space-time property, in opposition to a dynamical one, is independent of processes in space-time [14, pp. 62–64]. Such counterconcepts serve to distinguish between various aspects that remain interwoven and undifferentiated in the original notion. Introducing distinctions of this sort contributes to illuminating the meaning of spatiotemporal notions, as understood in general relativity.

A second dimension of philosophical reflection concerns *epistemological analysis*. Recall Ernst Mach's objections against Isaac Newton's absolute space that strongly shaped Albert Einstein's position on the subject. Mach criticized that absolute space was inaccessible to experience and that Newton's notion was not a legitimate part of scientific theory for this reason. A third branch is *ontological*: if the theory under consideration should turn out to be completely true, what is the ensuing appropriate picture of nature? Regarding space-time theory, one of the relevant issues is whether the spatiotemporal metric or the metric field is rather to be taken as a part of space-time or of matter-energy [8, pp. 28–31].

These considerations are intended to make plausible, in a preliminary fashion, that there are interesting and non-trivial interpretational questions that cannot be decided by recourse to experience alone. There are sensible challenges left to the philosophical reflection of physical theories. Conceptual clarification, epistemological analysis, and ontological exploration can contribute to a deeper understanding of what the relevant theory is all about.

I begin by recounting some traditional puzzles and their attempted resolutions in the philosophy of quantum theory. Then I turn to the EPR-correlations as a task for conceptual and ontological investigation. Afterward, I address the quantum measurement problem as an example of an epistemological challenge.

## 3.2 Traditional Puzzles and Positions in Quantum Philosophy

The Copenhagen Interpretation is at the origin of the philosophy of quantum mechanics. This interpretation takes macroscopic measuring instruments as being correctly describable by the concepts of classical physics. By contrast, quantum phenomena do not fit together so as to yield physical bodies in the familiar, macroscopic sense, bodies, that is, with a fixed set of properties and another set of properties that change in accordance with observable interactions with other bodies. The Copenhagen claim was that physical theory is prevented from getting any coherent picture of a quantum object, a picture that is capable of capturing the phenomena in causal and in spatiotemporal terms at the same time. Wave-particle duality or complementarity are indications of this inability to come up with a single, coherent notion that could make sense of quantum events [18, pp. 68–69, 87–94].

This failure prompted Copenhagen instrumentalism, according to which Schrödinger's equation does not represent physical processes directly but merely predicts the outcome of certain quantum measurements. Werner Heisenberg took an operationalist stance that was advertised as being analogous to Einstein's approach to distant simultaneity. The registered relationships are the only phenomena of relevance. As Heisenberg put it: "Physics ought to describe only the correlation of observations" (quoted after [4, p. 455]). A quantum phenomenon is always characterized by its relation to a particular experimental setup, and this essentially relational nature of quantum states is supposed to imply that there is no independent quantum reality behind these observations—at least no reality that is subject to human knowledge. By contrast, the measuring instruments were considered real, and this is why the need arose to draw a line between these two realms of the quantum phenomena, on the one hand, and classical objects or bodies, on the other. This line is known as "Heisenberg's cut." As a result, it is illicit to apply quantum theory to the measuring apparatus [19, p. 115].

Of course, this nomological split between parts of the world appeared unappealing to many. After all, the Scientific Revolution of the seventeenth century had come into being by removing the alleged nomological divide between the celestial and the sublunar spheres. Further, restricting oneself to correlations among observables seems to rule out any deeper understanding of nature's contrivances. In addition, no clear reason was given why quantum theory should not be applied to measuring instruments. After all, the latter consisted of atoms, too. To make things worse, Niels Bohr, the Copenhagen champion, had been inconsistent enough to apply quantum theory to measuring procedures in his debate with Einstein in the late 1920s about the coherence of the theory ([18, pp. 127–136]; [19, p. 110]; [5, pp. 294–295]). It was only natural that this road was explored more systematically. This is what John von Neumann achieved in 1932 with his approach, sometimes called the "orthodox interpretation." Von Neumann conceived of the measuring instrument as a quantum system and analyzed its interaction with the object in quantum theoretical terms. The result was the emergence of the "*quantum measurement problem.*" The interaction between apparatus and quantum object produces entangled states between the two, but no prediction of any observation. In other words, judged on the basis of quantum theory, no definite measuring values should ever occur. But they regularly do.

The so-called "collapse postulate" or the "reduction of the wave function" was introduced as a separate process in addition to the Schrödinger evolution. While the wave function evolved continuously in ways governed by Schrödinger's equation, an abrupt change was supposed to occur whenever a measurement was made. This collapse is not captured by this equation so that two categorically distinct kinds of behavior of the wave function were assumed to exist. It is only in virtue of this latter, non-Schrödinger-like type of evolution that definite measuring values turn up. The point is not that quantum mechanics fails to predict which particular measuring value emerges. This inability might simply arise from the indeterminate nature of quantum processes according to which a particular value is taken by chance. Which value turns up may be objectively uncertain and this indeterminacy is reflected by quantum mechanics. There is nothing mysterious about that. The point rather is that the Schrödinger evolution suggests that a superposition of states, rather than a particular state corresponding to a particular measuring value, results from coupling the observed particle to the measuring apparatus. Accordingly, what appears anomalous and mysterious from the point of view of Schrödinger's equation is that a definite measuring value ensues from the interaction between the particle and the apparatus in the first place. This is the gap the collapse postulate was supposed to bridge.

The Copenhagen instrumentalism is now considered obsolete in many quarters. In the physics community the view has gained prominence that quantum theory describes natural processes and not just human interventions in such processes. As a result, the distinguished notion of observation or measurement, that is characteristic of the Copenhagen approach, has lost its earlier appeal. Rather, observations are increasingly considered as a special kind of interaction and not as a categorically distinct sort of intervening in nature. In the Copenhagen vein, only observations and measurements, but not physical interactions in general, manage to produce the collapse of the wave packet. Accordingly, quantum theory was not supposed to describe nature objectively but rather the relationship between human interventions and nature's response. By contrast, beginning around 1980, a realist interpretation of quantum theory has increasingly gained acceptance. This attitude is also based on the fact that quantum theory has proven robust. During the early decades of the development of the theory, the general idea had been that all the mysteries and puzzles would be resolved by a future theory that would supersede and replace quantum theory in its present shape. Quantum theory is provisional and makeshift, and the most sensible strategy to follow is to extract robust results from this ramshackle scheme, results that need be and will be recovered by the expected future account. Yet now everybody believes that quantum theory is here to stay. No emergence of a new distinct physical theory is in the offing.

As a result, many physicists grant quantum theory explanatory or realist import. However, this is not tantamount to the realism suggested by classical physics. The latter had pursued the project of accounting for nature by abstracting from ourselves, as it were, by disregarding or correcting for human interaction with nature. This approach cannot be upheld in the quantum realm. Reference to human intervention and interaction with an apparatus is indispensable for marshaling the phenomena appropriately. Quantum realism needs to be of a perspectival character which is not the same thing as introducing subjectivity into our understanding of reality. Still, this perspectivalism involves a dependence of our understanding of nature on how we interact with nature. Even if observation and measurement lose their special position and are reintegrated into the realm of physical interaction, the ensuing realism is of a particularly relational character.

Such a relational view of physical reality is genuinely different from the realism suggested by classical physics. It is sometimes argued that prequantum approaches to nature should as well have given rise to a relational understanding of objects. After all, we come to know the properties of objects by entering into a relation with them. We need to connect the object under scrutiny to a measuring device. Accordingly, classical realism granted that measuring physical quantities involves an interaction with the pertinent quantity and is thus bound to interfere with its magnitude. However, first, this interaction can be made arbitrarily small in principle, and, second, it can possibly be accounted for by the physical theories of the measurement process and thus be corrected. Third, the readings of various instruments may agree with each other which suggests that these measuring devices register the same property consistently. Although it is true indiscriminately for the classical realm and the quantum realm that observing an object means interacting with this object, the coherence among classical measuring values suggests that the object and its properties exist independently of this interaction. The numerical agreement among the results of different methods of measurement suggests that the results do not essentially depend on these methods. Rather, this agreement indicates that the results are produced by intrinsic properties of the object under consideration. After all, this is how Jean Perrin argued for the reality of atoms in 1913. Since 13 dissimilar methods yield the same numerical value of Avogadro's constant, something real is referred to by this constant [21, pp. 217–220]. That is, although we need to rely on establishing a relation with an object in order to register its properties, the agreement and coherence among the outcome of diverse approaches of this sort suggest that the nature of these relations is inessential. This argument provides a basis for taking such measurement results as revealing intrinsic properties of the object.

Precisely this is different in quantum theory. For example, in many instances, changing the order of measurements affects the outcome. Under such circumstances it is clearly unjustified to abstract from the kinds of interactions used for exploring objects. This is why the relational interpretation of quantum states is non-trivial. It was not anticipated by classical physics, it was not an unrecognized aspect of classical physics. It is a feature truly different from classical physics.

Such considerations are often appealed to in favor of what is called "structural realism," or "ontological structural realism," to be more precise. The idea is that reality does not consist of separate objects with their intrinsic properties, but is rather made of relations. Bohr's notion of complementarity is frequently suggested in support of structural realism. The properties a quantum object exhibits depend essentially on the apparatus it interacts with. Complementarity in Bohr's sense means that two modes of description are appealed to that would contradict each other if they were applied at the same time. Yet both are necessary for a comprehensive elucidation of the phenomena. Bohr made this more concrete by saying that we can either give a spatiotemporal description by measuring positions at a certain time or a causal description by registering momentum and energy. Using the two in combination is ruled out by the indeterminacy relations. The bottom line is that there are two different accounts that are essentially tied to the use of specific apparatus. In other words, what the phenomena are is determined by how they interact with other objects. Quantum systems are characterized by their relations with other quantum systems [7, pp. 219–220].

## 3.3   Entanglement and the EPR-Correlations

One of the chief puzzles emerging against this background is entanglement. The philosophical challenge posed by entanglement is to understand which deeper property of nature it reveals. It is clear what entanglement, or the EPR-correlations at that, is not. The EPR-correlations are not produced by local preparation and distributed subsequently by usual signaling processes. In other words, the EPR-correlations are not brought about by common causation. It is more difficult to say how they are fashioned alternatively. It is helpful for any such venture to heed a distinction introduced by Jon Jarrett in 1983. John Bell's locality condition attempts to capture the intuition that the measurement results in each of the two correlated wings of an EPR-experiment depends only on the local circumstances. The empirical violation of this condition shows that quantum mechanics is non-local. However, it conduces to clarity to split this condition up into two. The distinction Jarrett drew is now mostly called *"parameter independence"* and *"outcome independence."* Parameter independence is intended to express the independence of a measurement result in one wing of an EPR-setup from the selection of the observable to be measured in the other wing. Outcome independence means the independence of a measurement result in one wing from the result obtained in the other wing. These two independence claims are sufficient for deriving Bell's theorem. This conceptual analysis suggests that the violation of Bell's theorem means that at least one of these claims needs to go. Parameter independence is satisfied in quantum mechanics while outcome independence is violated.

The philosophical impact comes out more clearly if Jarrett's two conditions are further sharpened by conceptual analysis. Parameter independence means that the measurement results in one wing are stochastically independent from the experimental setup at the second wing. This property is a consequence of special relativity and says that the state of a system is unaffected by events in spacelike regions. Accordingly, parameter independence captures a locality condition that is more specific than Bell-locality and entails that any causal influence spreads at most with the speed of light. Second, as Don Howard has pointed out, outcome independence is equivalent to saying that each of the two EPR-correlated systems has its own physical state and that the joint state is the product of these separate states. In other words, outcome independence is equivalent to the separability of the two systems. Jarrett's result can now be rephrased to the effect that any theory that satisfies locality and separability in this sense is in conformity with Bell's theorem. It follows from the violation of this theorem that one of these conditions must be given up. In fact, quantum mechanics *denies separability* by abandoning factorizability. In entangled states, it is the composite state that is primary since it cannot be neatly divided into two states that unambiguously pertain to the partial systems. Rather, superpositions between these partial states obtain that do not belong to either part and rather belong to both simultaneously. The formal treatment in quantum mechanics agrees with the experimental results in that quantum mechanics assumes physical states that are *local, but non-separable* [16, pp. 226–228].

It is striking that Einstein in his 1948 presentation of the EPR argument proceeded precisely along these lines. Einstein was not happy with the 1935 EPR paper. He thought that it buried the chief message by erudition. The approach he sketched in 1948, 13 years after the initial publication,had been, for all we know, the nucleus of the earlier EPR argument, that had been transformed and altered by Boris Podolsky and Nathan Rosen. This original Einsteinian thought involved the demonstration that the EPR-correlations meant either a violation of locality in the sense of limited speed of causal propagation or separability in the sense of factorizability or both. Einstein spoke of *"Nahewirkung"* and *"Trennbarkeit."* Yet Einstein thought that both conditions were prerequisites for doing science so that none of them could possibly be given up. In particular, he argued that separability is necessary for individuating systems and believed that individuation is a presupposition of testability. Individuating systems is necessary for testing, since testing presupposes the assigning of some property to a system. Without individuation, each property could only be assigned to the whole universe. Since Einstein believed that the spatiotemporal interval is the only objective basis for individuation, he took separability as imperative for testing theories. Yet he also thought that locality is necessary for securing the existence of closed systems. Closed systems are indispensable for testing as well. Otherwise, any discrepancy between theory and evidence could always be attributed to a distant, instantaneous influence [12, 15, 16].

Einstein's conclusion in his 1948 paper was that quantum mechanics is not in accordance with at least one of these two essential requirements, and he challenged the physics community by demanding a stark choice: either adhere to quantum mechanics in its present shape and give up separability or hold fast to separability and grant that quantum mechanics is incomplete. That is, the distant correlations are fixed by additional states in the quantum systems involved, but unknown to and unrecognized by quantum mechanics. In other words, Einstein opted for a local hidden variable theory [12]. We know today that this is of no avail. Today, the most prominent way out of this Einsteinian quandary is jettisoning separability.

But what does it mean to give up separability? Abandoning separability is tantamount to accepting entanglement as a basic trait of nature. If a quantum system consists of various parts, it is the total state that is primary. Two entangled electrons may have a definite total spin value in each direction, but no one of the two electrons has a definite spin value in itself in any direction. The components stand in a certain relation, but the total state of the composite system cannot be derived from the non-relational properties of these components. In contrast to classical objects, quantum objects are not characterized by intrinsic properties but by being part of more comprehensive systems. In other words, total states cannot be traced back to properties that the parts possess independently of each other; total states are not produced by an interaction among the parts. Rather, the parts are only created by an intervention in or outside interaction with the comprehensive system. In entangled systems, no such parts are realized. This is the precise opposite of separability ([13, p. 5657]; [3, pp. 129–130]).

    This feature can be interpreted in ontological terms as suggesting a *holist* picture
of nature. Holism involves the primacy of the whole over its parts. In entangled
systems, the total state determines the properties of the parts, while the reverse
is not true. The violation of Bell's inequalities demonstrates that locally prepared
properties of the parts are insufficient for fixing the total state. The correlations
are so strong that they cannot be produced by the pre-established properties of the
parts. Yet the correlations are generated by the total quantum state ([17, pp. 15–16];
[13, pp. 77–78]). This involves a reversal of traditional mereological supervenience.
The latter notion suggests that the properties of the parts define the properties
of the whole. This is expressed by the condition that there is no change in the
composite system without an accompanying change in at least one of the parts.
Further, supervenience entails that the converse is not true. It is considered well
possible that the properties of the parts change while the traits of the whole remain
invariant. This feature is usually called multiple realization; it means that the same
effect can be produced by different means [17]. Take statistical mechanics: the
same macroscopic or thermodynamic properties can be brought about by different
distributions of the relevant molecules across the pertinent configuration space. This
is precisely the contrary of what quantum theory entails. Quantum holism means a
primacy of the whole over its parts, and this is an asymmetric relation. Classical
thought, by contrast, is governed by the inverse primacy of the parts over the whole.
    Entanglement has become part of many attempts to capture the structure of
reality on the basis of physical theory. In this vein, entanglement is sometimes
included in the arguments for structural realism. As I mentioned in Sect. 3.2,
structural realism takes relations to be the stuff of which reality is made. Yet I
wonder whether entanglement really fits well with such a structuralist approach.
Entanglement produces a sort of union of the two original states that cannot be split
up into two separate quantum systems. After all, this is what the loss of separability
means in this context. Entangled states are undivided wholes that do not consist of
interacting objects. The total system is the basic entity. Yet this holism does not
seem to square well with structural realism and its emphasis on relations rather
than intrinsic properties. Holism is usually thought to be tantamount to a relational
interpretation of reality. But such a relational interpretation says that the parts are
tied to each other by relations that do not hook up with the properties that the parts
possess independently of each other. However, the fusion of initial objects into an
undivided whole suggests a more intimate connection than the relation between
parts, even if the parts are granted no separate existence and no intrinsic properties.
Accordingly, I suggest distinguishing more clearly between holism and a relationist
ontology, as supported by structural realism [7, p. 221].
    I mentioned that abandoning separability is often thought to be the right way
to go. Giving up separability and endorsing holism leads to the following account
of the EPR correlations. The two wings of an EPR system do not possess definite
values of measuring quantities, such as spin. There is only one comprehensive state.
This total state is changed by registering a spin value in one wing. This change
of the total state affects the probabilities of measuring values in other parts of the

comprehensive system. This change becomes manifest if a local interaction with a measuring device at the other wing occurs [13, p. 69].

## 3.4  Puzzling Features of Entanglement

I hope these considerations have added some plausibility to the supposition that conceptual clarification and ontological exploration provides a deeper understanding of what quantum mechanics is all about. The introduction of a sharper notion of locality and its distinction from separability suggests that, in contradistinction to a widespread sentiment, quantum mechanics respects locality and rather parts company with classical physics when it comes to separability. Further, and likewise in opposition to a widely received view, the EPR-correlations do not embody a causal interaction between the two systems at hand but rather flow from the holistic or inseparable nature of quantum states. I take it that such reconstructions are conducive to an improved understanding of quantum theory—in spite of the fact that no new predictions are made. However, it also belongs to the professional duties of philosophy to avoid glossing over difficulties and to prevent premature agreement. Philosophy is called upon to confess its confusion, to produce *taumazein*, and to insist that certain notions or ideas are still obscure.

Regarding entanglement, two such murky spots can be identified. The first one is connected to the question how the EPR-correlations can be understood. A prominent notion of understanding emphasizes causal processes. We understand a phenomenon if we know how it is brought about [21]. Along such lines, James Cushing argues that understanding a physical process must always rely on a mechanism or process that can be pictured. Yet the two major roads to a causal explanation of the EPR-correlations are blocked: neither can they be accounted for by a direct causal link, nor can they be traced back to common causation. So we need to face the possibility that quantum theory represents the "endgame for understanding" [9]. It is to be granted that we don't have the faintest idea as to how the EPR-correlations come about. There is no causal mechanism that could produce such correlations. However, there are other modes of producing understanding, and the challenge is to explore whether other avenues might be suitable for providing a non-causal understanding. The trouble is that a simple appeal to holism is of no help. Holism operates with extended wholes as primitive states that are destroyed if they are divided It is in virtue of their extended nature that distant states look as if they were adjusted by instantaneous action at a distance. Yet, the extended total state changes if some local alterations are performed. Two entangled electrons do not possess spin values separately, and this is why a measurement performed at one electron changes the total state and thereby the probability distributions of spin values measured at the other electron [13, p. 69]. Yet it is difficult to make sense of this non-causal instantaneous adjustment over arbitrary distances. Introducing relational states go some way in the desired direction. If an electron is moved by 1 m, the distance of a second electron to the one moved has increased as well. The problem is not that the relational properties of an object may change without intervening in this object. The trouble rather is that these changes also extend to relations that this untouched object bears to further objects such as measuring devices. Such puzzling features are responsible for the lack of any idea of a mechanism that could underlie the EPR-correlations.

Yet it is not written in stone that all explanation needs mechanisms. Explaining phenomena by appeal to conservation laws is a traditional avenue for producing understanding. Such pathways to understanding are captured by the unificatory account of understanding, advocated, among others, by Philip Kitcher. Understanding is produced by realizing that a variety of seemingly different phenomena instantiate a common explanatory pattern [20]. There is no doubt that entanglement serves to unify a variety of different effects and produces understanding in this sense. Yet another notion of understanding conforms well to the quantum mechanical practice of giving explanations. On the "contextual approach" entertained by Henk de Regt and Denis Dieks [10], understanding is the ability to apply scientific conceptions properly and to see their full implications. Understanding helps us recognize qualitatively characteristic consequences of a theory and thus facilitates the construction of test opportunities for this theory. This approach takes up Richard Feynman's quip that he understands an equation if he knows what the solutions are like without actually solving it.

One thing is worth emphasizing, when the nature of understanding is at issue. Understanding is in no way necessarily tied to familiarity. Exploring the kind of understanding that quantum mechanics can provide ought not to be abused for insisting on bringing back pictures and models familiar from classical physics. Quantum phenomena and processes are different from classical ones. Nature behaves in an unaccustomed way in the quantum realm, and philosophy can do nothing about it. The challenge is different: we want to realize how the different aspects of quantum phenomena hang together. We want to produce a coherent picture that expounds interconnections between the quantum phenomena in an orderly way and thus allows us to make sense of them—even if some premises and principles need to be granted that look foreign from the macroscopic point of view.

As a result, quantum mechanics is not in need of justifying the introduction of unfamiliar properties, such as relational or holistic ones. Fundamental properties can never be explained; this is what makes them fundamental. The principle of inertia was met with amazement in the seventeenth century, and it would not have been a legitimate question to ask René Descartes (who conceived rectilinear inertial motion) why bodies that are free of external influences continue to move uniformly. Alternatively, a legitimate question is what the basic states that a theory introduces and appeals to are like, and how the theory proceeds from these pure states to more complex ones that are subject to experience. Inertial motion is a basic state in classical mechanics, relational states and holistic properties are basic entities in quantum mechanics. The requirement only is to produce a coherent account on such grounds.

Another puzzling feature is the relation between the EPR-correlations and special relativity theory. It goes without saying that there is no formal contradiction between the two since no information can be transmitted by taking advantage of the correlations. Light rays remain the fastest signal. This is why Abner Shimony characterized the relationship between these two established parts of scientific knowledge as a "peaceful coexistence" [22]. However, on closer inspection this is not that obvious. The trouble is that the EPR-correlations (to all appearances) establish an instantaneous relationship between events located at arbitrary distances. But this suggests that quantum mechanical non-separability introduces a relation of absolute simultaneity. The correlations obtained between distant measurements seem to be simultaneous in a frame-independent sense; they seem to be located at a preferred spacelike hyperplane [2].

I take it that it does not really help to stress that no information can be transmitted along such absolute simultaneity planes. This would mean to stick to special relativity in letter, but abandon it in spirit. On that account, the speed of light would assume the character of a technical barrier of information transmission while nature at more profound levels remains unfettered by such limitations. At the level of ontology, such an account would rather resemble the one developed by Hendrik Lorentz and Henri Poincaré rather than by Einstein. Lorentz and Poincaré assumed absolute simultaneity, introduced an upper bound of signal transmission as a technical constraint and derived the principle of relativity for electrodynamics. The latter principle originated in these technical limitations, whereas at the level of the natural processes themselves the distinction between absolute rest and absolute motion was unscathed. By contrast, Einstein took the principle of relativity as being engrained in the workings of nature and built a new structure of space-time on this basis. It is not obvious whether this privileged position can be retained for the principle of relativity if the EPR-correlations establish a relation of absolute simultaneity behind our back, as it were.

This section is intended to show that philosophical attempts to produce additional clarity may end up with recognizing further obstacles to clarification. Although it may sound paradoxical, it means making progress in understanding when gaps in understanding are identified and when challenges are elaborated that still lie ahead.

## 3.5   Quantum Measurement and Decoherence

The quantum measurement problem is a serious one and has haunted quantum theory and quantum philosophy since von Neumann performed his now classic analysis of the measurement process in 1932. Von Neumann applied quantum mechanical principles to the system composed of the object under scrutiny and the measuring apparatus and obtained a series of terms, some of which referred to properties of the object, others to properties of the apparatus, and additional interference terms that involve a superposition of both entities. Two challenges arose from this analysis: First, quantum mechanics did not predict that any measuring value occurs in the first place; the measuring process gets stuck in unfolding a spectrum of possibilities. Yet we regularly obtain a measuring value (see Sect. 3.2). Second, the spectrum of possibilities contains superpositions of states of the object and the device. Yet such interferences are never observed as a measurement outcome. Accordingly, quantum mechanics is faced with a serious empirical problem or an anomaly.

Another way to present the problem is to emphasize the conceptual incoherence it involves. Consider a system consisting of a quantum object and a measurement apparatus. On the first approach, the apparatus is treated as a macroscopic object and as being subject to classical physics. The usual story runs that the apparatus serves to reduce the quantum mechanical state description. This collapse of the wave function makes a definite measuring value emerge (see Sect. 3.2). On the second approach, the same situation is redescribed by construing the apparatus as a quantum object, too, so that the composite system is subject to a quantum mechanical analysis. In this framework, the Schrödinger evolution should continue. No collapse occurs and no measuring value shows up. We get two different predictions depending on how we treat the composite system. As a result, the quantum measurement problem is an internal inconsistency of quantum theory [24, pp. 211–212].

Major progress has been made in the past decades by the development of the *decoherence approach*. Decoherence exclusively proceeds from the Schrödinger evolution. The treatment is based on nothing but ordinary quantum mechanics; no interpretation, no additions are necessary. Decoherence takes the fact into consideration that a measurement process does not only involve the object and the measuring apparatus. Rather, the two are part of the larger environment and enter into a relation with this environment. A measurement is never performed in isolation from surrounding objects, and yet these further relations have not been included in the traditional accounts of quantum measurements. As a result, the environmental objects also enter into a superposition with the quantum object under scrutiny and the measuring apparatus. Decoherence theory is able to show that these superposed states are inaccessible to local observers. For instance, a photon hits upon a quantum object, enters into an entangled state and moves away from the observer. The interference terms are still there, but do not become manifest in any measuring instrument. The superpositions are not destroyed, but they are moved aside and do not show up anywhere. Decoherence makes interference terms vanish not by suppressing them but by delocalizing them. The only states left pertain unambiguously to the object or the apparatus, respectively. This is how quantum objects assume their separate existence and their definite properties. For instance, a quantum state appears particle-like since the continual collisions with ambient molecules and photons amount to a series of position measurements [5, p. 296].

The decoherence approach provides an epistemological analysis by applying quantum theory to the process of observation and measurement. In general, this is a familiar scheme that has been subsumed under the label of "measuremental theory-ladenness." Physical theories are used for reconstructing the measuring process and for showing that this process is suitable for registering and representing the quantity at hand. As the case may be, an observation theory of this sort may coincide with the corresponding explanatory theory which is expected to account for the results obtained with the instrument. Consider measuring current intensity with an old-fashioned moving-coil galvanometer. The electromagnetic interactions that make this device a reliable indicator of current intensity are governed by the same theory of electromagnetism that also introduces and characterizes the notion of current intensity. A theory that is able to figure as its own observation theory is sometimes called "complete" [6, pp. 20–27]. The application of quantum mechanics to the process of quantum mechanical registration, as begun by von Neumann and heavily improved in the decoherence approach, shows that quantum mechanics is a complete theory (if understood in a sense different from the one employed in the EPR-debate).

This advanced analysis of the measuring process brings two things to the fore. First, the unobserved interference states are actually unobservable for a local observer. The observable states clearly belong either to the quantum object or the apparatus. This resolves an anomaly of quantum theory. Second, one item from this reduced set of properties will actually show up as a measurement reading. This resolves another anomaly of quantum theory (see Sect. 3.2). So it looks pretty clear at first sight that decoherence gives us a solution to the measurement problem.

Yet, in fact, this claim is highly contentious. Critics object that decoherence does not account for the transition from the superposition to one of its elements. Decoherence does not entail the collapse of the wave function. Instead, there is no collapse, and all the superpositions are still there—if unobservable. This is why critics have objected that, conceptually speaking, nothing has changed regarding the quantum measurement problem. We obtain a spectrum of options but none of these options is actually selected and turns up as measurement outcome. This is why early proponents of decoherence, such as H. Dieter Zeh, think that the true selection of a measuring value demands a branching of causal histories and thus leads to the "many-worlds interpretation."

Other pioneers of decoherence, such as Wojciech Zurek, are more optimistic regarding the explanatory potential of decoherence. The central questions to ask are: what is the evidence that a collapse occurs in the first place? How do we know that one item from the range of possible results is really picked by the measurement? Conversely, is it possible to rule out that the superpositions are never broken up and that the situation only looks as if a particular property emerged? Advocates of the view that decoherence solves the measurement problem indeed claim that it is sufficient to show that the measurement appears to disrupt superpositions. It is sufficient to expound that the world appears classical, although it is of a quantum nature all the way up ([24, pp. 213–215]; [5, p. 296]).

The objective of an epistemological analysis of quantum measurement is to clarify what the demands are for considering the quantum measurement problem settled. The decoherentist answer suggests that the problem is ill-posed in that in reality there is no selection of a measuring value from the set of alternatives. Yet, it looks as if one such value was selected. In this vein, the quantum measurement problem is rather dissolved than solved. The transition from the quantum world to the classical world, a cornerstone of the Copenhagen interpretation, does never occur, in fact. The world is and remains thoroughly a quantum world but looks classical because of human limitations to getting hold of quantum states. The human grip remains local and thus misses global interferences that extend through the universe. Objects possess definite properties because of human epistemic frailty. Critics demand the true emergence of classical objects.

I have tried to elaborate two questions of philosophical import in this section. The first one concerns the issue what is, precisely, the quantum measurement problem and what do we require considering it settled. Some critics suggest that a solution means to show which measurement outcome actually turns up. This seems clearly overdemanding since the quantum world may be truly indeterministic (and was assumed to be so for quite some time). Which measuring value will turn up may be due to genuine chance. Others insist that it needs to be shown that one such outcome is actually selected. This is tantamount to demanding that macroscopic objects truly possess definite properties. Most of the opponents to the claim that decoherence solves the measurement problem take this avenue and require that superpositions actually be destroyed (see [1]). A solution of the measurement problem needs to demonstrate how classical objects emerge from quantum states and how the collapse of the wave function actually proceeds. Advocates of the decoherentist solution respond that such requirements aim at the wrong explanandum: strictly speaking, no classical objects exist, and the attempt to derive their existence is ill-conceived and bound to fail. Dealing appropriately with such contentious issues obviously demands addressing conceptual questions and the in part normative issue what the appropriate explanandum and a satisfactory explanation is.

The second question to which philosophy may have something to contribute is the relation between decoherence and other interpretations of quantum mechanics, in particular, the many-worlds interpretation and the Copenhagen interpretation. The many-worlds interpretation claims that it is complete in the sense that no additional observation theory is needed to interpret quantum mechanical predictions [6, pp. 94–96]. Opponents reject the inference to many worlds and maintain that the mixture of states obtained by applying decoherence refers to different possible outcomes of a measurement on one and the same quantum system.[1] That is, decoherence is not necessarily tied up with the many-worlds interpretation. Another question at this juncture is the connection of decoherence to the Copenhagen view. On the one hand, decoherence can be credited with actually deriving the boundary between the classical and quantum areas that is pivotal for the Copenhagen view [5, p. 297]. On the other hand, this boundary is claimed to be deceptive and not to coincide with any demarcation in nature, which deviates from the Copenhagen view.

---

[1]It is granted within this approach that the Born rule needs to be postulated in addition in order to translate the quantum mechanical coefficients into relative frequencies of measurement results whereas many-worlds champions claim to be able to derive the Born rule.

## 3.6   What the Philosophical Interpretation of Quantum Theory Can Accomplish

Philosophical analysis addresses conceptual, ontological, and epistemological issues. Regarding conceptual analysis, the challenge is to contribute to a clarification of the questions, to analyze the requirements of a satisfactory answer and to suggest criteria for judging about the appropriateness of such requirements. Consider the following example of a conceptual clarification. In the early days of quantum mechanics, the presumption prevailed that the predictions of measuring values involved a particular relationship to experimental devices and observers. Consider a double-slit experiment. Only if the experiment was set up such that no information about the trajectory of a particle could be obtained so that the observer was prevented from telling through which of the slits the particle had passed, interference patterns and a wavelike behavior emerged. If, by contrast, an observer could possibly glean the particle's path from the experiment, then the results suggested a more localized, particle-like event. Decoherence theory has suggested a different picture in the past years according to which this dependence of the results on the observer is only apparent. The crucial point rather is whether the quantum system is sufficiently isolated to avoid the occurrence of decoherence. If the quantum system interacts with its environment, it is possible to obtain information about what is going on in the system. Photons are emitted and allow the observer to identify more specific features of the system. At the same time, this interaction produces decoherence which is in turn tantamount to make entanglement and interference terms invisible [23, p. 57]. As a result, underlying the seemingly subjective feature whether or not information about a particle trajectory can be obtained is the objective, process-related feature whether or not decoherence occurs in the system. Conceptual analysis serves to identify the impact of certain premises and principles.

Ontological reconstruction is another major field of philosophical analysis. Such reconstruction is shaped by what kind of understanding of the world quantum mechanics is able to supply. Different notions of understanding are invoked in the debate about achievements of quantum theory and these notions can be clarified by conceptual analysis. The various camps and opposing factions can be identified more easily if a distinction between different notions of understanding is introduced. In pragmatic or action-oriented approaches as well as in intuitive approaches, understanding has to do with the capacity to anticipate the impact of certain premises and principles. Understanding in this sense is connected to being able to foresee the outcome of certain actions or assumptions. By contrast, understanding in a coherentist sense refers to an overall account that serves to integrate the various aspects of a situation or a theory into an articulate whole. In particular, no gaps or lacunae distort the comprehensiveness of the picture; no murky parts leave room for conflicting interpretations. Understanding in this coherentist sense conveys the impression that things are marshaled in an ordered, transparent, and sensible fashion. This does not mean to feature principles familiar from classical physics;

the point rather is to realize how the different aspects of quantum phenomena are connected to each other. It appears that quantum theory has managed to produce understanding in pragmatic or intuitive respect but has some way to go to generate understanding in coherentist respect as well.

Developing such a coherentist understanding of quantum theory is an urgent challenge. If science is supposed to transcend the mere technical business of anticipating future experience and is also thought to provide some sort of grasping nature's workings, then we need a coherent picture of what physical theory tells us about the world. In particular, one of the relevant ambitions is to spell out the kind of holism that quantum theory instantiates. The ambition is to develop a coherent picture that allows us to make sense of quantum phenomena—even if some premises and principles need to be granted that are unfamiliar from the macroscopic point of view. The goal of fundamental research in the natural sciences is understanding nature, and this challenge is not appropriately met by merely relating equations to experimental results. We also want to relate the equations among themselves such that they form a coherent whole and reciprocally support each other. This is part of the philosophical endeavor to explore the broader consequences of physical theories and to trace their impact on what might be called a scientific worldview.

Third, epistemological analysis. Quantum theory has been plagued by anomalies and inconsistencies regarding the relation between the formalism and its predictions. The relation between theory and experience has been a particularly problematic one in this area, and many attempts to come to terms with this relation have been criticized as being nothing but hand-waving and sloganeering.[2] Yet in the past three decades this debate has gained in prominence and maturity. Philosophical analysis of the relationship between theoretical states and observed result in quantum theory may provide deeper insights into the epistemic potential and limitations of scientific research.

# References

1. Adler, S.L.: Why decoherence has not solved the measurement problem: a response to P.W. Anderson. Stud. Hist. Philos. Mod. Phys. **34**, 135–142 (2003)
2. Albert, D.Z., Galchen, R.: Was Einstein wrong?: a quantum threat to special relativity. Sci. Am. **300**, 32–39 (2009)
3. Audretsch, J.: Die sonderbare Welt der Quanten. Beck, München (2008)
4. Audretsch, J.: Causality and entanglement in the quantum world. Eur. Rev. **18**, 455–467 (2010)
5. Camilleri, K.: A history of entanglement: decoherence and the interpretation problem. Stud. Hist. Philos. Mod. Phys. **40**, 290–302 (2009)

---

[2]Dürr and Lazarovici [11, pp. 111–112] complain about philosophy and criticize it for not taking seriously such challenges. But their chief culprit is Bohr who was not affiliated with a philosophy department. So, a fairer judgment might be to attribute a lack of attentiveness to such problems to some physicists.

6. Carrier, M.: The Completeness of Scientific Theories. On the Derivation of Empirical Indicators within a Theoretical Framework: The Case of Physical Geometry. Kluwer, Dordrecht (1994)
7. Carrier, M.: Raum-Zeit. de Gruyter, Berlin (2009)
8. Carrier, M.: Die Struktur der Raum-Zeit in der klassischen Physik und der allgemeinen Relativitätstheorie. In: Esfeld, M. (ed.) Philosophie der Physik, pp. 13–31. Suhrkamp, Berlin (2012)
9. Cushing, J.T.: Quantum theory and explanatory discourse: endgame for understanding? Philos. Sci. **58**, 337–358 (1991)
10. De Regt, H.W., Dieks, D.: A contextual approach to scientific understanding. Synthese **144**, 137–170 (2005)
11. Dürr, D., Lazarovici, D.: Quantenphysik ohne Quantenphilosophie. In: Esfeld, M. (ed.) Philosophie der Physik, pp. 110–134. Suhrkamp, Berlin (2012)
12. Einstein, A.: Quanten-Mechanik und Wirklichkeit. Dialectica **2**, 320–324 (1948)
13. Esfeld, M.: Einführung in die Naturphilosophie. Wissenschaftliche Buchgesellschaft, Darmstadt (2002)
14. Friedman, M.: Foundations of Space-Time Theories. Relativistic Physics and Philosophy of Science. Princeton University Press, Princeton (1983)
15. Howard, D.: Einstein on locality and separability. Stud. Hist. Philos. Sci. **16**, 171–201 (1985)
16. Howard, D.: Holism, separability, and the metaphysical implications of the bell experiments. In: Cushing, J.T., McMullin, E. (eds.) Philosophical Consequences of Quantum Theory. Reflections on Bell's Theorem, pp. 224–253. University of Notre Dame Press, Notre Dame (1989)
17. Humphreys, P.: How properties emerge. Philos. Sci. **64**, 1–17 (1997)
18. Jammer, M.: The Philosophy of Quantum Mechanics. Wiley, New York (1974)
19. Kiefer, C.: Der Quantenkosmos. S. Fischer, Frankfurt (2008)
20. Kitcher, P.: Explanatory unification. Philos. Sci. **48**, 507–531 (1981)
21. Salmon, W.C.: Scientific Explanation and the Causal Structure of the World. Princeton University Press, Princeton (1984)
22. Shimony, A.: Metaphysical problems in the foundations of quantum mechanics. Int. Philos. Q. **18**, 3–17 (1978)
23. Zeilinger, A.: Die Wirklichkeit der Quanten. Spektrum der Wissenschaft **11**, 54–63 (2008)
24. Zwirn, H.: Can we consider quantum mechanics to be a description of reality?. In: Soler, L., Sankey, H., Hoyningen-Huene, P. (eds.) Rethinking Scientific Change and Theory Comparison: Stabilities, Ruptures, Incommensurabilities?, pp. 209–217. Springer, Dordrecht (2008)

# Chapter 4
# On the Sufficiency of the Wavefunction

**Roger Colbeck and Renato Renner**

## 4.1 Introduction

> Theorem IV: Whoever endows $\Psi$ with more meaning than is needed for computing
> observable phenomena is responsible for the consequences. [1]

Physical theories allow us to make predictions about future observations from preexisting data. For example, we may have data about the configuration $C$ of a collection of particles at time $t_0$, based on which we can predict the particles' positions $X$ at some later time $t > t_0$.[1] Some of our theories are deterministic, such as classical mechanics. In this case, if $C$ consists of a complete description of all relevant particle positions and momenta, so that the system's state is completely determined, the corresponding prediction for $X$ would be a certain one.

What about quantum theory? We could use the data $C$ to infer the particles' (joint) state at time $t_0$, which may be represented as a wavefunction $\Psi$. Then, employing Schrödinger's equation and the Born rule, we could establish a probability distribution over $X$, telling us how likely we are, at time $t$, to find the particles at certain positions. However, due to the probabilistic nature of the Born rule, even when $\Psi$ is known to arbitrary accuracy, the corresponding prediction for $X$ is in general uncertain.

---

[1] We may of course also employ physical theories to infer past events, corresponding to a retrodiction (rather than a prediction).

R. Colbeck (✉)
Department of Mathematics, University of York, YO10 5DD, UK
e-mail: roger.colbeck@york.ac.uk

R. Renner
Institute for Theoretical Physics, ETH Zurich, 8093 Zurich, Switzerland

© Springer-Verlag Berlin Heidelberg 2015
P. Blanchard, J. Fröhlich (eds.), *The Message of Quantum Science*, Lecture Notes
in Physics 899, DOI 10.1007/978-3-662-46422-9_4

The question then arises as to whether the quantum-mechanical wavefunction $\Psi$ is "sufficient" to generate the most precise predictions. This question is closely related to that raised in 1935 by Einstein, Podolsky, and Rosen who asked whether the quantum-theoretic description of physical reality can be considered complete [2] (cf. Footnote 2). A longstanding debate was initiated [3–6] that is still ongoing today. This matter, besides its relevance for the foundations of quantum theory, has recently gained practical importance. For example, the sufficiency of the wavefunction is a prerequisite for standard security proofs in quantum cryptography, or for establishing that quantum random number generators are "truly random" [7].

The aim of this chapter is to review a recent line of work [8–11] in which we presented arguments supporting the sufficiency of the wavefunction for optimal predictions—and hence that quantum theory is complete. More precisely, we show that any extension of quantum theory yielding predictions that are more informative than those based on the wavefunction necessarily has a rather undesirable feature: it is incompatible with a natural notion of "free" choice.

The chapter is organised as follows. In Sect. 4.2 we describe the main claims and their underlying assumptions on an informal level. We then continue with a more formal treatment. Section 4.3 presents a simple mathematical framework that allows us to study and compare the predictions made by different physical theories. The notion of free choice, which is central to our considerations, is defined and discussed in Sect. 4.4. Sections 4.5–4.9 are devoted to the mathematical formulation and proofs of (some of) our claims, first in a basic (Sects. 4.7 and 4.8) and then in a generalised form (Sect. 4.9). The concluding Sect. 4.10 discusses limitations of the approach and its relation to similar results.

## 4.2 Overview of the Main Claims

Suppose (as above) that we would like to predict a future observation, $X$, based on available data, $C$. Most generally, $C$ and $X$ may be modelled as random variables, with joint probability distribution $P_{CX}$. That the setup admits a quantum-mechanical description means that there exists a wavefunction, $\Psi$ (which we model as another random variable that may be correlated with $C$) such that the probability distribution of $X$ conditioned on $\Psi$, $P_{X|\Psi}$, obeys the Born rule. For our considerations, we will usually assume that the available data $C$ allows us to infer the wavefunction $\Psi$ to arbitrary precision, so that $\Psi$ is a function of $C$, i.e., $\Psi = \Psi(C)$.

To formulate our claims, we need to compare the quantum-mechanical predictions to those obtained from possible alternative theories. For any alternative theory, we denote by $\Lambda$ the collection of variables that it uses to describe a given experimental setting. One may think of $\Lambda$ as a "state variable" within the alternative theory, analogously to the wavefunction $\Psi$ in quantum theory. In our framework, $\Lambda$ is simply a random variable, and the prediction of the alternative theory for a future observation $X$ corresponds to the conditional distribution $P_{X|\Lambda}$.

We now describe a few basic properties that an alternative theory may (or may not) have. They play a crucial role for our considerations and will be defined more formally in the subsequent sections.

**(QM) Compatibility with Quantum Theory** The alternative theory can be applied to any quantum-mechanical experiment. More precisely, any experimental data that admits a quantum-mechanical description (in the sense that it is compatible with the predictions based on $\Psi$) also admits a description using the alternative theory (with predictions based on $\Lambda$).

**(FR) Compatibility with Free Choice** If an experimental parameter can be chosen independently of certain other values according to quantum theory, then this should also be allowed according to the alternative theory.

**($\Psi$-S) Sufficiency of $\Psi$** No prediction of quantum theory (based on $\Psi$) is improved by the alternative theory (i.e., when taking into account $\Lambda$).[2]

These properties are all we need to describe our main claim.

**Claim 1** *Properties* (QM) *and* (FR) *imply* ($\Psi$-S).

It can be argued that Properties (QM) and (FR) are quite reasonable requirements in light of experiments. Property (QM) merely demands that the alternative theory can be used whenever quantum theory can (without contradicting it), and hence can be considered an "extension" of quantum theory. Given that quantum theory has so far not been falsified by any experiment, there is at least no experimental evidence that speaks against this assumption. We also note that in a recent experiment [12], the validity of Property (QM) has been verified to good accuracy for a particular setup relevant to our claims. Furthermore, Property (FR) means that if, according to quantum theory, we can choose an experimental parameter at random then this should also be the case according to the extended theory.

Theories that satisfy Properties (QM) and (FR) may be termed *free extensions of quantum theory*. Claim 1 then reads as follows.

**Claim 1'** *The wavefunction* $\Psi$ *is sufficient within any free extension of quantum theory.*

An interesting corollary of this result is that the wavefunction is determined uniquely by the state variable $\Lambda$ of any free extension of quantum theory, provided $\Lambda$ is sufficiently informative. To make this more precise, consider any alternative

---

[2]In previous work, we have sometimes used the term "completeness" rather than "sufficiency" to describe Property ($\Psi$-S). The notion is indeed closely related to what Einstein, Podolsky and Rosen (EPR) [2] called "completeness of quantum theory". They characterised this by the requirement that "every element of the physical reality must have a counterpart [in quantum theory]", where the "elements of physical reality" correspond to quantities that can be predicted with certainty. This means that, if the prediction based on $\Lambda$ is deterministic then also the prediction based on $\Psi$ must be deterministic. Hence, Property ($\Psi$-S) implies completeness of quantum theory according to EPR.

theory with state variable $\Lambda$. The theory may have the following property, which is the counterpart to ($\Psi$-S).

($\Lambda$-S) **Sufficiency of** $\Lambda$  No prediction of the alternative theory (based on $\Lambda$) is improved by quantum theory (i.e., when taking into account $\Psi$).

The corollary may now be phrased as follows.

**Claim 2**  *Properties* (QM), (FR), *and* ($\Lambda$-S) *imply that* $\Psi$ *is determined by* $\Lambda$.

Remarkably, this claim is in some sense a converse to Claim 1. To see this, we reformulate it analogously to Claim 1'.

**Claim 2'**  *Any sufficient free extension of quantum theory contains the wavefunction* $\Psi$.

The claim is closely related to a recent result by Pusey, Barrett, and Rudolph [13], who showed that the wavefunction $\Psi$ is uniquely determined by the "real physical state" of a system (and is therefore also "real"). Indeed, the same conclusion can be obtained from Claim 2, if one interprets the "real physical state" of a system as a variable $\Lambda$ that satisfies Properties (QM), (FR), and ($\Lambda$-S). Further discussion on this point is deferred to Sect. 4.10.

We conclude this introductory section by noting that Claims 1' and 2', taken together, imply that any sufficient free extension of quantum theory is essentially equivalent to quantum theory. More precisely, the variable $\Lambda$, on which the predictions of the extended theory is based, is in one-to-one correspondence with the quantum-mechanical wavefunction $\Psi$ (up to possible redundancies).

## 4.3  Predictions

It should be clear from the informal description above that our approach is operational. To illustrate this, it is useful to imagine an experimental setup where each of the components (e.g., sources and measurement devices) is equipped with a printer that continuously prints all relevant information (such as its configuration or the result of a measurement) on a slip of paper. Our arguments may then be formulated in terms of statements about the printed values. Mathematically, the printed values are modelled as random variables, and our technical claims will refer to their joint probability distribution.[3]

When we analyse a given experiment, there is usually a well-defined set of quantities, in the following denoted by $\Upsilon$, about which we want to make statements.

---

[3]We note that there is some tension between quantum systems undergoing unitary evolution and the existence of classical random variables describing print-outs. We take the view that random variables are defined from the point of view of an observer and that it makes sense for an observer to assign random variables to distant devices prior to the outcome reaching him. (See also Sect. 4.10.)

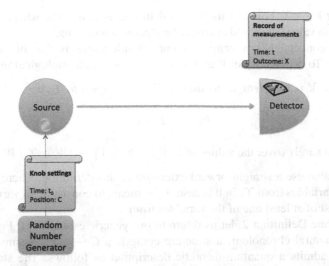

**Fig. 4.1** Basic experiment. Physical theories can be used to predict the outcome of a measurement $X$ based on data $C$. The figure shows a simple example, where $X$ is the measurement output of a detector at time $t$, and where $C$ is the configuration of a source at some earlier time $t_0$. For our argument, we will often assume that this configuration is chosen at random

As a simple (but rather generic) example, consider a setup consisting of a particle source and a detector, as depicted by Fig. 4.1. Here $\Upsilon$ could comprise two quantities, $C$ and $X$, where $C$ corresponds to a print-out of the source's configuration at time $t_0$, when the particle is emitted, and $X$ is the print-out of the detector at time $t$, after the particle has interacted with it. We could then ask how well we can predict $X$ from $C$.

In general, when talking about predictions, it is convenient to distinguish between "future" observations, which we want to predict, and "past" ones, on which our predictions are based. This time ordering could be obtained naturally from our operational interpretation, by associating to each value in $\Upsilon$ the point in spacetime at which it is printed. For our purposes, the exact location of the values is not relevant, and we merely need to know their relative order in time. This motivates the following definition.

**Definition 1** A *chronological structure* $\Upsilon$ is a set of random variables equipped with a binary relation, denoted by $\rightarrow$, such that[4]

- $X \rightarrow X$ (reflexivity);
- $X \rightarrow Y$ and $Y \rightarrow Z$ imply $X \rightarrow Z$ (transitivity).

The statement $X \rightarrow Y$ should be interpreted as "$X$ was printed before $Y$". Note that two elements of $\Upsilon$ may be incomparable, i.e., it could be that neither

---

[4]In mathematics, such a binary relation is called a *preorder* or *quasiorder*.

$X \to Y$ nor $Y \to X$ holds. This is useful to model situations where $X$ and $Y$ correspond to values observed at spacelike separated locations.[5]

Another concept that is central to our considerations is that of "sufficient" information. To define this, let $V$ and $X$ be elements of a chronological structure $\Upsilon$.

**Definition 2**  $V$ is *sufficient for predicting* $X$ *(with respect to* $\Upsilon$*)* if

$$P_{X|V} = P_{X|VV_{\downarrow}} \tag{4.1}$$

holds almost surely (over the values of $V$, $V_{\downarrow}$), where $V_{\downarrow} \equiv \{W \in \Upsilon : W \to V\}$.

We will also use a straightforward extension of this definition, where $X$ and $V$ are sets of variables from $\Upsilon$. In this case, $V_{\downarrow}$ is meant to consist of all variables that lie in the past of at least one of the variables from $V$.

To illustrate Definition 2, let us return to our generic example (cf. Fig. 4.1) for which the natural chronological structure comprises $C \to X$. Assume that the experiment admits a quantum-theoretic description as follows. The state of the particle emitted at time $t_0$ is given by a wavefunction $\Psi$ on a Hilbert space $\mathcal{H}$. Furthermore, the evolution of the particle's state from time $t_0$ (when it leaves the source) to $t$ (when it reaches the detector) is given by a unitary $U$ on $\mathcal{H}$. Finally, the measurement carried out by the detector may be modelled by a family $\{\Pi_x\}_{x \in \mathcal{X}}$ of projectors on $\mathcal{H}$. The probability distribution $P_{X|\Psi}$ of the measurement result $X$, conditioned on any value of the wavefunction $\Psi = \psi$, is then given by the Born rule,[6]

$$P_{X|\Psi}(x|\psi) = \langle \psi | U^{\dagger} \Pi_x U | \psi \rangle . \tag{4.2}$$

To treat this quantum-mechanical description within our framework, it is convenient to consider extending the chronological structure $\Upsilon$ to $\Upsilon^{QM}$, which additionally includes the random variable $\Psi$ together with the relations[7]

$$C \leftrightarrow \Psi \tag{4.3}$$

The latter reflects that we are interested in the quantum state $\Psi$ that the system has at the time when $C$ is chosen.[8] That the wavefunction $\Psi$ is sufficient for predicting

---

[5]While it is natural to use a chronological structure compatible with relativistic spacetime, we stress that this is not necessary for our technical claims.

[6]Note that we do not consider more general states (described by density operators) or measurements (described by POVMs) here—such generalised states and measurements already have "extensions" within quantum theory as a consequence of Naimark's theorem.

[7]We use $C \leftrightarrow \Psi$ as a shorthand for the two relations $C \to \Psi$ and $C \leftarrow \Psi$.

[8]To comply with our operational approach, one may imagine that the source, upon printing the value $C$ at time $t_0$, also prints $\Psi$.

$X$ w.r.t. $\Upsilon^{\text{QM}}$ then corresponds to the condition

$$P_{X|\Psi} = P_{X|\Psi C} .$$

Sufficiency of $\Psi$ w.r.t. $\Upsilon^{\text{QM}}$ thus means that $\Psi$ contains all information about the configuration $C$ of the particle source that is relevant for predicting the measurement outcome $X$.

Analogously to quantum theory, in which the predictions are based on the wavefunction $\Psi$, an alternative theory may give predictions based on other variables, which we collectively denote by $\Lambda$. To study the alternative theory, we consider an extended chronological structure $\Upsilon^{\text{Ext}}$ that, in addition to the variables of $\Upsilon^{\text{QM}}$, contains $\Lambda$, as well as the relations $C \leftrightarrow \Lambda$, generalising (4.3). Property ($\Psi$-S), specified informally in Sect. 4.2, then corresponds to the requirement that $\Psi$ is sufficient for predicting $X$ w.r.t. the extended chronological structure $\Upsilon^{\text{Ext}}$ ($P_{X|\Psi} = P_{X|\Psi \Lambda C}$ and, in particular, $P_{X|\Psi} = P_{X|\Psi \Lambda}$). Similarly, Property ($\Lambda$-S) means that $\Lambda$ is sufficient for predicting $X$ w.r.t. $\Upsilon^{\text{Ext}}$ (in particular, $P_{X|\Lambda} = P_{X|\Lambda \Psi}$).

## 4.4   Free Choice

It is difficult to imagine talking about physics without free choice. It is a notion that finds itself embedded within the usual language we use to describe physical scenarios. We ask questions of the form "What would happen if ..." and reason about the consequences. For example, we may ask what peak altitude a ball obtains if we throw it at different angles and speeds. Clearly, this is only meaningful under the premise that it is possible to set up, at least in principle, the different scenarios in question.

Thus, free choice is a property of the way we describe the world, with an interventionist picture. We would not find a theory satisfying if it were unable to compute the future evolution for certain choices of the initial conditions. Sometimes free choice has been called a no-conspiracy assumption, the idea being that if free choice were not to hold—i.e., were it impossible to set up a particular scenario—it would be a conspiracy on the part of nature preventing certain experiments being performed in certain situations.

The concept of free choice also plays an important role for the formulation of our main claims. We view it as a property of a theory, capturing the idea that certain parameters of the theory can be chosen independently. While, as argued above, any (reasonable) physical theory should have this property (in some form), it is usually not described explicitly. One of the notable exceptions is the work of Bell [14], who characterised the "free" variables of a theory as follows:

> For me this means that the values of such variables have implications only in their future light cones.

The definition we use here (Definition 3 below) may be seen as a formal version of this criterion. Although some bespoke definitions have been used in particular scenarios, as far as we are aware, no other consistent criterion has been proposed in the literature that is applicable to a general scenario (see also [15] for a discussion).[9]

Consider, once again, our generic setting consisting of a particle source and a detector (cf. Fig. 4.1). Suppose that the source has a knob that allows selection among a set of possible configurations, and that, at time $t_0$, its setting $C$ is determined by a random number generator. Note that $C$ may determine the properties of the emitted particle, which could in turn influence the detector output $X$ obtained at time $t$. Hence, despite $C$ being chosen "freely" at random, it will in general be correlated with $X$. Crucially, however, $C$ can only be correlated with things generated after it is chosen, and if $X$ was obtained at a time $t < t_0$, before $C$ is randomly generated, we would expect them to be independent, i.e.,[10]

$$P_{CX} = P_C \times P_X . \qquad (4.4)$$

This motivates the following definition. Let $\Upsilon$ be a chronological structure that contains $C$ as an element.

**Definition 3** $C$ is *free (with respect to $\Upsilon$)* if

$$P_{CC_\uparrow} = P_C \times P_{C_\uparrow} \qquad (4.5)$$

where $C_\uparrow \equiv \{X \in \Upsilon : C \nrightarrow X\}$.

In words, $C$ is free if it is statistically independent of all other variables in $\Upsilon$, except those that lie in its future.

Note that because in general the chronological structure contains incomparable elements, $C_\uparrow$ is a larger set of variables than those that lie in the past (see Fig. 4.2 for examples). One may therefore wonder whether a weakened variant of Definition 3, where $C_\uparrow$ was replaced by the set $C_\downarrow = \{X \in \Upsilon : X \rightarrow C\}$, would be sensible, i.e., whether the independence criterion (4.5) could be replaced by

$$P_{CC_\downarrow} = P_C \times P_{C_\downarrow} . \qquad (4.6)$$

To illustrate that it cannot, consider a set $\Upsilon$ consisting of two variables, $C$ and $C'$, and assume that their chronological structure is trivial, i.e., $C \nrightarrow C'$ and $C' \nrightarrow C$.

---

[9]The definitions given in the literature usually refer to specific scenarios, e.g., where the measurements applied to a fixed initial state are chosen freely. However, it is unclear how to generalise these definitions, for instance, to the case where the initial state may also be chosen freely. Indeed, while these definitions ensure that a freely chosen measurement setting is independent of certain hidden variables (such as the particle positions according to Bohmian mechanics), this requirement would make little sense for the free choice of an initial state.

[10]In relativistic spacetime, we would expect the independence condition (4.4) to be satisfied whenever $t < t_0$ holds in *some* inertial frame.

**Fig. 4.2** Two examples of a chronological structure. In (**a**), $F$ being free implies $P_{F|EG} = P_F$, while in (**b**), if $A$ is free then $P_{A|BYZ} = P_A$, for example

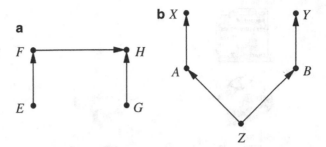

Specifically, one may imagine that $C$ and $C'$ correspond to the configuration of two distant devices, and that $C$ and $C'$ are chosen simultaneously at time $t_0$ (w.r.t. some inertial frame) by two experimentalists sitting next to the devices. Suppose that we ask the two experimentalists to choose the respective configurations at random, but that we later find that $C$ and $C'$ are perfectly correlated. Then it would be natural to conclude that (at least) one of the choices was not "free" (for any reasonable interpretation of this term). Indeed, neither $C$ nor $C'$ meet the requirement of Definition 3, which demands that $P_{CC'} = P_C \times P_{C'}$. Nevertheless, both $C$ and $C'$ would meet criterion (4.6), because $C_\downarrow$ is the empty set. This weaker criterion is therefore not sufficient for characterising free variables.

We also note that, although Definition 3 refers to probability distributions, it is equally applicable to deterministic theories. For example, we may use classical mechanics to compute the trajectory of a particle even if its initial conditions, $C$, were chosen using a random number generator. As long as we only use the output of the random number generator, but do not require that its internal workings be described within classical mechanics, we can safely consider a chronological structure that does not include any variables correlated to $C$ that lie in its past. Hence, $C$ could still be considered free according to Definition 3.

## 4.5 Bipartite Measurement Scenario

To formulate our Claims 1 and 2 on a technical level, we will refer to a particular experimental setup, similar to the one considered by Einstein et al. [2] and by Bell [3]. It consists of a source, which we call *Charlie*, and two detectors, called *Alice* and *Bob*, arranged symmetrically around Charlie (see Fig. 4.3). Charlie emits a signal (e.g., a photon pulse) at time $t_0$. Later, at time $t$ when the signal reaches Alice and Bob, they carry out measurements, whose outcomes we denote by $X$ and $Y$, respectively. Furthermore, each of the devices is equipped with a knob to choose between different configurations, and the knobs are set using random number generators at time $t_0$.

The experiment may be described within the framework introduced above. For this, we consider a chronological structure $\Upsilon$ (cf. Fig. 4.4) that contains the

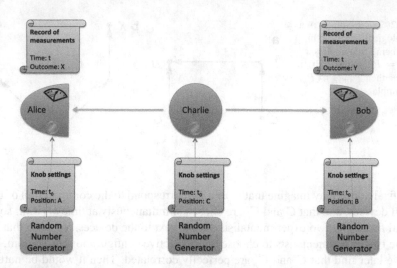

**Fig. 4.3** Bipartite measurement scenario. A source, Charlie, sends signals to detectors, Alice and Bob. The configurations $A$, $B$, and $C$ of the detectors and the source are randomly chosen at time $t_0$, and the detectors yield measurement outcomes $X$ and $Y$ at time $t$. Under the condition that $c(t - t_0)$ is smaller than the distance between Alice and Bob, the two measurement processes are spacelike separated

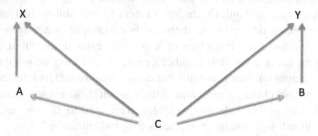

**Fig. 4.4** Bipartite chronological structure. The bipartite chronological structure $\Upsilon$ consists of all values relevant to the bipartite measurement scenario depicted in Fig. 4.3. The properties we need for our statements are (4.7) and (4.8); the relations $C \to A$ and $C \to B$ may or may not hold

random variables $C$, $A$, $B$, $X$, and $Y$, corresponding to the knob settings and measurement outcomes. $\Upsilon$ is equipped with the relations

$$A \to X, \quad B \to Y, \quad C \to X, \quad C \to Y, \tag{4.7}$$

which reflect the fact that $C$ and $A$ are chosen before outcome $X$ is obtained, and that $C$ and $B$ are chosen before outcome $Y$ is obtained. Furthermore, we will use that

$$A \nrightarrow Y, \quad B \nrightarrow X, \tag{4.8}$$

i.e., there is no definite time-ordering between the two measurements by Alice and Bob. One may think of the devices being located far enough apart that the two measurement processes are spacelike separated. In the following we will refer to this particular $\Upsilon$ as the *bipartite chronological structure*.

Note that we have, up to this point, not said anything about the probability distribution of the random variables of $\Upsilon$. To specify this distribution, we will from now assume that the experiment admits a particular description within quantum theory. More precisely, we assume that the state of the signal emitted at time $t_0$ is given by a wavefunction $\Psi$, i.e., a unit vector in a Hilbert space $\mathcal{H}$, which may depend on $C$. Furthermore, for any possible knob setting $A = a$, the measurement carried out by Alice at time $t$ is represented by a family $\{\Pi_x^a\}_{x \in \mathcal{X}}$ of projectors on a Hilbert space $\mathcal{H}$ with $\sum_x \Pi_x^a = \mathrm{id}_{\mathcal{H}}$. Similarly, for any $B = b$, Bob's measurement is given by a family $\{\Pi_y^b\}_{y \in \mathcal{Y}}$ of projectors on $\mathcal{H}'$. Finally, we assume that the evolution of the system from time $t_0$ to time $t$ corresponds to an isometry $U$ from $\mathcal{H}$ to $\mathcal{H} \otimes \mathcal{H}'$. We can then apply the Born rule, according to which the conditional probability distribution $P_{XY|AB\Psi}$ is given by

$$P_{XY|AB\Psi}(x, y|a, b, \psi) = \langle \psi | U^\dagger (\Pi_x^a \otimes \Pi_y^b) U | \psi \rangle \ . \tag{4.9}$$

As in Sect. 4.3 [see the text around Eq. (4.3)], we denote by $\Upsilon^{\mathrm{QM}}$ the chronological structure obtained by adding $\Psi$ to $\Upsilon$, together with the relations $C \leftrightarrow \Psi$ (cf. Fig. 4.5).

The variables defined within the bipartite measurement scenario have natural properties w.r.t. the chronological structure $\Upsilon^{\mathrm{QM}}$: the wavefunction $\Psi$ is sufficient for predictions within quantum theory (Remark 1), and $A$, $B$, and $C$ can be chosen freely (Remark 2). These properties hold under the assumption that the Born rule (4.9) is valid conditioned on any value of the source configuration $C$, i.e.,

$$P_{XY|AB\Psi C} = P_{XY|AB\Psi} \ . \tag{4.10}$$

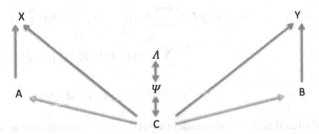

**Fig. 4.5** Extended bipartite chronological structure. The bipartite chronological structure $\Upsilon$ (see Fig. 4.4) may be supplemented with additional random variables. We denote by $\Upsilon^{\mathrm{QM}}$ the chronological structure that includes the wavefunction $\Psi$, and by $\Upsilon^{\mathrm{Ext}}$ the one that also contains the state variable $\Lambda$ of an extended theory

Because $\Psi$ may depend arbitrarily on $C$, this assumption is fulfilled whenever the experiment admits a quantum-mechanical description.

*Remark 1* $\{\Psi, A, B\}$ is sufficient for predicting $\{X, Y\}$ w.r.t. $\Upsilon^{QM}$. In particular, $\Psi$ is sufficient for predicting $X$ and $Y$ conditioned on any $A = a$ and $B = b$.

*Proof* By the definition of sufficiency (cf. Definition 2 as well as the extension described thereafter) and because $\{\Psi, A, B\}_\downarrow = \{C\}$ the claim is equivalent to (4.10). $\qquad\square$

Note that sufficiency of $\Psi$ w.r.t. $\Upsilon^{QM}$ is a strictly weaker condition than Property $\Psi$S introduced in Sect. 4.2. The latter corresponds to the requirement that $\Psi$ is not only sufficient w.r.t. $\Upsilon^{QM}$, but also w.r.t. any extended chronological structure, denoted $\Upsilon^{Ext}$, which may include additional variables defined within an alternative theory (see Fig. 4.5). Proving this stronger sufficiency condition ($\Psi$-S) is one of the main aims of this chapter (cf. Claim 1).

Before stating the next property, we remark that quantum theory does not prescribe how $A$, $B$, and $C$ are chosen. We may thus, in particular, choose $A$ and $B$ independently of $C$ and $\Psi$, i.e., such that

$$P_{ABC\Psi} = P_A \times P_B \times P_{C\Psi} . \tag{4.11}$$

*Remark 2* $A$, $B$, and $C$ are free w.r.t. $\Upsilon^{QM}$ whenever they are chosen according to (4.11).

*Proof* That $C$ is free follows directly from the fact that $C_\uparrow$ can only contain $A$ and $B$, and that $P_{ABC} = P_{AB} \times P_C$, according to (4.11). Furthermore, we have $A_\uparrow = \{B, C, \Psi, Y\}$. Hence, to verify that $A$ is free it suffices to check the condition

$$P_{ABC\Psi Y} = P_A \times P_{BC\Psi Y} . \tag{4.12}$$

But using (4.9), (4.10), and $\sum_x \Pi_x^a = \mathrm{id}_{\mathcal{H}}$ we find that

$$
\begin{aligned}
P_{Y|ABC\Psi}(y|a, b, c, \psi) &= \sum_x P_{XY|ABC\Psi}(x, y|a, b, c, \psi) \\
&= \sum_x \langle \psi | U^\dagger (\Pi_x^a \otimes \Pi_y^b) U | \psi \rangle \\
&= \langle \Psi | U^\dagger (\mathrm{id}_{\mathcal{H}} \otimes \Pi_y^b) U | \Psi \rangle .
\end{aligned}
$$

Since the right hand side is independent of $A = a$, we conclude that $P_{Y|ABC\Psi} = P_{Y|BC\Psi}$. Furthermore, (4.11) implies that $P_{BC\Psi|A} = P_{BC\Psi}$. We thus find that $P_{YBC\Psi|A} = P_{Y|ABC\Psi} P_{BC\Psi|A} = P_{Y|BC\Psi} P_{BC\Psi}$ is independent of $A = a$, which proves (4.12). That $B$ is free follows by symmetry. $\qquad\square$

Remark 2 explains our formulation of Property (FR) given in Sect. 4.2. There we refer to the variables that can be chosen freely within quantum theory. Following Remark 2, these include $A$, $B$, and $C$. Property (FR) demands that these can also be chosen freely within an alternative theory. Technically, this means that $A$, $B$, and $C$ should not only be free w.r.t. $\Upsilon^{QM}$, but also w.r.t. the extended chronological structure, $\Upsilon^{Ext}$, which includes all variables used by the alternative theory.

We conclude this section with the remark that the bipartite nature of the chronological structure, in particular condition (4.8), is crucial for the formulation of our technical claims and for their operational interpretation. For example, if we used a chronological structure that includes the relations $A \leftrightarrow B$, the assumption that $A$ and $B$ are free (w.r.t. this structure) would not suffice to guarantee that they are uncorrelated.

## 4.6   Free Choice and Local Causality

The notion of "local causality", introduced formally by Bell [16], captures the idea that physical influences propagate continuously in spacetime. It is implicitly used in Einstein, Podolsky, and Rosen's argument [2], and it is an assumption of Bell's theorem, which asserts that quantum correlations cannot be reproduced by a realistic model [3]. While our claims do not rely on this assumption, there is a connection between local causality and the concept of free choice, as defined in Sect. 4.4. In the following we briefly discuss this.[11] In addition, note that Lemma 1 will play a role in the proof of our main claims.

Consider the bipartite measurement scenario described in Sect. 4.5 and let $\Upsilon$ be the bipartite chronological structure, which consists of the choices $A$, $B$, and $C$ as well as the measurement outcomes $X$ and $Y$ (cf. Fig. 4.4). Bell argued that, if the latter are correlated then they must have a common cause that lies in their past. We may model this common cause by an additional variable $\Lambda$, which is added to $\Upsilon$, together with the relations

$$\Lambda \to X, \quad \Lambda \to Y . \tag{4.13}$$

The latter reflect the idea that $\Lambda$ is in the common past of $X$ and $Y$. The resulting extended chronological structure then corresponds to $\Upsilon^{Ext}$ defined above (cf. Fig. 4.5).[12]

Local causality demands that, given the information in the past of a measurement, there is no other information that can be correlated to the outcome. Applied to the

---

[11]For a general discussion of local causality we refer to [17].

[12]In this section we don't make reference to the variables $C$ and $\Psi$ (which are also part of $\Upsilon^{Ext}$). However, one may include them by replacing $\Lambda$ with the triple $(\Lambda, C, \Psi)$.

two measurements in our scenario, this criterion corresponds to the conditions

$$P_{X|ABY\Lambda} = P_{X|A\Lambda} \quad \text{and} \quad P_{Y|ABX\Lambda} = P_{Y|B\Lambda} \,. \tag{4.14}$$

Note that these imply

$$P_{X|AB\Lambda} = P_{X|A\Lambda} \quad \text{and} \quad P_{Y|AB\Lambda} = P_{Y|B\Lambda} \,, \tag{4.15}$$

as well as

$$P_{X|ABY\Lambda} = P_{X|AB\Lambda} \quad \text{and} \quad P_{Y|ABX\Lambda} = P_{Y|AB\Lambda} \,. \tag{4.16}$$

The term "parameter independence" is often used for the conditions (4.15), while "outcome independence" is used for (4.16).[13] Bell's theorem can be summarised as saying that quantum theory is incompatible with local causality and free choice in the bipartite measurement scenario [6]. We discuss how this follows from our result in Sect. 4.10.

It turns out that, in this scenario, parameter independence (4.15) is implied by free choice w.r.t. the bipartite chronological structure.[14] This follows directly from Lemma 1 below. Local causality and free choice are thus connected in that both have parameter independence as a consequence.

**Lemma 1** *If $A$ and $B$ are free w.r.t. $\Upsilon^{\text{Ext}}$ then $P_{X\Lambda|AB} = P_{X\Lambda|A}$ and $P_{Y\Lambda|AB} = P_{Y\Lambda|B}$ (wherever these conditional distributions are defined).*

*Proof* The distribution $P_{X\Lambda AB}$ can be decomposed as

$$P_{X\Lambda AB} = P_{X\Lambda|AB} P_A P_{B|A}$$

$$P_{X\Lambda AB} = P_{X\Lambda|A} P_A P_{B|X\Lambda A} \,.$$

By definition, if $B$ is free then $P_{B|A\Lambda X} = P_B$. This implies that the last factors on the right hand side of the two equalities, $P_{B|A}$ and $P_{B|X\Lambda A}$, are equal. Since the left hand side is also equal, we find $P_{X\Lambda|AB} = P_{X\Lambda|A}$ as desired. The second part of the claim follows by symmetry.                                                                    □

Although this argument shows that free choice implies parameter independence (4.15), it does not imply outcome independence (4.16), and hence is strictly weaker than local causality. Indeed, while "plain" quantum theory (where the common cause $\Lambda$ consists only of the wavefunction $\Psi$) is perfectly compatible with

---

[13]Parameter independence (4.15) is also related to "no-signalling" in the sense that if Bob (for example) has access to $\Lambda$ then the condition $P_{Y|AB\Lambda} = P_{Y|B\Lambda}$ follows from the assumption that Alice cannot signal to Bob by choosing different values for $A$.

[14]We do not have a general definition of parameter independence, hence we speak only of the bipartite scenario here.

free choices (see Remark 2), it violates outcome independence.[15] This also points to an important difference between the usual formulation of Bell's theorem [6] (as well as similar results, such as Leggett's theorem [18]) and Claim 1. The assumptions of the latter (in particular freedom of choice) are perfectly compatible with quantum theory. In contrast, Bell's theorem is based on local causality, so that plain quantum theory is already excluded by assumption.

In general, if a particular theory violates outcome independence (4.16), this need not imply that no locally causal explanation exists. Instead, the violation could be due to the insufficiency of $\Lambda$, i.e., there could be correlations between $X$ and $Y$ beyond those mediated by $\Lambda$. If so, one possibility is that introduction of additional variables restores outcome independence. However, in the case of quantum theory, Bell's theorem implies that no such additional variables can exist.

We also note that certain "non-local" extensions of quantum theory violate the parameter independence conditions (4.15). This is for example the case for Bohmian mechanics, if we include the hidden particle positions in the variable $\Lambda$. For these theories, it follows directly from Lemma 1 that they cannot be compatible with free choice w.r.t. the bipartite chronological structure.

Possibly due to such considerations, the definition of free choice we use has sometimes been mistaken for a locality assumption. However, as explained above, it is strictly weaker than local causality. Indeed, because the free choice assumption is satisfied by plain quantum theory, it may be motivated by experimental observations: in our bipartite measurement scenario, for example, one would always find that the statistics obeys the condition $P_{X|AB} = P_{X|A}$. Furthermore, this observation holds whether or not the measurements are spacelike separated. This motivates taking this as a feature of an alternative theory, the rationale being that it would be strange if $P_{X|AB\Lambda}$ were not equal to $P_{X|A\Lambda}$, but that averaging over $\Lambda$ happens in just the right way to ensure $P_{X|AB} = P_{X|A}$.

## 4.7   Basic Claims

The goal of this section is to provide formal versions of Claims 1 and 2 (cf. Sect. 4.2). For this, we consider the bipartite measurement scenario described in Sect. 4.5, for specific choices of states prepared by the source and measurements carried out by the detectors.

Let $\mathcal{H}$ and $\mathcal{H}'$ be $d$-dimensional Hilbert spaces with orthonormal bases $\{|z\rangle\}_{z\in[d]}$, where $[d] = \{0,\ldots,d-1\}$. Let $\psi_d$ be an arbitrary state on $\mathcal{H}$ and let $U$ be an

---

[15]The outcomes $X$ and $Y$ of a measurement on an entangled state $\Psi$ are generally correlated and thus violate outcome independence (4.16) for $\Lambda = \Psi$.

isometry from $\mathcal{H}$ to $\mathcal{H} \otimes \mathcal{H}'$ such that

$$U |\psi_d\rangle = \sqrt{1/d} \sum_{z \in [d]} |z\rangle \otimes |z\rangle \in \mathcal{H} \otimes \mathcal{H}' . \qquad (4.17)$$

We denote by $\hat{X}_d = \sum_{z \in [d]} |z\rangle\langle z \oplus 1|$ the generalised Pauli operator (where $\oplus$ means addition modulo $d$). For any $n \in \mathbb{N}$ and for any rational values[16] $a, b \in [0,1]_\mathbb{Q} \equiv [0,1] \cap \mathbb{Q}$ we define the families of projectors $\{\Pi_x^a\}_{x \in [d]}$ on $\mathcal{H}$ and $\{\Pi_y^b\}_{y \in [d]}$ on $\mathcal{H}'$ by

$$\Pi_x^a = (\hat{X}_d)^a |x\rangle\langle x| (\hat{X}_d^\dagger)^a \qquad (4.18)$$

$$\Pi_y^b = (\hat{X}_d)^b |y\rangle\langle y| (\hat{X}_d^\dagger)^b . \qquad (4.19)$$

Finally, let $\Upsilon^{\text{Ext}}$ be the extended bipartite chronological structure defined in Sect. 4.5 (see Fig. 4.5), with random variables $A, B, C, X, Y, \Lambda$, and $\Psi$.

Theorem 1 below is a formal version of Claim 1. This version is however restricted, as it applies only to local measurements on the particular state defined by (4.17). Later, in Sect. 4.9, we will provide a generalisation of the theorem to arbitrary measurements on arbitrary states.

**Theorem 1 (Restricted Version)** *If*

- (QM) *conditioned on $\Psi = \psi_d$, the Born rule (4.9) holds;*
- (FR) $A$ *and* $B$ *are free (w.r.t. $\Upsilon^{\text{Ext}}$) and their support contains* $[0,1]_\mathbb{Q}$

*then, conditioned on $A = 0$ and $\Psi = \psi_d$,*

- ($\Psi$-S) $\Psi$ *is sufficient for predicting $X$ (w.r.t. $\Upsilon^{\text{Ext}}$).*

Claim 2 can be obtained as a corollary from this theorem. We first provide a restricted version where the wavefunction $\Psi$ is chosen from a set of two different wavefunctions, $\psi_d$ and $\psi_r$, both of which satisfy (4.17) for the same, appropriately chosen isometry $U$.

**Corollary 1 (Restricted Version)** *Assume that $\Psi$ takes values only from the set $\{\psi_d, \psi_r\}$ for some $r, d \in \mathbb{N}$, with $r < d$. If*

- (QM) *the Born rule (4.9) holds;*
- (FR) $A$ *and* $B$ *are free (w.r.t. $\Upsilon^{\text{Ext}}$) and their support contains* $[0,1]_\mathbb{Q}$;
- ($\Lambda$-S) *conditioned on $A = 0$, $\Lambda$ is sufficient for predicting $X$ (w.r.t. $\Upsilon^{\text{Ext}}$)*

*then there exists a function $f$ such that $f(\Lambda) = \Psi$ holds almost surely.*

Note that for any two wavefunctions $\psi_d$ and $\psi_r$ on $\mathcal{H}$ with overlap $\langle \psi_d | \psi_r \rangle = \sqrt{r/d}$ there exists an isometry $U$ such that (4.17) holds for both $d$

---

[16]We restrict to the set of rational values since these are the only ones required for the proof.

and $r$. Since, in addition, $d$ and $r$ can be any positive integers, Corollary 1 is applicable to any two wavefunctions whose overlap is the square root of a rational number. In Sect. 4.9 we will provide a more general version of the corollary which extends the claim to any countable set of wavefunctions.

## 4.8   Proof of the Basic Claims

In this section we give an overview of the proofs of Theorem 1 and Corollary 1. For more details, we refer to [9] and [19]. The argument uses the idea of "non-signalling correlations", which we explain first.

Let $d, n \in \mathbb{N}$, let $X$ and $Y$ be random variables that take values from the set $[d] = \{0, 1, \ldots, d - 1\}$, and suppose that $A$ and $B$ take values $A \in \mathcal{A}_n \equiv \{0, \frac{2}{2n}, \ldots, \frac{2n-2}{2n}\}$ and $B \in \mathcal{B}_n \equiv \{\frac{1}{2n}, \frac{3}{2n}, \ldots, \frac{2n-1}{2n}\}$. Define the quantity $I_{n,d}(P_{XY|AB})$, a function of the conditional distribution $P_{XY|AB}$, as follows

$$I_{n,d}(P_{XY|AB}) \equiv P(X \oplus 1 \neq Y | A = 0, B = \tfrac{2n-1}{2n})$$

$$+ \sum_{\substack{a,b \\ |a-b|=1/2n}} P(X \neq Y | A = a, B = b),$$

where the addition $\oplus$ is modulo $d$, and where

$$P(X \neq Y | A = a, B = b) \equiv 1 - \sum_x P_{XY|AB}(x, x | a, b).$$

The quantity $I_{n,d}$ is an extension of a quantity used to formulate chained Bell inequalities [20, 21]. That a large violation of these implies low correlation with other variables was first shown in [22], where it was used for a novel key distribution scheme.

Consider an arbitrary additional random variable $\Lambda$. As we shall see (Lemma 2 below), if $P_{XY\Lambda|AB}$ satisfies the *non-signalling conditions*[17]

$$P_{X\Lambda|AB} = P_{X\Lambda|A} \quad \text{and} \quad P_{Y\Lambda|AB} = P_{Y\Lambda|B} \tag{4.20}$$

then the value of $I_{n,d}(P_{XY|AB})$ gives an upper bound on the maximum information that $\Lambda$ can provide about $X$ (or $Y$). Note that $I_{n,d}$ is a function of the distribution without $\Lambda$, so can be estimated from the correlations between $X$ and $Y$ even if $\Lambda$ is unknown.

---

[17]Whereas the proof does of course not depend on the interpretation of these conditions, we note that the term "non-signalling" originates from the operational meaning described in Footnote 13. Alternatively, the conditions could be interpreted as mirroring an analogous property of the wavefunction $\Psi$ (namely $P_{X\Psi|AB} = P_{X\Psi|A}$) to the variable $\Lambda$.

To illustrate the idea behind the proof of Lemma 2, let $d = n = 2$ and consider the distribution $P_{XY|AB} = P^{\text{NL}}_{XY|AB}$, known as the "non-local box distribution" [23, 24], defined by

$$P^{\text{NL}}_{XY|AB}(x, y|a, b) = \begin{cases} 1/2\delta_{x,y} & \text{if } (a,b) \in \{(0, \frac{1}{4}), (\frac{2}{4}, \frac{1}{4}), (\frac{2}{4}, \frac{3}{4})\} \\ 1/2\delta_{x,1\oplus y} & \text{if } (a,b) = (0, \frac{3}{4}). \end{cases}$$

The distribution satisfies

$$P(X \neq Y|0, \tfrac{1}{4}) = P(X \neq Y|\tfrac{2}{4}, \tfrac{1}{4}) = P(X \neq Y|\tfrac{2}{4}, \tfrac{3}{4}) = P(X = Y|0, \tfrac{3}{4}) = 0\,,$$

which is equivalent to $I_{2,2}(P^{\text{NL}}_{XY|AB}) = 0$. Now let $\Lambda$ be such that the non-signalling conditions (4.20) hold and let $Z = Z(\Lambda)$ be an arbitrary binary value computed from $\Lambda$ (one may think of $Z$ as a guess for $X$). We then have $P_{XZ|AB} = P_{XZ|A}$ and $P_{YZ|AB} = P_{YZ|B}$. Using this, we find

$$P(Z = X|A = 0, B = \tfrac{3}{4}) = \sum_z P_{XZ|AB}(z, z|0, \tfrac{3}{4}) = \sum_z P_{XZ|AB}(z, z|0, \tfrac{1}{4})$$

$$= \sum_z P_{YZ|AB}(z, z|0, \tfrac{1}{4}) = \sum_z P_{YZ|AB}(z, z|\tfrac{2}{4}, \tfrac{1}{4})$$

$$= \sum_z P_{XZ|AB}(z, z|\tfrac{2}{4}, \tfrac{1}{4}) = \sum_z P_{XZ|AB}(z, z|\tfrac{2}{4}, \tfrac{3}{4})$$

$$= \sum_z P_{YZ|AB}(z, z|\tfrac{2}{4}, \tfrac{3}{4}) = \sum_z P_{YZ|AB}(z, z|0, \tfrac{3}{4})$$

$$= \sum_z P_{XZ|AB}(1 \oplus z, z|0, \tfrac{3}{4}) = P(Z \neq X|A = 0, B = \tfrac{3}{4})$$

$$= 1 - P(Z = X|A = 0, B = \tfrac{3}{4})\,.$$

The equality implies that $P(Z = X|A = 0, B = \tfrac{3}{4}) = 1/2$.[18] This implies that $Z$ is uncorrelated to $X$. Since $Z$ is an arbitrary function of $\Lambda$, the same holds for $\Lambda$, that is, $\Lambda$ cannot be used to predict $X$.

Although this is for a specific case, there is a general connection between $I_{n,d}$ and the ability to make improved predictions. The latter is measured in terms of the distance between the conditional distribution $P_{X\Lambda|a} \equiv P_{X\Lambda|A}(\cdot, \cdot|a)$ and the product distribution $\bar{P}_X \times P_{\Lambda|a}$, where $\bar{P}_X$ is the uniform distribution. The following lemma, whose proof can be found in [11], provides an upper bound on this distance.[19]

---

[18] An analogous argument works for any $a$ and $b$, so $P(Z = X|A = a, B = b) = 1/2$.

[19] Note that for two distributions, $P_X$ and $Q_X$, defined on a set $\mathcal{X}$, $\|P_X - Q_X\|_1 \equiv \sup_{\mathcal{X}' \subseteq \mathcal{X}}(P_X(\mathcal{X}') - Q_X(\mathcal{X}'))$.

**Lemma 2** *Let $P_{XY\Lambda|AB}$ be defined for any $X \in [d]$, $Y \in [d]$, $A \in \mathcal{A}_n$, and $B \in \mathcal{B}_n$. If (4.20) holds then*

$$\left\| P_{X\Lambda|a} - \bar{P}_X \times P_{\Lambda|a} \right\|_1 \leq \frac{d}{2} I_{n,d}(P_{XY|AB}),$$

*for any $a \in \mathcal{A}_n$, with $\bar{P}_X(x) = 1/d$ for $x \in [d]$.*

We saw that $I_{2,2}(P_{XY|AB}^{NL}) = 0$ for the non-local box distribution $P_{XY|AB}^{NL}$ and the lemma thus implies that $X$ is uniform and independent of $\Lambda$, in agreement with our derivation above. More generally, for any $n$ and $d$, there exist distributions $P_{XY|AB}$ for which $I_{n,d}(P_{XY|AB}) = 0$. However, such distributions cannot be realised by local measurements on two parts of a quantum system in the bipartite measurement scenario described in Sect. 4.5. Nevertheless, for the particular state and measurements defined in Sect. 4.7, for any $d$, $I_{n,d}$ approaches 0 for large $n$.

**Lemma 3** *Let $P_{XY|AB\psi_d} \equiv P_{XY|AB\Psi}(\cdot, \cdot | \cdot, \cdot, \psi_d)$ be the probability distribution given by the Born rule (4.9), for $U$, $\psi = \psi_d$, $\{\Pi_x^a\}_{x \in [d]}$, and $\{\Pi_y^b\}_{y \in [d]}$ defined by (4.17)–(4.19), $a \in \mathcal{A}_n$ and $b \in \mathcal{B}_n$. Then*

$$I_{n,d}(P_{XY|AB\psi_d}) \leq \frac{\pi^2}{6n}.$$

The statement follows from a relatively straightforward calculation, whose details are given in [11].

We are now ready to prove the claims formulated in Sect. 4.7.

*Proof of Theorem 1* Note first that distinction between the random variables $\Lambda$ and $C$ is irrelevant for the statement of the theorem. We can therefore without loss of generality think of $C$ as being included in $\Lambda$, which simplifies the notation. Furthermore, since the claim only needs to hold conditioned on $\Psi = \psi_d$, it suffices to consider the conditional probability distribution $P_{XYAB\Lambda|\psi_d}$. Because of Property (FR), Lemma 1 implies that this distribution satisfies the non-signalling conditions (4.20) (see also Footnote 12). Furthermore, because of Property (QM), we have by Lemma 3 that for any $\varepsilon > 0$

$$I_{n,d}(P_{XY|AB\psi_d}) \leq \frac{2\varepsilon}{d}$$

for sufficiently large $n \in \mathbb{N}$. Hence, by Lemma 2 we find that

$$\left\| P_{X\Lambda|a\psi_d} - \bar{P}_X \times P_{\Lambda|a\psi_d} \right\|_1 \leq \varepsilon$$

for any $a \in \mathcal{A}_n$. Note that, for the particular choice $a = 0$, this upper bound holds for any $\varepsilon > 0$, as $a = 0$ is an element of $\mathcal{A}_n$ for any $n$. We thus have

$$P_{X\Lambda|a\psi_d} = \bar{P}_X \times P_{\Lambda|a\psi_d}$$

for $a = 0$. The only variable in the past of $\Psi$ is $\Lambda$ (as well as $C$, which we assume to be included in $\Lambda$). As we have shown that $\Lambda$ is independent of $X$, conditioned on $A = a = 0$, Property ($\Psi$-S) follows. $\square$

*Proof of Corollary 1* If one of the events $\Psi = \psi_d$ and $\Psi = \psi_r$ has probability 0 then the statement is trivially true. We therefore assume in the following that both events have strictly positive probability. In particular, the conditional distribution $P_{X\Lambda|\Psi}(\cdot, \cdot|\psi_d)$ is well defined and hence, because of Assumption (FR), also $P_{X\Lambda|A\Psi}(\cdot, \cdot|a, \psi_d)$ for $a \in [0, 1]_{\mathbb{Q}}$.

Let $\mathcal{L}$ be the set of values $\lambda$ in the range of $\Lambda$ such that the conditional probability $P_{X|\Lambda\Lambda\Psi}(d - 1|0, \lambda, \psi_d)$ is defined and equal to $^1/_d$. By Theorem 1, if we condition on $\Psi = \psi_d$ and $A = 0$, $\Psi$ is sufficient for predicting $X$, which means that

$$P_{X|\Lambda\Lambda\Psi}(d - 1|0, \lambda, \psi_d) = P_{X|A\Psi}(d - 1|0, \psi_d)$$

holds almost surely over the values $\lambda$ for $\Lambda$. Furthermore, it follows from Property (QM) that $X$ is uniformly distributed over $[d]$, so that

$$P_{X|\Lambda\Lambda\Psi}(d - 1|0, \lambda, \psi_d) = {}^1/_d$$

holds almost surely. This implies that the set $\mathcal{L}$ must have weight 1, i.e., $P_{\Lambda|\Psi}(\mathcal{L}|\psi_d) = P_{\Lambda|A\Psi}(\mathcal{L}|0, \psi_d) = 1$.

Likewise, Property ($\Lambda$-S) implies the existence of a set $\tilde{\mathcal{L}}$ such that $P_{\Lambda|A\Psi}(\tilde{\mathcal{L}}|0, \psi_d) = 1$, and, for $\lambda \in \tilde{\mathcal{L}}$,

$$P_{X|\Lambda\Lambda\Psi}(d - 1|0, \lambda, \psi_d) = P_{X|\Lambda\Lambda}(d - 1|0, \lambda). \tag{4.21}$$

Defining $\bar{\mathcal{L}} = \mathcal{L} \cap \tilde{\mathcal{L}}$, we have that for $\lambda \in \bar{\mathcal{L}}$, $P_{X|\Lambda\Lambda}(d - 1|0, \lambda) = {}^1/_d$ with

$$P_{\Lambda|A\Psi}(\bar{\mathcal{L}}|0, \psi_d) = 1. \tag{4.22}$$

Using Property (QM), for $\Psi = \psi_r$ we have $P_{X|A\Psi}(d - 1|0, \psi_r) = 0$ (since $r < d$). From ($\Lambda$-S), it follows that

$$P_{X|A\Psi}(d - 1|0, \psi_r) = \int dP_{\Lambda|A\Psi}(\lambda|0, \psi_r) P_{X|\Lambda\Lambda}(d - 1|0, \lambda), \tag{4.23}$$

and hence that there exists a set $\bar{\mathcal{L}}'$ with

$$P_{\Lambda|A\Psi}(\bar{\mathcal{L}}'|0, \psi_r) = 1, \tag{4.24}$$

such that for $\lambda \in \bar{\mathcal{L}}'$, $P_{X|A\Lambda}(d-1|0,\lambda) = 0$. It follows that $\bar{\mathcal{L}} \cap \bar{\mathcal{L}}' = \emptyset$, and hence we can define the function $f$ such that

$$f(\lambda) = \begin{cases} \psi_d & \text{if } \lambda \in \bar{\mathcal{L}} \\ \psi_r & \text{if } \lambda \in \bar{\mathcal{L}}'. \end{cases}$$

It follows from (4.22) and (4.24) that $f(\Lambda) = \Psi$ holds with probability 1.

$\square$

## 4.9 Generalised Claims

Theorem 1, as stated in Sect. 4.7, holds for local measurements on a bipartite maximally entangled state, but is silent about other measurements. In this section we provide a more general formulation of the theorem, which applies to any projective measurement, in the following denoted $\{\bar{\Pi}_x\}_{x\in\bar{\mathcal{X}}}$, on any pure state $\psi$ in a finite dimensional Hilbert space $\mathcal{H}$.

To formulate the assumptions of the generalised theorem, it is no longer sufficient to assume validity of the Born rule for the particular measurements considered in Sect. 4.7. Instead we need to construct additional measurements, depending on the measurement of interest, $\{\bar{\Pi}_x\}_{x\in\bar{\mathcal{X}}}$, and on $\psi$. The additional measurements are defined on an extended Hilbert space, which contains $\mathcal{H}$ as a subsystem. Property (QM) of the generalised theorem then demands that the Born rule is valid for the additional measurements on the extended space.

To construct the additional measurements, we first note that, according to quantum theory, a projective measurement $\{\bar{\Pi}_x\}_{x\in\bar{\mathcal{X}}}$ corresponds to a physical evolution specified by a trace-preserving completely positive mapping (TPCPM) that takes any density operator $\rho$ on $\mathcal{H}$ to a convex combination of the (mutually orthogonal) post-measurement states. This may be seen as part of a reversible evolution on a larger space, known as the *Stinespring dilatation*. Formally, the latter is given by an isometry $U$ from $\mathcal{H}$ to a product space $\mathcal{H} \otimes \mathcal{H}'$ such that[20]

$$\text{tr}_{\mathcal{H}'}(U\rho U^\dagger) = \sum_{x\in\bar{\mathcal{X}}} \bar{\Pi}_x \rho \bar{\Pi}_x \quad (\forall \rho), \tag{4.25}$$

---

[20]One may argue that *any* implementation of the measurement defined by $\{\bar{\Pi}_x\}_{x\in\bar{\mathcal{X}}}$ admits such a description, as the only condition on $U$ is that it is a Stinespring dilation of the projection operation, cf. (4.25). The physical reason for this is that the measurement process (at least for a short time) corresponds to a unitary evolution of an extended system, which may include part of the measurement apparatus. We also remark that this claim only refers to standard quantum theory and is (at least in principle) experimentally testable.

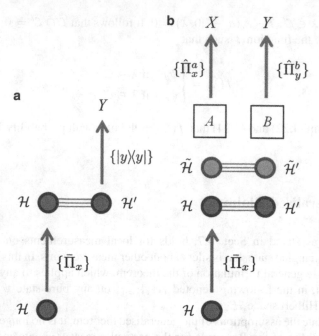

**Fig. 4.6** Illustration of the general argument. (**a**) The measurement of interest, defined by projectors $\bar{\Pi}_x$, entangles the system with another system such that a subsequent measurement on that system, with projectors $|y\rangle\langle y|$, reveals the outcome of the original measurement. (**b**) After the interaction between the two systems corresponding to the quantum description of the measurement, one can consider sets of local measurements, one set that act on the first system and one half of an embezzling state, and the other set that act on the second system and the other half of the embezzling state. These measurements are chosen such that $P_{XY|AB}$ is the same as if the measurements given in (4.18) and (4.19) are applied to a state of the form given in (4.17). In addition, a measurement labelled $B = 0$ is included, which corresponds to measuring the second system with projectors $|y\rangle\langle y|$ (and ignoring the second half of the embezzling state), in other words, when $B = 0$, the outcome of the original measurement is read out

and we can take $U$ to be

$$U = \sum_{x \in \bar{\mathcal{X}}} \bar{\Pi}_x \otimes |x\rangle \ , \tag{4.26}$$

where $\{|x\rangle\}_{x \in \bar{\mathcal{X}}}$ is a family of orthonormal vectors on $\mathcal{H}'$. Thus the original measurement can be thought of as the concatenation of this isometry with a measurement defined by the projectors $\{|x\rangle\langle x|\}_{x \in \bar{\mathcal{X}}}$ on $\mathcal{H}'$ (see also Fig. 4.6a).[21]

For our construction, we will further expand this space by adding a bipartite system $\tilde{\mathcal{H}} \otimes \tilde{\mathcal{H}}'$ initialised in a specific state $|\theta\rangle$. We then consider measurements

---

[21]One may think of $\mathcal{H}'$ as representing (part of) the measurement device used to measure w.r.t. $\{\bar{\Pi}_x\}_{x \in \bar{\mathcal{X}}}$ and the subsequent measurement on $\mathcal{H}'$ as reading the value from the device.

given by families of projectors $\{\hat{\Pi}_x^a\}_{x \in \mathcal{X}}$ and $\{\hat{\Pi}_y^b\}_{y \in \mathcal{Y}}$ on $\mathcal{H} \otimes \tilde{\mathcal{H}}$ and $\mathcal{H}' \otimes \tilde{\mathcal{H}}'$, respectively. The Born rule, adapted to this situation, reads

$$P_{XY|AB\Psi}(x, y|a, b, \psi) = \left(\langle\psi|U^\dagger \otimes \langle\theta|\right)\left(\hat{\Pi}_x^a \otimes \hat{\Pi}_y^b\right)\left(U|\psi\rangle \otimes |\theta\rangle\right). \tag{4.27}$$

With these preparatory remarks, we are ready to state the general version of Claim 1. As before, the chronological structure $\Upsilon^{\text{Ext}}$ is the one defined in Sect. 4.5. The theorem applies to any random variables $A$, $B$, $C$, $X$, $Y$, $\Lambda$, and $\Psi$.

**Theorem 1 (General Version)** *Let* $|\psi\rangle \in \mathcal{H}$, $\{\bar{\Pi}_x\}_{x \in \bar{\mathcal{X}}}$ *be a projective measurement on* $\mathcal{H}$ *and let* $U$ *be an isometry such that (4.26) holds (i.e., $U$ is a Stinespring dilatation). Then there exists* $|\theta\rangle$ *and projective measurements* $\{\hat{\Pi}_x^a\}_{x \in \mathcal{X}}$ *and* $\{\hat{\Pi}_y^b\}_{y \in \mathcal{Y}}$ *parameterised by* $a, b \in [0, 1]_{\mathbb{Q}}$ *such that*

$$\hat{\Pi}_y^{b=0} = |y\rangle\langle y| \otimes \mathrm{id}_{\tilde{\mathcal{H}}'} \quad (\forall y \in \bar{\mathcal{X}}) \tag{4.28}$$

*and such that the following holds. If*

- *(QM) conditioned on* $\Psi = \psi$, *the Born rule (4.27) holds;*
- *(FR) $A$ and $B$ are free (w.r.t. $\Upsilon^{\text{Ext}}$) and their support contains* $[0, 1]_{\mathbb{Q}}$;

*then, conditioned on $B = 0$ and $\Psi = \psi$,*

- *($\Psi$-S) $\Psi$ is sufficient for predicting $Y$ (w.r.t. $\Upsilon^{\text{Ext}}$).*

Note that, for the case $B = b = 0$, which is relevant for Property ($\Psi$-S), the outcome $Y$ is obtained by a measurement that corresponds to the measurement of interest, i.e., the one defined by the projectors $\{\bar{\Pi}_x\}_{x \in \bar{\mathcal{X}}}$. Indeed, it follows from (4.26) and (4.28) that, for any $y \in \bar{\mathcal{X}}$,

$$\sum_x \left(U^\dagger \otimes \langle\theta|\right)\left(\hat{\Pi}_x^a \otimes \hat{\Pi}_y^{b=0}\right)\left(U \otimes |\theta\rangle\right) = U^\dagger(\mathrm{id}_{\mathcal{H}} \otimes |y\rangle\langle y|_{\mathcal{H}'})U = \bar{\Pi}_y.$$

$$\tag{4.29}$$

The theorem is proved in [19]. It is not our intention to repeat the argument here, but rather to explain the ideas behind the result at a more informal level through two examples.

The first simple example is the projective measurement $\{\bar{\Pi}_0, \bar{\Pi}_1\} = \{|0\rangle\langle0|, |1\rangle\langle1|\}$ on the state $|\psi\rangle = \sqrt{1/2}(|0\rangle + |1\rangle)$. For this measurement the isometry $U$ can be taken to be

$$U = \left(|0\rangle_{\mathcal{H}} \otimes |0\rangle_{\mathcal{H}'}\right)\langle0|_{\mathcal{H}'} + \left(|1\rangle_{\mathcal{H}} \otimes |1\rangle_{\mathcal{H}'}\right)\langle1|_{\mathcal{H}'} \tag{4.30}$$

so that

$$U|\psi\rangle = \sqrt{1/2}\left(|0\rangle_{\mathcal{H}} \otimes |0\rangle_{\mathcal{H}'} + |1\rangle_{\mathcal{H}} \otimes |1\rangle_{\mathcal{H}'}\right). \tag{4.31}$$

It is easy to see that this $U$ satisfies (4.26).

We now show that in this case the claim of the generalised version of Theorem 1 follows from its restricted version stated in Sect. 4.7 for $d = 2$. We do not need $|\theta\rangle$, i.e., we can set $\tilde{\mathcal{H}} = \tilde{\mathcal{H}}' = \mathbb{C}$. Furthermore, we take $\hat{\Pi}_x^a = \Pi_x^a$ and $\hat{\Pi}_y^b = \Pi_y^b$ to be the projectors defined by (4.18) and (4.19), respectively. Note that this choice is compatible with condition (4.28). Furthermore, (4.31) implies that (4.17) holds for $U$ as defined above and $\psi_d = \psi$. With (4.17)–(4.19) satisfied, all requirements of the restricted form of Theorem 1 are fulfilled and we can conclude that, provided $A$ and $B$ can be chosen freely and that the Born rule holds, $\Psi$ is sufficient for predicting the measurement outcome $X$ in the case $A = 0$.

In order to extend this conclusion to the case $B = 0$, note that the Born rule gives $P_{XY|AB\psi}(x, y|0, 0, \psi) = 0$ whenever $x \neq y$, in other words, when $A = 0$ and $B = 0$ the outcomes $X$ and $Y$ are always equal. It follows that $\Psi$ is also sufficient for predicting the outcome $Y$ when $B = 0$.

For more general states and projective measurements, using an isometry $U$ with the same form as (4.30) won't necessarily lead to a maximally entangled state of the form (4.17), and hence won't directly allow the restricted form of Theorem 1 to be applied. Hence, to derive the claim we need a slightly stronger argument. This involves some pre-shared entanglement, corresponding to the state $|\theta\rangle$ in (4.27). The next example illustrates this idea (see also Fig. 4.6).

Suppose that the state $|\psi\rangle = \frac{1}{2}(|0\rangle + \sqrt{3}|1\rangle)$ is measured with projectors $\{\bar{\Pi}_0, \bar{\Pi}_1\} = \{|0\rangle\langle 0|, |1\rangle\langle 1|\}$. In this case, the isometry $U$ can be chosen as in the previous example [see Eq. (4.30)] and gives rise to the state

$$U|\psi\rangle = \frac{1}{2}(|0\rangle_{\mathcal{H}} \otimes |0\rangle_{\mathcal{H}'} + \sqrt{3}|1\rangle_{\mathcal{H}} \otimes |1\rangle_{\mathcal{H}'}),$$

and, as before, a subsequent measurement of $\{|0\rangle\langle 0|, |1\rangle\langle 1|\}$ on $\mathcal{H}'$ reads out the outcome. Suppose now that an additional entangled state, $|\theta\rangle_{\tilde{\mathcal{H}}\tilde{\mathcal{H}}'}$, is available. This is taken to be in a special state, called an *embezzling state* [25]. The key feature of this state is that for all $m \in \mathbb{N}$ there exist isometries $V_m : \tilde{\mathcal{H}} \mapsto \hat{\mathcal{H}} \otimes \tilde{\mathcal{H}}$ and $W_m : \tilde{\mathcal{H}}' \mapsto \hat{\mathcal{H}}' \otimes \tilde{\mathcal{H}}'$ such that

$$(V_m \otimes W_m)|\theta\rangle_{\tilde{\mathcal{H}}\tilde{\mathcal{H}}'} = \frac{1}{\sqrt{m}} \sum_{i=0}^{m-1} |m\rangle_{\hat{\mathcal{H}}} \otimes |m\rangle_{\hat{\mathcal{H}}'} \otimes |\theta\rangle_{\tilde{\mathcal{H}}\tilde{\mathcal{H}}'}.$$

Using these, we can define $V : \mathcal{H} \otimes \tilde{\mathcal{H}} \mapsto \mathcal{H} \otimes \hat{\mathcal{H}} \otimes \tilde{\mathcal{H}}$ and $W : \mathcal{H}' \otimes \tilde{\mathcal{H}}' \mapsto \mathcal{H}' \otimes \hat{\mathcal{H}}' \otimes \tilde{\mathcal{H}}'$ via $V = |0\rangle\langle 0| \otimes V_1 + |1\rangle\langle 1| \otimes V_3$ and $W = |0\rangle\langle 0| \otimes W_1 + |1\rangle\langle 1| \otimes W_3$, so that the state $(V \otimes W)(U|\psi\rangle \otimes |\theta\rangle)$ has the form

$$\frac{1}{2}(|00\rangle_{\mathcal{H}\hat{\mathcal{H}}} |00\rangle_{\mathcal{H}'\hat{\mathcal{H}}'} + |10\rangle_{\mathcal{H}\hat{\mathcal{H}}} |10\rangle_{\mathcal{H}'\hat{\mathcal{H}}'}$$
$$+ |11\rangle_{\mathcal{H}\hat{\mathcal{H}}} |11\rangle_{\mathcal{H}'\hat{\mathcal{H}}'} + |12\rangle_{\mathcal{H}\hat{\mathcal{H}}} |12\rangle_{\mathcal{H}'\hat{\mathcal{H}}'}) \otimes |\theta\rangle_{\tilde{\mathcal{H}}\tilde{\mathcal{H}}'},$$

which is isomorphic to

$$\tfrac{1}{2}\big(|0\rangle_{\bar{\mathcal{H}}}\,|0\rangle_{\bar{\mathcal{H}}'} + |1\rangle_{\bar{\mathcal{H}}}\,|1\rangle_{\bar{\mathcal{H}}'} + |2\rangle_{\bar{\mathcal{H}}}\,|2\rangle_{\bar{\mathcal{H}}'} + |3\rangle_{\bar{\mathcal{H}}}\,|3\rangle_{\bar{\mathcal{H}}'}\big) \otimes |\theta\rangle_{\hat{\mathcal{H}}\hat{\mathcal{H}}'}$$

(i.e., $\bar{\mathcal{H}} \cong \mathcal{H} \otimes \hat{\mathcal{H}}$ with the isometry $|00\rangle_{\mathcal{H}\hat{\mathcal{H}}} \mapsto |0\rangle_{\bar{\mathcal{H}}}$, $|10\rangle_{\mathcal{H}\hat{\mathcal{H}}} \mapsto |1\rangle_{\bar{\mathcal{H}}}$, $|11\rangle_{\mathcal{H}\hat{\mathcal{H}}} \mapsto |2\rangle_{\bar{\mathcal{H}}}$, $|12\rangle_{\mathcal{H}\hat{\mathcal{H}}} \mapsto |3\rangle_{\bar{\mathcal{H}}}$ and, similarly, $\bar{\mathcal{H}}' \cong \mathcal{H}' \otimes \hat{\mathcal{H}}'$).

Take $\hat{\Pi}_y^{b=0} = |y\rangle\langle y| \otimes \mathrm{id}_{\bar{\mathcal{H}}'}$, and use $V$ and $W$ to construct the measurements $\hat{\Pi}_x^a = V^\dagger \Pi_x^a V$ and $\hat{\Pi}_y^b = W^\dagger \Pi_y^b W$ ($b \neq 0$) in terms of the projectors $\Pi_x^a$ and $\Pi_y^b$ defined as in (4.18) and (4.19) (except that they are now defined on $\bar{\mathcal{H}}$ and $\bar{\mathcal{H}}'$). Note that the isometries $V$ and $W$ are controlled on $\mathcal{H}$ and $\mathcal{H}'$ respectively, and hence the outcome of the measurement corresponding to $A = 0$ (which is equivalent to a measurement with projectors $\{|x\rangle\langle x|_{\bar{\mathcal{H}}}\}$ on $(V \otimes W)(U|\psi\rangle \otimes |\theta\rangle))$) allows direct determination of the outcome of a measurement on $U|\psi\rangle$ with projectors $\{|y\rangle\langle y|_{\mathcal{H}'}\}$. In other words, when both $A = 0$ and $B = 0$ the outcome $Y$ is a function of $X$ (in our example, $Y = 0$ if $X = 0$ and $Y = 1$ if $X \in \{1,2,3\}$).

The restricted form of Theorem 1 applies to $|\psi_d\rangle = |\psi\rangle \otimes |\theta\rangle$ using the Born rule (4.27), and hence in the case $A = 0$, $\Psi$ is sufficient for predicting $X$. Since for $A = 0$ and $B = 0$, $Y$ is a function of $X$, it follows that $\Psi$ is also sufficient for predicting the outcome $Y$ when $B = 0$. (Somewhat remarkably, using the same embezzling state but varying the isometries $V$ and $W$ always allows us to generate a state arbitrarily close to one for which the restricted form of Theorem 1 applies.)

Finally, we give a generalised version of Corollary 1, which in Sect. 4.7 is stated for the case where $\Psi$ is chosen from a set of only two wavefunctions.

**Corollary 1 (General Version)** *Let $\mathcal{S}$ be a countable set of wavefunctions on $\mathcal{H}$ such that $|\langle \psi | \psi' \rangle| < 1$ for any $|\psi\rangle, |\psi'\rangle \in \mathcal{S}$. Then there exist measurements $\{\Pi_x^a\}_{x \in \mathcal{X}}$ and $\{\Pi_y^b\}_{y \in \mathcal{Y}}$ parameterised by $a, b \in [0,1]_{\mathbb{Q}}$ such that the following holds whenever $\Psi$ takes values from the set $\mathcal{S}$. If*

- *(QM) the Born rule (4.9) holds;*
- *(FR) $A$ and $B$ are free (w.r.t. $\Upsilon^{\mathrm{Ext}}$) and their support contains $[0,1]_{\mathbb{Q}}$;*
- *($\Lambda$-S) conditioned on $A = 0$, $\Lambda$ is sufficient for predicting $X$ (w.r.t. $\Upsilon^{\mathrm{Ext}}$)*

*then there exists a function $f$ such that $f(\Lambda) = \Psi$ holds almost surely.*

For a proof of this extended version of the corollary we refer to [11].

## 4.10  Discussion

We begin this section by briefly discussing the relation between our work and others in the literature.

One of the most prominent results of this type, Bell's theorem [6], follows as a corollary of Theorem 1 (the restricted version given in Sect. 4.7 is sufficient). To see this, let us first consider the original version of Bell's theorem [3]. The theorem refers to "hidden variables" $\Lambda$ that determine the measurement outcome $X$ for any fixed choice of measurement $A$. In our notation, this means that $P_{X|A\Lambda} \in \{0, 1\}$. But since the outcomes of quantum measurements are generally not deterministic, i.e., $P_{X|A\Psi} \notin \{0, 1\}$, this contradicts Property ($\Psi$-S). Theorem 1 thus implies that, provided the Born rule holds and the measurement settings can be chosen freely, there cannot exist any hidden variable $\Lambda$ that determines $X$. This is exactly the statement of Bell's theorem [3].

In later variants of Bell's theorem, the requirement of determinism was weakened to "local causality", as defined in Sect. 4.6, cf. (4.14). In fact, in our framework we can obtain this more general variant from Theorem 1 using only outcome independence (4.16), which is a consequence of local causality. One way to do this was sketched in [19]. However, a more direct way was recently given by Forster [26], who proved the following lemma.

**Lemma 4** *Properties* (QM), ($\Psi$-S) *and outcome independence* (4.16) *are incompatible.*

*Proof* According to Property (QM), the outcomes of the measurements (4.18) and (4.19) on entangled states are generally correlated, so that

$$P_{XY|AB} \neq P_{X|AB}P_{Y|AB} \tag{4.32}$$

(we consider the state, $\Psi$, to be fixed so suppress it in the distributions here). If Property ($\Psi$-S) holds then $P_{X|AB\Lambda} = P_{X|AB}$, and similarly for $Y$. Using outcome independence (4.16) followed by this relation, and writing $P_{XY|AB\lambda}$ for $P_{XY|AB\Lambda}(\cdot, \cdot|\cdot, \cdot\lambda)$, we have

$$P_{XY|AB} = \int dP_{\Lambda|AB}(\lambda) P_{XY|AB\lambda}$$

$$= \int dP_{\Lambda|AB}(\lambda) P_{X|AB\lambda} P_{Y|AB\lambda}$$

$$= \int dP_{\Lambda|AB}(\lambda) P_{X|AB}P_{Y|AB} = P_{X|AB}P_{Y|AB} ,$$

which is in contradiction with (4.32).                                                                   □

Since Theorem 1 asserts that Properties (QM) and (FR) imply ($\Psi$-S), it follows that (QM), (FR) and outcome independence are incompatible, which is another way to state Bell's theorem.

Our second claim, Corollary 1, has also been established by Pusey et al. [13] using a different argument. Instead of being based on free choice, i.e., Property (FR), they assume "preparation independence". The latter means that in situations where quantum theory assigns a product state, so does the alternative theory. More precisely, if independent preparations of $N$ quantum systems are made, so that the joint quantum state is $\rho_1 \otimes \rho_2 \otimes \cdots \otimes \rho_N$, then the "real physical state" $\Lambda$

[in our language, the state of an alternative theory that satisfies Property (Λ-S)] takes a similar product form, i.e., Λ can be divided into $\Lambda_1$, $\Lambda_2$, etc., where $\Lambda_i$ represents the physical state of the $i$th system. As discussed in [11, 19], preparation independence and free choice are related. Furthermore, it has been shown that a setup consisting of at least two systems (as in Sect. 4.5) is necessary to arrive at the claim [27].

One may also ask whether it is possible to further strengthen the claims of Theorem 1 and Corollary 1 by relaxing the assumptions. In particular, it would be interesting to weaken assumption (FR), as this assumption cannot be experimentally verified (without access to Λ). This has led to the question of "randomness amplification": given a random value that satisfies (FR) only approximately, is it possible to generate a random value that satisfies (FR) to arbitrarily good accuracy [28]? This question has been studied recently in a series of work. The conclusion is that randomness amplification is indeed possible, even if the initial randomness satisfies (FR) only with arbitrarily small probability [29–31].

Although this chapter is about quantum theory, we have talked about the predictions it makes from the point of view of a classical observer. For example, we considered the maximum knowledge we can have about the state of a system within quantum theory to be a (classical) description of a pure state. Likewise, in a higher theory we have also considered the predictions based on an additional classical parameter (or set of parameters), Λ.[22] In fact, we can make an analogous argument in the case that the predictions of the higher theory are made based on a hypothetical additional system that has both an input and an output [9]. Such a system can be modelled in terms of its (classical) input-output behaviour (this is the higher theory analogue of modelling a quantum system by its behaviour under measurement).

The use of classical random variables also appears to be in tension with our modelling of a quantum measurement as a unitary interaction. If all quantum processes are unitary, what is the meaning of a random variable? This is not an easy question, and is related to the question of why we don't experience superpositions. If all processes are quantum, then when we read out a measurement outcome, our brain will become entangled with the measured particle and the measurement device. However, we experience a single outcome occurring.

Thus, any random variables are defined from an observer's perspective. Being observer-centred is the way most theories evolved, their main purpose being to enable observers to make predictions. In quantum theory, the random variables do not have universal meaning: in principle, the process of making an observation could be undone if the unitary enacting that observation is reversed. After the reversal, the "outcome" would no longer have any meaning, and the observer would have no

---

[22]The parameters are classical in the sense of being modelled as random variables. However, they may be derived from measurements on a quantum (or more general) system, so the correlations between them need not have a locally causal explanation (which is often taken to be a property of classicality).

memory of it. Thus, the notion of a measurement having taken place only makes sense from the point of view of an observer. In addition, the question of *when* an outcome occurred is observer-dependent. (An observer may not be able to give a precise moment, but can at least give a time after which it makes sense (from their perspective) to say that the measurement has taken place.)

A more general theory may dispense with such classical notions and consider the predictions of a quantum observer (one could think of this as a machine that processes information in a quantum way). This goes beyond the scope of this chapter.

# References

1. van Kampen, N.G.: Ten theorems about quantum mechanical measurements. Physica A **153**, 97–113 (1988)
2. Einstein, A., Podolsky, B., Rosen, N.: Can quantum-mechanical description of physical reality be considered complete? Phys. Rev. **47**, 777–780 (1935)
3. Bell, J.S.: On the Einstein Podolsky Rosen paradox. Physics **1**, 195–200 (1964)
4. Kochen, S., Specker, E.P.: The problem of hidden variables in quantum mechanics. J. Math. Mech. **17**, 59–87 (1967)
5. Bell, J.S.: On the problem of hidden variables in quantum mechanics. Rev. Mod. Phys. **38**, 447–452 (1966)
6. Bell, J.S.: Speakable and Unspeakable in Quantum Mechanics. Cambridge University Press, Cambridge (1987)
7. Frauchiger, D., Renner, R., Troyer, M.: True randomness from realistic quantum devices (2013). arXiv:1311.4547
8. Colbeck, R., Renner, R.: Hidden variable models for quantum theory cannot have any local part. Phys. Rev. Lett. **101**, 050403 (2008)
9. Colbeck, R., Renner, R.: No extension of quantum theory can have improved predictive power. Nat. Commun. **2**, 411 (2011)
10. Colbeck, R., Renner, R.: Is a system's wave function in one-to-one correspondence with its elements of reality? Phys. Rev. Lett. **108**, 150402 (2012)
11. Colbeck, R., Renner, R.: A system's wave function is uniquely determined by its underlying physical state (2013). arXiv:1312.7353
12. Stuart, T.E., Slater, J.A., Colbeck, R., Renner, R., Tittel, W.: Experimental bound on the maximum predictive power of physical theories. Phys. Rev. Lett. **109**, 020402 (2012)
13. Pusey, M.F., Barrett, J., Rudolph, T.: On the reality of the quantum state. Nat. Phys. **8**, 476–479 (2012)
14. Bell, J.S.: Free variables and local causality. In: Speakable and Unspeakable in Quantum Mechanics, Chap. 12. Cambridge University Press, Cambridge (1987)
15. Colbeck, R., Renner, R.: A short note on the concept of free choice (2013). arXiv:1302.4446
16. Bell, J.S.: La nouvelle cuisine. In: Speakable and Unspeakable in Quantum Mechanics, Chap. 24, 2nd edn. Cambridge University Press, Cambridge (2004)
17. Norsen, T.: J.S. Bell's concept of local causality. Am. J. Phys. **79**, 1261 (2011)
18. Leggett, A.J.: Nonlocal hidden-variable theories and quantum mechanics: an incompatibility theorem. Found. Phys. **33**, 1469–1493 (2003)
19. Colbeck, R., Renner, R.: The completeness of quantum theory for predicting measurement outcomes (2012). arXiv:1208.4123
20. Pearle, P.M.: Hidden-variable example based upon data rejection. Phys. Rev. D **2**, 1418–1425 (1970)

21. Braunstein, S.L., Caves, C.M.: Wringing out better Bell inequalities. Ann. Phys. **202**, 22–56 (1990)
22. Barrett, J., Hardy, L., Kent, A.: No signalling and quantum key distribution. Phys. Rev. Lett. **95**, 010503 (2005)
23. Tsirelson, B.: Some results and problems on quantum Bell-type inequalities. Hadronic J. Suppl. **8**, 329–345 (1993)
24. Popescu, S., Rohrlich, D.: Quantum nonlocality as an axiom. Found. Phys. **24**, 379–385 (1994)
25. van Dam, W., Hayden, P.: Universal entanglement transformations without communication. Phys. Rev. A **67**, 060302(R) (2003)
26. Forster, M.R.: How the quantum sorites phenomenon strengthens Bell's argument (2014). e-print arXiv:1403.1598
27. Lewis, P.G., Jennings, D., Barrett, J., Rudolph, T.: Distinct quantum states can be compatible with a single state of reality. Phys. Rev. Lett. **109**, 150404 (2012)
28. Colbeck, R., Renner, R.: Free randomness can be amplified. Nat. Phys. **8**, 450–453 (2012)
29. Gallego, R., et al.: Full randomness from arbitrarily deterministic events. Nat. Commun. **4**, 2654 (2012)
30. Brandão, F.G.S.L., et al.: Robust device-independent randomness amplification with few devices (2013). e-print arXiv:1310.4544
31. Chung, K.-M., Shi, Y., Wu, X.: Physical randomness extractors: generating random numbers with minimal assumptions (2014). e-print arXiv:1402.4797

# Chapter 5
# The Role of the Probability Current for Time Measurements

Nicola Vona and Detlef Dürr

## 5.1 Introduction

Think of a very simple experiment, in which a particle is sent towards a detector.

*When will the detector click?*

Imagine to repeat the experiment many times, starting a stopwatch at every run. The instant at which the particle hits the detector will be different each time, forming a statistics of *arrival times*. Experiments of this kind are routinely performed in almost any laboratory, and are the basis of many common techniques, collectively known as time-of-flight methods (TOF). In spite of that, how to theoretically describe an arrival time measurement is a very debated topic since the early days of quantum mechanics [22]. It is legitimate to wonder why it is so easy to speak about a position measurement at a fixed time, and so hard to speak about a time measurement at a fixed position. An overview of the main attempts and a discussion of the several difficulties they involve can be found in [18–20].

In the following, we will discuss the theoretical description of time measurements with particular emphasis on the role of the probability current.

N. Vona (✉) • D. Dürr
Mathematisches Institut, Ludwigs-Maximilians-Universität München, Theresienstraße 39, D-80333 München, Germany
e-mail: vona@math.lmu.de

© Springer-Verlag Berlin Heidelberg 2015
P. Blanchard, J. Fröhlich (eds.), *The Message of Quantum Science*, Lecture Notes in Physics 899, DOI 10.1007/978-3-662-46422-9_5

## 5.2  What is a Measurement?

We will start recalling the general description of a measurement in quantum mechanics in terms of *positive operator valued measures* (POVMs). This framework is less common than the one based on self-adjoint operators, but is more general and more explicit than the latter.

### 5.2.1  Linear Measurements: POVMs

When we speak about a *measurement*, what are we speaking about?

A measurement is a situation in which a physical system of interest interacts with a second physical system, the apparatus, that is used to inquire into the former. In general, we are interested in those cases in which the experimental procedure is fixed and independent of the state of the system to be measured given as input; these cases are called *linear measurements*. The meaning of this name will be clarified in the following. The analysis of the general properties of a linear measurement, and of the general mathematical description of such a process, has been carried out mostly by Ludwig [16], and finds a natural completion within Bohmian mechanics [8]. In the following, we will present a simplified form of this analysis [6, 21].

We will denote by $x$ the configuration of the system and by $\psi_0$ its initial state, element of the Hilbert space $L^2(\mathbb{R}^{3n})$, while we will use $y$ for the configuration of the apparatus and $\Phi_0 \in L^2(\mathbb{R}^{3N})$ for its ready state; moreover, we will denote by $(0, T)$ the interval during which the interaction constituting the measurement takes place. The evolution of the composite system is a usual quantum process, so the state at time $T$ is

$$\Psi_{T(x,y)} = (U_T \Psi_0)_{(x,y)} = U_T (\psi_0 \Phi_0)_{(x,y)}, \tag{5.1}$$

where $U_T$ is a unitary operator on $L^2(\mathbb{R}^{3(N+n)})$. We call such an interaction a *measurement* if for every initial state $\psi_0$ it is possible to write the final state $\Psi_T$ as

$$\Psi_{T(x,y)} = \sum_{\alpha} \psi_\alpha(x) \, \Phi_\alpha(y), \tag{5.2}$$

with the states $\Phi_\alpha$ normalized and clearly distinguishable, i.e. with supports $G_\alpha = \{y \mid \Phi_\alpha(y) \neq 0\}$ macroscopically separated. This means that after the interaction it is enough to "look" at the position of the apparatus pointer to know the state of the apparatus. Each support $G_\alpha$ corresponds to a different result of the experiment, that we will denote by $\lambda_\alpha$. One can imagine each support to have a label with the value $\lambda_\alpha$ written on it: if the position of the pointer at the end of the measurement is inside the region $G_\alpha$, then the result of the experiment is $\lambda_\alpha$. The probability of getting the outcome $\lambda_\alpha$ is

$$\mathbb{P}_\alpha = \int dx \int_{G_\alpha} dy \, |\Psi_{T(x,y)}|^2 = \int dx \int_{G_\alpha} dy \, |\psi_\alpha(x) \, \Phi_\alpha(y)|^2 = \int |\psi_\alpha(x)|^2 \, dx, \tag{5.3}$$

indeed $\Phi_{\alpha'}(y) = 0 \; \forall y \in G_\alpha, \alpha' \neq \alpha$, and the $\Phi_\alpha$ are normalized. Consider now the projectors $P_\alpha$ that act on the Hilbert space $L^2(\mathbb{R}^{3(N+n)})$ of the composite system and project to the subspace $L^2(\mathbb{R}^{3n} \times G_\alpha)$ corresponding to the pointer in the position $\alpha$, i.e. in particular

$$P_\alpha \Psi_T = \psi_\alpha \, \Phi_\alpha. \tag{5.4}$$

Through the projectors $P_\alpha$ we can define the operators $R_\alpha$ such that

$$P_\alpha \Psi_T = \psi_\alpha \, \Phi_\alpha = (R_\alpha \psi_0) \, \Phi_\alpha, \tag{5.5}$$

that means $R_\alpha \psi_0 = \psi_\alpha$. Finally, we can also define the operators $O_\alpha = R_\alpha^\dagger R_\alpha$. These operators are directly connected to the probability (5.3) of getting the outcome $\alpha$

$$\mathbb{P}_\alpha = \|\psi_\alpha\|^2 = \langle \psi_0 | O_\alpha \psi_0 \rangle. \tag{5.6}$$

Therefore, *the operators $O_\alpha$ together with the set of values $\lambda_\alpha$ are sufficient to determine any statistical quantity related to the experiment.* The fact that any experiment of the kind we have considered can be completely described by a set of linear operators, explains the origin of the name *linear measurement*. Equation (5.6) implies also that the operators $O_\alpha$ are *positive*, i.e.

$$\langle \psi_0 | O_\alpha \psi_0 \rangle \geq 0 \qquad \forall \psi_0 \in L^2(\mathbb{R}^{3n}). \tag{5.7}$$

In addition, they constitute a *decomposition of the unity*, i.e.

$$\sum_\alpha O_\alpha = 1, \tag{5.8}$$

as a consequence of the unitarity of $U_T$ and of the orthonormality of the states $\Phi_\alpha$, that imply

$$1 = \|\psi_0 \Phi_0\|^2 = \|\Psi_T\|^2 = \sum_\alpha \|\psi_\alpha\|^2 = \sum_\alpha \langle \psi_0 | O_\alpha \psi_0 \rangle \qquad \forall \psi_0 \in L^2(\mathbb{R}^{3n}). \tag{5.9}$$

A set of operators with these features is called discrete *positive operator valued measure*, or simply POVM. It is a measure on the discrete set of values $\lambda_\alpha$. In case the value set is a continuum, the POVM is a Borel-measure on that continuum, taking values in the set of positive linear operators.

It is important to note that in the derivation of the POVM structure the orthonormality of the states $\Phi_\alpha$ and the unitarity of the overall evolution play a crucial role, while in general, the states $\psi_\alpha$ do not need to be neither orthogonal nor distinct.

In case the operators $O_\alpha$ happen to be orthogonal projectors, then the usual measurement formalism of standard quantum mechanics is recovered by defining the selfadjoint operator $\hat{A} = \sum_\alpha \lambda_\alpha O_\alpha$. Physically, this condition is achieved for example in a reproducible measurement, i.e. one in which the repetition of the measurement using the final state $\psi_\alpha$ as input, gives the result $\alpha$ with certainty.

We remark that calculating the action of a POVM on a given initial state requires that the initial state is evolved for the duration of the measurement together with an apparatus, and therefore its evolution in general differs from the evolution of the system alone. This circumstance is evident if one thinks that the state of the system after the measurement will depend on the measurement outcome.[1] Usually, if the measurement is not explicitly modeled, this evolution is considered as a black box that takes a state as input and gives an outcome and another state as output. It is important to keep in mind that the measurement formalism always entails such a departure from the autonomous evolution of the system, even if not explicitly described.

### 5.2.2   Not Only POVMs

Although a linear measurement is a very general process, there are many quantities that are not measurable in this sense. An easy example is the probability distribution of the position $|\psi|^2$. Indeed, suppose to have a device that shows the result $\lambda_1$ if the input is a particle in a state for which the position is distributed according to $|\psi_1|^2$, and $\lambda_2$ if it is in a state with distribution $|\psi_2|^2$. If the process is described by a POVM, the linearity of the latter requires that when the state $\psi_1 + \psi_2$ is given as input, the result is *either* $\lambda_1$ *or* $\lambda_2$, as for example the result of a measurement of spin on the state $|up\rangle + |down\rangle$ is either "up" or "down". On the contrary, if the device was supposed to measure the probability distribution of the position, the result had to be $\lambda_+$, corresponding to $|\psi_1 + \psi_2|^2$, possibly distinct both from $\lambda_1$ and from $\lambda_2$.

To overcome a limitation of this kind, the only possibility is to give up on linearity, accepting as measurement also processes different than the one devised in the previous section. These processes use additional information about the $x$-system, for example giving a result dependent on previous runs, or adjusting the interaction according to the state of the $x$-system. In particular, to measure

---

[1] It will be an eigenstate of the selfadjoint operator corresponding to the measurement, in case it exists.

the probability distribution of the position one exploits the fact that $|\psi(x)|^2 = \langle\psi|O_x|\psi\rangle$, where $O_x = |x\rangle\langle x|$ is the density of the POVM corresponding to a position measurement. Instead of measuring directly $|\psi|^2$, one measures $x$, and repeats the measurement on many systems prepared in the same state $\psi$. The distribution $|\psi|^2$ is then recovered from the statistics of the results of the position measurements. The additional information needed in this case is that all the $x$-systems used as input were prepared in the same state. The outcome shown by the apparatus depends then on the preparation procedure of the input state: if we change it, we have to notify the change to the apparatus, that needs to know how to collect together the single results to build the right statistics.

For other physical quantities not linearly measurable, like for example the wave function, a similar, but more refined strategy is required. This strategy is known as *weak measurement* [1]. An apparatus to perform a weak measurement is characterized first of all by having a very weak interaction with the $x$-system; loosely speaking, we can say that the states $\psi_\alpha$ are very close to the initial state $\psi_0$. As a consequence of such a small disturbance, the information conveyed to the $y$-system by the interaction is very little. The departure from linearity is realized in a way similar to that of the measurement of $|\psi|^2$: the single run does not produce any useful information because of the weak coupling, therefore the experiment is repeated many times on many $x$-systems prepared in the same initial state $\psi_0$; the result of the experiment is recovered from a statistical analysis of the collected data.

The advantage of this arrangement is that the output state $\psi_\alpha$ can be used as input for a following linear measurement of usual kind (*strong*), whose reaction is almost as if its input state was directly $\psi_0$. In this case the experiment yields a joint statistics for the two measurements, and it is especially interesting to *postselect* on the value of the strong measurement, i.e. to arrange the data in sets depending on the result of the strong measurement and to look at the statistics of the outcomes of the weak measurement inside each class. For example, a weak measurement of position followed by a strong measurement of momentum, postselected on the value zero for the momentum, allows to measure the wave function [17].

The nonlinear character of weak measurements becomes apparent if one understands the many repetitions they involve in terms of a calibration. Indeed, one can think of the last run as the actual measurement, and of all the previous runs as a way for the apparatus to collect information about the $x$-system used in the last run, profiting from the knowledge that it was prepared exactly as the $x$-systems of the previous runs. The $x$-systems used in the preliminary phase can be then considered part of the apparatus, used to build the joint statistics needed to decide which outcome to attribute to the last strong measurement. For example, the result of the experiment could be the average of the previous weak measurements postselected on the strong value obtained in the last run. If we then change the initial wave function $\psi_0$ to some $\psi_0'$, the calibration procedure has to be repeated. In this case, the apparatus itself depends on the state of the $x$-system to be measured, breaking linearity.

## 5.3 Time Statistics

Now we finally come to our topic: time measurements. At first, we have to note that there are several different experiments that can be called time measurements: measurements of dwell times, sojourn times, and so on. We will refer in the present discussion exclusively to *arrival times*, although it is possible to recast everything to fit any other kind of time measurement. More precisely, we will consider the situation described at the beginning: a particle is prepared in a certain initial state and a stopwatch is set to zero; the particle is left evolving in presence of a detector at a fixed position; the stopwatch is read when the detector clicks. The time read on the stopwatch is what we call *arrival time*.

A measurement of this kind is necessarily linear, and we can ask for the statistics of its outcomes given the initial state of the particle. If, for example, we measure the position at the fixed time $t$, then we can predict the statistics of the results by calculating the quantity

$$\langle \psi_t | x \rangle \langle x | \psi_t \rangle. \tag{5.10}$$

Which calculation do we have to perform to predict the statistics of the stopwatch readings with the detector at a fixed position?

### 5.3.1 The Semiclassical Approach

Arrival time measurements are routinely performed in actual experiments, and they are normally treated semiclassically: essentially, they are interpreted as momentum measurements. The identification with momentum measurements is motivated by the fact that the detector is at a distance $L$ from the source usually much bigger than the uncertainty on the initial position of the particles, so one can assume that each particle covers the same length $L$. Hence, the randomness of the arrival time must be a consequence of the uncertainty on the momentum, and the time statistics must be given by the momentum statistics. For a free particle in one dimension, the connection between time and momentum is provided by the classical relation $p(t) = m L/t$. By a change of variable, this relation implies that the probability density of an arrival at time $t$ is

$$|\tilde{\psi}(p(t))|^2 \left| \frac{dp(t)}{dt} \right| = |\tilde{\psi}(p(t))|^2 \frac{mL}{t^2}, \tag{5.11}$$

where $\tilde{\psi}$ is the Fourier transform of the wave function $\psi$.

This semiclassical approach is justified by the distance $L$ being very big, that is true for most experiments so far performed. On the other hand, we tacitly assumed that the particle moves on a straight line with constant velocity $v$, whose ignorance is the source of the arrival time randomness: such a classical picture is inadequate to describe the behavior of a quantum particle in general conditions, and is expected to fail in future, near-field experiments. A deeper analysis is needed.

## 5.3.2   An Easy but False Derivation

Consider that the particle crosses the detector at time $t$ with certainty. This implies that the particle is on one side of the detector before $t$, and on the other side after $t$. One can therefore think that it is possible to connect the statistics of arrival time to the probability that the particle is on one side of the detector at different times. Because the latter is known, this seems like a good strategy.

For simplicity we will consider only the one dimensional case, that already entails all the relevant features that we want to discuss.[2] The detector is located at the origin; we will assume the evolution of the particle in presence of the detector to be very close to that of the particle alone. We consider the easiest possible case: a free particle, initially placed on the negative half-line and moving towards the origin, i.e. prepared in a state $\psi_0$ such that

$$\psi_0(x) \approx 0 \qquad \forall x \geq 0; \tag{5.12}$$

$$\tilde{\psi}(p) = 0 \qquad \forall p \leq 0, \tag{5.13}$$

where $\tilde{\psi}$ denotes the Fourier transform of $\psi_0$, and Eq. (5.12) is a shorthand for $\int_0^\infty \psi_0(x) dx \ll 1$. The particle can only have positive momentum, therefore it will get at some time to the right of the origin and thus it has to cross the detector from the left to the right.

One might think that the probability to have a crossing at a time $\tau$ later than $t$ is equal to the probability that the particle at $t$ is still in the left region,

$$\mathbb{P}(\tau \geq t) = \mathbb{P}(x \leq 0; t) = \int_{-\infty}^0 dx \, |\psi_t(x)|^2. \tag{5.14}$$

Conversely, the probability that the particle arrived at the detector position before $t$ is

$$\mathbb{P}(\tau < t) = 1 - \mathbb{P}(\tau \geq t) = \int_0^\infty dx \, |\psi_t(x)|^2. \tag{5.15}$$

Therefore, the probability density $\Pi(t)$ of a crossing at $t$ is

$$\Pi(t) = \frac{d}{dt}\mathbb{P}(\tau < t) = \int_0^\infty dx \, \partial_t |\psi_t(x)|^2. \tag{5.16}$$

---

[2]The same treatment is possible in three dimensions, provided that the detector is sensitive only to the arrival time and not to the arrival position, and that the detecting surface divides the whole space in two separate regions (i.e. it is a closed surface or it is unbounded).

We can now make use of the continuity equation for the probability

$$\partial_t \left( |\psi_t(x)|^2 \right) + \partial_x j(x,t) = 0, \tag{5.17}$$

that is a consequence of the Schrödinger equation, with the probability current

$$j(x,t) := \frac{\hbar}{m} \Im \psi_t^*(x) \, \partial_x \psi_t(x). \tag{5.18}$$

Substituting,

$$\Pi(t) = -\int_0^\infty dx \, \partial_x j(x,t) = j(x=0,t). \tag{5.19}$$

Thus, the probability density $\Pi(t)$ of an arrival at the detector at time $t$ is equal to the probability current $j(x=0,t)$, *provided everything so far has been correct*. Well, it hasn't. Equation (5.14) is problematical. It is only correct if the right hand side is a monotonously decreasing function of time, or, equivalently, if the current in (5.19) is always positive. But that is in general not the case and it is most certainly not guaranteed by asking that the momentum be positive. Indeed, even considering only free motion and positive momentum, there are states for which the current is not always positive, a circumstance known as *backflow* (for an example, see the Appendix). But a probability distribution must necessarily be positive, hence, *the current can not be equal to the statistics of the results of any linear measurement*, i.e. there is no POVM with density $O_t$ such that

$$\langle \psi_0 | O_t | \psi_0 \rangle = j(x=0,t). \tag{5.20}$$

This problem is well known [2] and has given rise to a long debate, aiming at finding a quantum prediction for the arrival time distribution with the needed POVM structure [18].

One might wonder: How can it be that the momentum is only positive, and yet the probability that the particle is in the left region is not necessarily decreasing? A state with only positive momentum is such that, if we *measure* the momentum, then we find a positive value with certainty. This is not the same as saying that the particle *moves* only from the left to the right when we do not measure it. Actually, in strict quantum-mechanical terms, it does not even make sense to speak about the momentum of the particle when it is not measured, as it does not make sense to speak about its position if we do not measure it, and therefore there is no way of conceiving how the particle moves in this framework. Think for example of a double slit setting: we can speak about the position of the particle at the screen, but we can not say through which slit the particle went.

Although the quantum-mechanical momentum is only positive, the conclusion that the particle moves only once from the left to the right is unwarranted. Even more: it simply does not mean anything.

### 5.3.3 The Moral

The problem with the simple derivation of the arrival time statistics is quite instructive, indeed it forces us to face the fact that quantum mechanics is really about measurement outcomes, and therefore it is a mistake to think of quantum-mechanical quantities as of quantities intrinsic to the system under study and independent of the measurement apparatus.

## 5.4 The Bohmian View

Bohmian mechanics is a theory of the quantum phenomena alternative to quantum mechanics, but giving the same empirical predictions (see [6, 10]). The two theories share at their foundation the Schrödinger equation. Quantum mechanics complements it by some further axioms like the collapse postulate, and describes all the objects around us only in terms of wave functions. On the contrary, according to Bohmian mechanics the world around us is composed by actual point particles moving on continuous paths, that are determined by the wave function. The Schrödinger equation is in this case supplemented by a guiding equation that specifies the relation between the wave function and the motion of the particles. The usual quantum mechanical formalism is recovered in Bohmian mechanics as an effective description of measurement situations (see [6]).

The main difference between quantum and Bohmian mechanics is that the first one is concerned only with measurement outcomes, while the second one gives account of the *physical reality* in any situation. Although every linear experiment corresponds to a POVM according to quantum mechanics as well as to Bohmian mechanics [8], for the former POVMs are the fundamental objects the theory is all about, while for the latter they are only very convenient tools that occur when the theory is used to make predictions.

We saw already how interpreting quantum-mechanical quantities as intrinsic properties of a system is mistaken, and how the framework of quantum mechanics is limited to measurement outcomes. In Bohmian mechanics the particle has a definite trajectory, so it makes perfectly sense to speak about its position or velocity also when they are not measured, and it is perfectly meaningful to argue about the way the particle *moves*. In doing so, one has just to mind the difference between the outcomes of hypothetical (quantum) measurements, and actual (Bohmian) quantities.

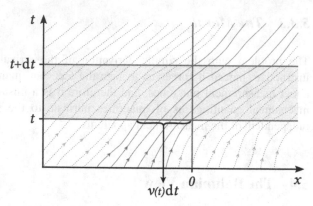

**Fig. 5.1** Bohmian trajectories in the vicinity of the detector, placed at $x = 0$. The trajectories, that cross the detector between the times $t$ and $t + \mathrm{d}t$, are those that at time $t$ have a distance from the detector smaller than the distance they cover during the interval $\mathrm{d}t$, that is $v_{(t)} \, \mathrm{d}t$

### 5.4.1 The Easy Derivation Again...

Let's review the derivation of Sect. 5.3.2 from the point of view of Bohmian mechanics.

To find out the arrival time of a Bohmian particle it is sufficient to literally follow its motion and to register the instant when it actually arrives at the detector position. A Bohmian trajectory $Q_{(t)}$ is determined by the wave function through the equation

$$\dot{Q}_{(t)} = \frac{j(Q(t),t)}{|\psi(Q(t),t)|^2}, \tag{5.21}$$

with $j$ defined in Eq. (5.18). Hence, the Bohmian velocity, that is the actual velocity with which the Bohmian particle moves, is not directly related to the quantum-mechanical momentum, that rather encodes only information about the possible results of a hypothetical momentum measurement. Even if the probability of finding a negative momentum in a measurement is zero, the Bohmian particle can still have negative velocity and arrive at the detector from behind, or even cross it more than once.[3] It is in these cases that the current becomes negative.

We can now repeat the derivation of Sect. 5.3.2 using the Bohmian velocity instead of the quantum-mechanical momentum. We consider again an initial state $\psi_0$ such that $\psi_0(x) \approx 0$ if $x \geq 0$, but we do not ask anymore the momentum to be positive: we rather ask the Bohmian velocity to stay positive for every time after the initial state is prepared. The particle crosses the detector between the times $t$ and $t + \mathrm{d}t$ if at time $t$ they are separated by a distance less than $v_{(x=0,t)} \, \mathrm{d}t$ (cf. Fig. 5.1). The probability that at time $t$ the particle is in this region is $v_{(0,t)} |\psi_{(0,t)}|^2 \, \mathrm{d}t$, thus the probability density of arrival times is simply

$$\Pi_{(t)} = v_{(0,t)} |\psi_{(0,t)}|^2 = j_{(0,t)}. \tag{5.22}$$

---

[3] Note that the notion of *multiple crossings* of the same trajectory is genuinely Bohmian, with no analog in quantum mechanics.

If the velocity does not stay positive, it is still true that the particle crosses the detector during $(t, t + dt)$ if at $t$ they are closer than $v_{(x=0,t)} dt$, but now this distance can also be negative. In this case the current $j_{(0,t)}$ still entails information about the crossing probability, but it also contains information about the direction of the crossing. To get a probability distribution from the current we have to clearly specify how to handle the crossings from behind the detector and the multiple crossings of the same trajectory. For example, one can count only the first time that every trajectory reaches the detector position, disregarding any further crossing, getting the so-called *truncated current* [4, 13].

The Bohmian analysis is readily generalized to three dimensions with an arbitrarily shaped detector, in which case also the arrival position is found. More complicated situations, like the presence of a potential, or an explicit model for the detector, can be easily handled too. Note that the presence of the detector can in principle be taken into account by use of the so-called *conditional wave function* [7, 23], that allows to calculate the actual Bohmian arrival time in exactly the same way as described in this section, although the apparatus needs to be explicitly considered.

### 5.4.2   Is the Bohmian Arrival Time Measurable in an Actual Experiment?

Any distribution calculated from the trajectories conveys some aspects of the actual motion of the Bohmian particle. Such a distribution does not need in principle to have any connection with the results of a measurement, similarly to the Bohmian velocity that is not directly connected to the results of a momentum measurement. The Bohmian level of the description is the one we should refer to when arguing about intrinsic properties of the system rather than measurement outcomes. Since, in the framework of Bohmian mechanics, an intrinsic arrival time exists, namely that of the Bohmian particle, one should ask the intrinsic question that constitutes the title of this section rather than asking the apparatus dependent question

*When will the detector click?*

We do not mean that the latter question is irrelevant, to the contrary, it points towards the prediction of experimental results, that is of course of high value. We shall continue the discussion of the latter topic in Sect. 5.5.

#### 5.4.2.1   Linear Measurement of the Bohmian Arrival Time

We now ask if a linear measurement exists, such that its outcomes are the first arrival times of a Bohmian particle. For sure, this can not be exactly true, indeed, if this was the case, then the outcomes of such an experiment would be distributed according to the truncated current, that depends explicitly on the trajectories and is not sesquilinear with respect to the initial wave function as needed for a POVM.

However, it is reasonable to expect it to be approximately correct for some set of "good" wave functions. That is motivated by the following considerations. A typical position detector is characterized by a set of sensitive regions $\{A_i \subset \mathbb{R}\}_{i=0,...,N}$, each triggering a different result. If the measurement is performed at a fixed time $t$, and if we get the answer $i$, then the Bohmian particle is at that time somewhere inside the region $A_i$. A time measurement is usually performed with a very similar set up: one uses a position detector with just one sensitive region $A_0$ (in our case located around the origin) and waits until it fires. In the ideal case, the reaction time of the detector is very small, and we can consider that the click occurs right after the Bohmian particle entered the sensitive region. As a consequence, if the Bohmian trajectories cross the detector region only once and do not turn back in its vicinity, then we can expect the response of the actual detector to be very close to the quantum current. This puts forward the set of wave functions such that the Bohmian velocity stays positive as a natural candidate for the set of good wave functions. Surprisingly, it can be shown that *there exists no POVM which approximates the Bohmian arrival time statistics on all functions in this set* [25].

On the other hand, it is easy to see that the Bohmian arrival time is approximately given by a measurement of the momentum for all *scattering states*, i.e. those states that reach the detector only after a very long time, so that they are well approximated by local plane waves. Numerical evidence for a similar statement for the states with positive Bohmian velocity and high energy was also produced [25], but a precise determination of the set of good wave functions on which the Bohmian arrival time can be measured is still missing.

An explicit example of a model detector whose outcomes in appropriate conditions approximate the Bohmian arrival time can be found in [3].

### 5.4.2.2 Nonlinear Measurement

An alternative to a linear measurement that directly detects the arrival time of a Bohmian particle is the reconstruction of its statistics from a set of measurements by a nonlinear procedure.

A first possibility in this direction starts by rewriting the probability current (5.18) as

$$j(x,t) = \langle \psi_t | \frac{1}{2m} \left( |x\rangle \langle x| \hat{p} + \hat{p} |x\rangle \langle x| \right) |\psi_t\rangle, \tag{5.23}$$

where $\hat{p} = -i\hbar \, \partial_x$ is the momentum operator. The operator $\hat{j} := \frac{1}{2m} \left( |x\rangle \langle x| \hat{p} + \hat{p} |x\rangle \langle x| \right)$ is selfadjoint, therefore it could be possible to measure the current at the position $x$ and at time $t$ by measuring the average value at time $t$ of the operator $\hat{j}$. Unfortunately, the operational meaning of this operator is unclear.

A viable solution is offered by weak measurements. As showed by Wiseman [27], it is possible to measure the Bohmian velocity, and therefore the current, by a sequence of two position measurements, the first weak and the second strong, used for postselection. Wiseman's proposal has been implemented with small modifications in an experiment with photons[4] [15]. A detailed analysis of the weak measurement of the Bohmian velocity and of the quantum current has been carried out by Traversa et al. [24].

It is worth noting that the weak measurement of the Bohmian velocity, if intended as a calibration of a non-linear measurement as explained in Sect. 5.2.2, gives rise to a genuine measurement, i.e. one whose outcome reveals the actual velocity possessed by the particle in that run [9].

## 5.5 When Will the Detector Click?

We still have to answer the question we posed at the beginning:

*When will the detector click?*

Surely, for any given experiment there is a POVM that describes the statistics of its outcomes. Such an object will depend on the details of the specific physical system and of the measurement apparatus used for the experiment. That is true not only for time measurements, but for any measurement, and for quantum mechanics as for Bohmian mechanics. Yet, we can speak for example of the position measurement in general terms, with no reference to any specific setting, as it was disclosing an intrinsic property of the system. How can that be?

One can speak of the position measurement and of its POVM in general terms because a POVM happens to exist, that has all the symmetry properties expected for a position measurement and that does not depend on any external parameter. That suggests that some kind of intrinsic position exists independently of the measurement details. Recalling how the POVMs have been introduced in Sect. 5.2.1, it is readily clear that they inherently involve an external system (the apparatus) in addition to the system under consideration, and therefore they encode the results of an interaction rather than the values of an intrinsic property. We also saw in Sect. 5.3.2 how interpreting quantum-mechanical statistics as intrinsic objects leads to a mistake. It is therefore very important to keep in mind that all POVMs describe the interaction with an apparatus. Having this clear, it still makes sense to look for a POVM that does not explicitly depend on any external parameter, meaning with this simply that one does not want to give too much importance to the details of the apparatus. Such a POVM may be regarded for example as the limiting element of a sequence of finer and finer devices, and it does not necessarily correspond to

---

[4]This experiment did not, of course, show the existence of a pointlike particle actually moving on the detected paths, but only the measurability of the Bohmian trajectories for a quantum system.

any realizable experiment. Nevertheless, the fortunate circumstance that occurs for position measurements, for which such an idealized POVM exists, does not need to come about for all physical quantities one can think of.

For the arrival time it is possible to show that some POVMs exist that have the transformation properties expected for a time measurement [16], but in three dimensions it is not possible to arrive at a unique expression in the general case, i.e. to something independent of any external parameter. To do so, one needs to restrict the analysis to detectors shaped as infinite planes, or similarly to restrict the problem to one dimension ([14, 26]; see also [11, 12, 18]). In this case, for arrivals at the origin, one finds the POVM

$$K_{(t_1, t_2)} = \sum_{\alpha=\pm 1} \int_{t_1}^{t_2} dT \ |T, \alpha\rangle \langle T, \alpha| \,, \tag{5.24}$$

$$\text{with} \quad \langle p|T, \alpha\rangle = \sqrt{\frac{|p|}{mh}} \ \theta_{(\alpha p)} \ e^{\frac{ip^2 T}{2m\hbar}} \,, \tag{5.25}$$

that corresponds to the probability density of an arrival at time $t$

$$\sum_{\alpha=\pm 1} |\langle t, \alpha|\psi_0\rangle|^2 = \sum_{\alpha=\pm 1} \frac{1}{mh} \left| \int_0^{\alpha\infty} dp \ \sqrt{|p|} \ \langle p|\psi_t\rangle \right|^2 \,. \tag{5.26}$$

Note that $K$ is not a projector valued measure because $\langle T, +|T, -\rangle \neq 0$. For scattering states $K$ becomes proportional to the momentum operator, and the density (5.26) gets well approximated by the probability current [5]. The general conditions under which this approximation holds are still not clear.

### 5.5.1  The Easy Derivation, Once Again

The analysis of Sect. 5.4.2.1 of the measurability of the Bohmian arrival time translates quite easily in an approximate derivation of the response of a detector: essentially what we tried to do in Sect. 5.3.2, just right.

Consider again the setting described in Sect. 5.3.2, but with an initial state such that the Bohmian velocity stays positive. That is equivalent to ask that the probability current stays positive, and therefore that the probability that the particle is on the left of the detector decreases monotonically in time. As described in Sect. 5.4.2.1, thinking of the arrival time detector as of a position detector with only one sensitive region $A_0$ around the origin, it is reasonable to expect that for some set of good wave functions the detector will click right when the particle enters $A_0$. Hence, the probability of a click at time $t$ is approximately equal to the increase of the probability that the particle is inside $A_0$ at that time, i.e. to the probability current through the detector. Therefore, for the good wave functions, the probability current is expected to be a good approximation of the statistics of the clicks of an arrival time detector. As remarked in Sect. 5.4.2.1 the set of the good wave functions is not exactly known, although it is clear that the scattering states are among its elements, and possibly also the states with positive probability current and high energy.

## Appendix: Example of Backflow

We mentioned that, even for states freely evolving and with support only on positive momenta, the quantum current can become negative. We provide now a simple example of this circumstance, depicted in Fig. 5.2. We use units such that $\hbar = 1$, and choose the mass to be one.

We consider the superposition of two gaussian packets, both with initial variance of position equal to 3, corresponding to a variance of momentum of $1/6$. The first packet is initially centered in $x = -10$ and moves with average momentum $p = 2$, while the second packet is centered in $x = -34$ and has momentum $p = 6$. The probability of negative momentum is in this case negligible. The second packet overcomes the first when they are both in the region around the origin, where the detector is placed. In this area the two packets interfere, but then they separate again (cf. Fig. 5.2a).

In Fig. 5.2d the Bohmian trajectories are shown on a big scale. One can see that they never cross, but rather switch from one packet to the other. Moreover, they are almost straight lines, except for the interference region. In that region, it is interesting to look at a higher number of trajectories, making apparent that the trajectories bunch together, resembling the interference fringes (cf. Fig. 5.2b, e).

Looking at the trajectories more in detail (Fig. 5.2f), one can see that they suddenly jump from one fringe to the next, somewhen even inverting the direction of their motion. In this case, it can happen that the particle crosses the detector backwards, leading to a negative current, as shown in Fig. 5.2c.

One could argue that gaussian packets always entail negative momenta, and that this could be the cause of the negative current. To show that this is not the case, we can compare the probability to have negative momentum

$$\mathbb{P}(p < 0) = \int_{-\infty}^{0} |\tilde{\psi}(p)|^2 \, dp \approx 10^{-33} \tag{5.27}$$

with the probability to have a negative Bohmian velocity

$$\mathbb{P}(v(t) < 0) = \int_{K_t} \rho(x,t) \, dx, \tag{5.28}$$

where $K_t := \{x \in \mathbb{R} \mid j(x,t) < 0\}$. For instance, at time $t = 5.2$ this probability is 0.008 (numerically calculated), therefore the negative current can not be caused by the negative momenta.

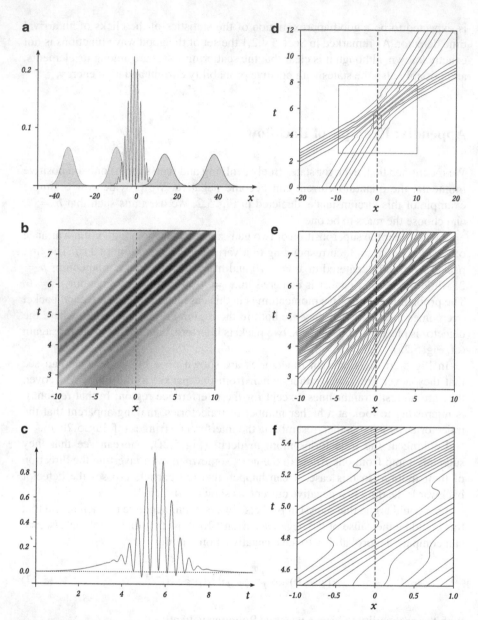

**Fig. 5.2** Example of backflow: superposition of two gaussian packets (for the parameters see text). The *dashed line* always represents the detector. (**a**) Probability density of the position at time $t = 0$ (*gray*), $t = 5.2$ (*blue*), and $t = 12$ (*red*). (**b**) Probability density of the position as a function of position and time. (**c**) Probability current at the screen as a function of time. (**d**) Overall structure of the Bohmian trajectories. The *blue* and the *red rectangles* are magnified in (**e, f**) (Color figure online)

**Acknowledgements** The determination of the Bohmian trajectories shown in Fig. 5.2 is based on the code by Klaus von Bloh for the double slit available at *http://demonstrations.wolfram.com/CausalInterpretationOfTheDoubleSlitExperimentInQuantumTheory/*.
Nicola Vona gratefully acknowledges the financial support of the Elite Network of Bavaria.

# References

1. Aharonov, Y., Albert, D.Z., Vaidman, L.: How the result of a measurement of a component of the spin of a spin-1/2 particle can turn out to be 100. Phys. Rev. Lett. **60**(14), 1351–1354 (1988). doi:10.1103/PhysRevLett.60.1351
2. Allcock, G.R.: The time of arrival in quantum mechanics iii. The measurement ensemble. Ann. Phys. **53**(2), 311–348 (1969). doi:10.1016/0003-4916(69)90253-X. ISSN 0003-4916. http://www.sciencedirect.com/science/article/pii/000349166990253X
3. Damborenea, J.A., Egusquiza, I.L., Hegerfeldt, G.C., Muga, J.G.: Measurement-based approach to quantum arrival times. Phys. Rev. A **66**, 052104 (2002). doi:10.1103/PhysRevA.66.052104. http://link.aps.org/doi/10.1103/PhysRevA.66.052104
4. Daumer, M., Dürr, D., Goldstein, S., Zanghì, N.: On the quantum probability flux through surfaces. J. Stat. Phys. **88**, 967–977 (1997). ISSN 0022-4715. http://dx.doi.org/10.1023/B:JOSS.0000015181.86864.fb. Reprinted in [10]
5. Delgado, V.: Probability distribution of arrival times in quantum mechanics. Phys. Rev. A **57**, 762–770 (1998). doi:10.1103/PhysRevA.57.762. http://link.aps.org/doi/10.1103/PhysRevA.57.762
6. Dürr, D., Teufel, S.: Bohmian Mechanics: The Physics and Mathematics of Quantum Theory. Fundamental Theories of Physics. Springer, New York (2009). ISBN 9783540893431
7. Dürr, D., Goldstein, S., Zanghì, N.: Quantum equilibrium and the origin of absolute uncertainty. J. Stat. Phys. **67**, 843–907 (1992). doi:10.1007/BF01049004. Reprinted in [10]
8. Dürr, D., Goldstein, S., Zanghì, N.: Quantum equilibrium and the role of operators as observables in quantum theory. J. Stat. Phys. **116**, 959–1055 (2004). doi:10.1023/B:JOSS.0000037234.80916.d0. Reprinted in [10]
9. Dürr, D., Goldstein, S., Zanghì, N.: On the weak measurement of velocity in Bohmian mechanics. J. Stat. Phys. **134**, 1023–1032 (2009). doi:10.1007/s10955-008-9674-0. Reprinted in [10]
10. Dürr, D., Goldstein, S., Zanghì, N.: Quantum Physics Without Quantum Philosophy. Springer, New York (2013). ISBN 9783642306907
11. Egusquiza, I.L., Muga, J.G.: Free-motion time-of-arrival operator and probability distribution. Phys. Rev. A **61**, 012104 (1999). doi:10.1103/PhysRevA.61.012104. http://link.aps.org/doi/10.1103/PhysRevA.61.012104
12. Giannitrapani, R.: Positive-operator-valued time observable in quantum mechanics. Int. J. Theor. Phys. **36**(7), 1575–1584 (1997). http://dx.doi.org/10.1007/BF02435757
13. Grübl, G., Rheinberger, K.: Time of arrival from bohmian flow. J. Phys. A **35**(12), 2907 (2002). http://stacks.iop.org/0305-4470/35/i=12/a=313
14. Kijowski, J.: On the time operator in quantum mechanics and the heisenberg uncertainty relation for energy and time. Rep. Math. Phys. **6**(3), 361–386 (1974). doi:10.1016/S0034-4877(74)80004-2. ISSN 0034-4877. http://www.sciencedirect.com/science/article/pii/S0034487774800042
15. Kocsis, S., Braverman, B., Ravets, S., Stevens, M.J., Mirin, R.P., Shalm, L.K., Steinberg, A.M.: Observing the average trajectories of single photons in a two-slit interferometer. Science **332**(6034), 1170–1173 (2011). doi:10.1126/science.1202218. http://www.sciencemag.org/content/332/6034/1170.abstract
16. Ludwig, G.: Foundations of Quantum Mechanics. Texts and Monographs in Physics. Springer, New York (1983, 1985)

17. Lundeen, J.S., Sutherland, B., Patel, A., Stewart, C., Bamber, C.: Direct measurement of the quantum wavefunction. Nature **474**(7350), 188–191 (2011). http://dx.doi.org/10.1038/nature10120
18. Muga, J.G., Leavens, C.R.: Arrival time in quantum mechanics. Phys. Rep. **338**(4), 353–438 (2000). doi:10.1016/S0370-1573(00)00047-8. ISSN 0370-1573. http://www.sciencedirect.com/science/article/pii/S0370157300000478
19. Muga, J., Mayato, R., Egusquiza, I. (eds.): Time in Quantum Mechanics—Vol. 1. Lecture Notes in Physics, vol. 734. Springer, Berlin/Heidelberg (2008)
20. Muga, J., Ruschhaupt, A., del Campo, A. (eds.): Time in Quantum Mechanics—Vol. 2. Lecture Notes in Physics, vol. 789. Springer, Berlin/Heidelberg (2009)
21. Nielsen, M.A., Chuang, I.L.: Quantum Computation and Quantum Information. Cambridge Series on Information and the Natural Sciences. Cambridge University Press, Cambridge (2000). ISBN 9780521635035. http://books.google.de/books?id=65FqEKQOfP8C
22. Pauli, W.: In: Flugge, S. (ed.) Encyclopedia of Physics, vol. 5/1, p. 60. Springer, Berlin (1958)
23. Pladevall, X.O., Oriols, X., Mompart, J.: Applied Bohmian Mechanics: From Nanoscale Systems to Cosmology. Pan Stanford, Singapore (2012). ISBN 9789814316392. http://books.google.it/books?id=mnqNx66amcIC
24. Traversa, F.L., Albareda, G., Di Ventra, M., Oriols, X.: Robust weak-measurement protocol for bohmian velocities. Phys. Rev. A **87**(5), 052124 (2013)
25. Vona, N., Hinrichs, G., Dürr, D.: What does one measure, when one measures the arrival times of a quantum particle? ArXiv e-prints (2013). http://arxiv.org/abs/1307.4366
26. Werner, R.: Screen observables in relativistic and nonrelativistic quantum mechanics. J. Math. Phys. **27**(3), 793–803 (1986). doi:10.1063/1.527184. http://link.aip.org/link/?JMP/27/793/1
27. Wiseman, H.M.: Grounding bohmian mechanics in weak values and bayesianism. New J. Phys. **9**(6), 165 (2007). http://stacks.iop.org/1367-2630/9/i=6/a=165

# Chapter 6
# Quantum Field Theory on Curved Spacetime and the Standard Cosmological Model

Klaus Fredenhagen and Thomas-Paul Hack

## 6.1 Introduction

The attempt to incorporate gravity into quantum theory meets great conceptual difficulties. The main reason for these problems seems to be the rather different roles played by space and time in quantum theory and in Einstein's theory of gravity. In quantum theory, an a priori notion of space and time enters the formulation and the interpretation of the theory in a crucial way. In Einstein's theory of gravity, on the other side, the structure of space and time is dynamical and strongly influenced by the distribution of matter which is treated classically.

These severe conceptual problems are accompanied by hard technical problems, hence testing ideas for solving the problem turns out to be extremely time consuming, and it is difficult to obtain reliable conclusions. In despair, rather radical approaches have been proposed as e.g. string theory and loop quantum gravity, but we think that it is fair to say that none of these approaches has reached its goal, up to now, nor could either of them be ruled out, neither by empirical results nor by inner theoretical reasons.

If one is less ambitious and takes into account, that gravitational forces tend to be very small compared to other forces, one may consider, in a first step, gravity as an external field, producing a curved spacetime, and treat quantum matter by quantum field theory on such a background. One may then, in a second step, treat quantum gravity as a quantum field fluctuating around a given background.

K. Fredenhagen (✉)
II. Institut für Theoretische Physik, Universität Hamburg, Luruper Chaussee 149, D-22761 Hamburg, Germany
e-mail: klaus.fredenhagen@desy.de

T.-P. Hack
Department of Mathematics, University of Genova, via Dodecaneso, 35, I-16146 Genova, Italy

© Springer-Verlag Berlin Heidelberg 2015
P. Blanchard, J. Fröhlich (eds.), *The Message of Quantum Science*, Lecture Notes in Physics 899, DOI 10.1007/978-3-662-46422-9_6

113

The second step meets severe problems: the arising theory is nonrenormalizable, which means that in every order of perturbation theory new interaction terms appear whose coupling constants have to be determined by experiments. Moreover, the causal structure of the theory is determined by the background metric, whereas physics would require that it depends only on the full metric, including the quantum fluctuations. Nevertheless, a consistent perturbative formulation was recently presented by Brunetti, Rejzner and one of us in [7].

Surprisingly, already the first step is by no means trivial. The reason is, that quantum field theory in its standard formulation heavily depends on the symmetries of Minkowski space. These symmetries are used to define the vacuum and the concept of a particle, and one can then, under quite general conditions, derive the existence of scattering states and of an S-matrix.

But on a generic Lorentzian spacetime, no nontrivial symmetries exist, and as a consequence, neither the concept of a vacuum state nor that of particles can be intrinsically introduced. In particular, the classical picture of particles moving in an empty spacetime is not supported by quantum field theory. The most spectacular consequence of this fact is the evaporation of black holes as predicted by Hawking.

The problems of quantum field theory on a given curved back ground have been solved within the last 20 years by using the concepts of algebraic quantum field theory and by replacing techniques of operators on Fock space by methods from microlocal analysis [25]. A compilation of references on algebraic quantum field theory on curved spacetimes can be found in [3].

Algebraic quantum field theory was originally developed in order to understand the relation between the local degrees of freedom of quantized fields and the observed multi-particle states [17]. It was then observed by Dimock and Kay that it provides a good starting point for formulating a theory on a curved spacetime [13, 29]. The absence of a distinguished Hilbert space representation, however, was a severe obstacle for extending the theory to nonlinear fields, the most prominent being the energy momentum tensor.

For this purpose it was necessary to understand the singularities of correlation functions. There was overwhelming evidence that the so-called Hadamard states yield a class of states with the correct singularity structure. A direct characterization of Hadamard states turned out to be rather complicated [30], and its use for the determination of correlation functions of nonlinear fields seemed to be extremely cumbersome.

The situation changed completely when Radzikowski discovered that the Hadamard condition could equivalently be replaced by a positivity condition on the wave front set of the 2-point function [41, 42]. This marked the breakthrough for the modern theory of quantum fields on curved back grounds, and within a few years it was possible to construct all kinds of composite fields [6] and to prove the existence of renormalized time ordered products [5].

Renormalization, however, had still the problem that renormalization conditions at different points of spacetime could not be compared with each other in the absence of nontrivial symmetries. A new principle was needed, the principle of local covariance [8]. This principle says that it is not meaningful to do physics on a special spacetime; instead all structures should depend only on the local geometry. Based on this principle, Hollands and Wald were able to finish the renormalization program [22, 23], which had been started by Brunetti and one of us [5]. One of the outcomes of this generalization of algebraic quantum field theory is that it is meaningful to consider the same field on different spacetimes.

A direct application of this fact is the use of the energy momentum tensor as a source term for Einstein's equation. But as long as gravity itself is not quantized one has the problem to compare a quantum object with a classical object. On a pragmatic level this may be solved by using the expectation value of the energy momentum tensor. This might be reasonable as long as the fluctuations are small enough. But here new problems arise. One is the fact that the correlation functions of the energy momentum tensor diverge at coinciding points. One therefore looks at appropriate averages; this, however, introduces a new parameter into the theory. The other problem is even worse: whereas fields exist which can be considered to be the same on different spacetimes, a corresponding identification of states on different spacetimes does not exist.

The latter problem can presumably only be treated in a theory containing quantized gravitational and matter fields. One may, however, restrict oneself to situations with higher symmetries, as they arise in cosmological spacetimes of the Friedmann–Robertson–Walker type. There, one may admit only states which are invariant under the spatial symmetries. Still, this does not fix the states uniquely, hence additional choices have to be introduced. Nevertheless, one can in this way reproduce the standard cosmological model from first principles, by modelling the matter-energy content of the universe entirely in terms of quantum fields rather than effectively by means of a classical perfect fluid [18].

## 6.2  The Free Scalar Field and Its Normal Ordered Products

Classically, a configuration of a scalar field may be understood as a smooth function on spacetime. Let $C^\infty(M)$ be the set of all smooth functions on a spacetime $M$, and let $\mathrm{Sol}(M)$ be the subset of smooth solutions of the Klein–Gordon equation. Classical observables are functions on $C^\infty(M)$ modulo functions which vanish on solutions. The observables of the quantum theory form a suitable subspace on which the algebraic structures of quantum theory can be defined. This subspace can be characterized in the following way.

We consider a globally hyperbolic time oriented spacetime. On such a spacetime the Klein–Gordon equation

$$P\phi = \left(\nabla^a \nabla_a + \xi R + m^2\right)\phi = 0,$$

with curvature scalar $R$, curvature coupling parameter $\xi$ and mass $m$, possesses unique retarded and advanced Green's functions $\Delta_{R,A}$ considered as maps from compactly supported densities to smooth functions. Their difference is the commutator function $\Delta$. A Hadamard solution of the Klein–Gordon operator $P$ is a distributional bisolution $h$ with the properties:

1. $h(x, y) - h(y, x) = i(\Delta(x, y))$.
2. $\mathrm{WF}(h) = \{(x, x'; k, k') \in \mathrm{WF}(\Delta) | k \in V_x^+\}$ where $V_x^+$ is the closed forward lightcone in $T_x^* M$.
3. $h$ is a distribution of positive type.

We want to introduce an associative product $\star_h$ on a subspace $\mathcal{F}(M)$ of the space of maps $\{F : C^\infty(M) \to \mathbb{C}\}$ by setting

$$(F \star_h G)(\phi) = \sum_{n=0}^{\infty} \frac{\hbar^n}{n!} \left\langle \frac{\delta^n F}{\delta \phi^n}, h^{\otimes n} \frac{\delta^n G}{\delta \phi^n} \right\rangle (\phi) . \tag{6.1}$$

In order to make this definition meaningful we require for $F \in \mathcal{F}(M)$:

1. $F$ is polynomial, therefore the sum over $n$ is finite.
2. $F$ is smooth in the sense of the calculus on locally convex spaces, where $C^\infty(M)$ is equipped with its standard topology (uniform convergence of all derivatives on any compact set). From these two conditions it follows that $F$ is of the form

$$F(\phi) = \sum_{n=0}^{N} \langle f_n, \phi^{\otimes n} \rangle$$

with compactly supported distributional densities $f_n$ on $M^n$.
3. The wave front set of $f_n$ does not intersect $(V^+)^n$ nor $(V^-)^n$. This condition guarantees by Hörmander's theorem on the multiplicability of distributions, that, in view of the wave front set of the Hadamard solution, the summands in the definition of the product are well defined.

The product is associative. Complex conjugation induces an involution on $\mathcal{F}(M)$,

$$\overline{F} \star_h \overline{G} = \overline{G \star_h F} ,$$

hence $\mathcal{F}(M)$ gets the structure of a unital *-algebra, where the unit is the constant function $F(\phi) \equiv 1$. The subspace $\{F \in \mathcal{F}(M) | F(\phi) = 0 \text{ for } \phi \in \mathrm{Sol}(M)\}$ is an ideal, and the quotient is the enlarged CCR-algebra. It contains as a subalgebra the CCR-algebra generated by linear functionals of the form $F(\phi) = \langle f, \phi \rangle$ with a smooth density $f$ on $M$ and in addition all local polynomials in the field and its derivatives,

$$F(\phi) = \int f(j_x(\phi)) d\mathrm{vol}(x)$$

where $x \mapsto j_x(\phi) = \{\phi + \psi | \psi \in C^\infty(M) \text{ with } \partial^\alpha \psi(x) = 0 \text{ for all multiindices } \alpha\}$ is the jet prolongation of $\phi$, and $f$ is a smooth function on the jet bundle which is a polynomial in $\phi$ and its derivatives at every point $x \in M$, and which has compact spacetime support

$$\mathrm{supp} F = \bigcup_\phi \mathrm{supp}(x \mapsto f(j_x(\phi))) .$$

The definition of the enlarged CCR-algebra depends on the choice of the Hadamard solution $h$. Since two Hadamard solutions differ by a smooth symmetric and real valued bisolution $w$, the arising algebras are isomorphic with the isomorphism

$$\Gamma_w = \exp \frac{1}{2}\hbar \left\langle w, \frac{\delta^2}{\delta\phi^2} \right\rangle .$$

Every Hadamard solution $h + w$ induces a family of coherent states by

$$\omega_{w,\phi}(F) = (\Gamma_w F)(\phi)$$

with $\phi \in \mathrm{Sol}(M)$. According to a result of Verch, the arising GNS-representations are locally equivalent [49].

A further crucial ingredient for the interpretation of the theory are locally covariant fields $A$. These are, for every spacetime $M$, linear maps $A_M$ from the space of (compactly supported) test tensors to the algebra $\mathcal{F}(M)$ such that, for every isometric, time orientation and causality preserving embedding $\chi : M \to N$ into a larger spacetime $N$ one has the relation

$$A_M(f)(\phi \circ \chi) = A_N(\chi_* f)(\phi)$$

where $\chi_*$ denotes the push forward of test tensors. In other words, a locally covariant field is a natural transformation between the functor $\mathcal{D}$ of test tensor spaces and the functor $\mathcal{F}$ of observable algebras, both based on the category of globally hyperbolic spacetimes with isometric, time orientation and causality preserving embeddings as morphisms.

In a first attempt one may look at a polynomial $p(\partial^\alpha \phi, \alpha \in \mathbb{N}_0^d)$ in $\phi$ and its derivatives and set

$$A_M(f)(\phi) = \int f(x) p(\partial^\alpha \phi(x)) d\,\mathrm{vol}(x) .$$

But this definition violates the naturality condition for locally covariant fields since there is no natural choice for the Hadamard solution, i.e. no choice which is compatible with all possible embeddings of a spacetime into another one, a fact which is responsible for the nonexistence of a vacuum state.

Let $p(\nabla)$ be a polynomial in covariant derivatives (with respect to the Levi-Civita connection) and consider the functionals

$$A(x)(\phi) = e^{p(\nabla)\phi(x)} .$$

Under the isomorphism $\Gamma_w$, $A(x)$ transforms as

$$\Gamma_w A(x) = e^{\frac{1}{2}p(\nabla) \otimes p(\nabla)w(x,x)} A(x) .$$

We now use the fact, that $w$ is the difference of 2 Hadamard solutions. Hadamard solutions admit an asymptotic expansion

$$h(x, y) = \frac{u(x, y)}{\sigma(x, y)} + \sum_{n=0}^{N} v_n(x, y)\sigma(x, y)^n \ln(\mu^2\sigma(x, y)) + w_N^h(x, y)$$

$$= h_N^{\text{sing}}(x, y) + w_N^h(x, y) \, .$$

Here $x, y$ are points in a geodesically convex open set, $\sigma(x, y)$ is the signed square of the geodesic distance between $x$ and $y$, the functions $u$ and $v_n$ are solutions of the so-called transport equations and are uniquely determined by the local geometry. $\mu$ is a free parameter with the dimension of inverse length. $w_N^h$ is an $2N + 1$ times continuously differentiable function which depends on the choice of $h$. We omit the $\epsilon$-prescription necessary for $h_N^{\text{sing}}$ to be well-defined, see [30].

We now set

$$A_h(x) = e^{\frac{1}{2}p(\nabla)\otimes p(\nabla)w_N^h(x,x)} A(x)$$

where $N$ is larger than or equal to twice the degree of $p$, and find

$$\Gamma_{h-h'} A_{h'}(x) = A_h(x) \, .$$

By expanding the exponential series we obtain a large class of locally covariant fields. These correspond to Wick powers of the scalar field and its derivatives regularised by point-splitting and suitable subtractions of derivatives of $h_N^{\text{sing}}$. This class may be enlarged by the $\phi$-independent locally covariant fields constructed from the metric. Further details may be found e.g. in [16].

A locally covariant field of particular interest is the energy momentum tensor $T_{ab}(x)$. However, it is by no means intrinsically clear which locally covariant field is the observable whose expectation value is the "correct" source term for Einstein's equation. Essentially this is due to the fact that gravity is sensitive to absolute energy densities rather than energy density differences. Wald [52] and later Hollands and Wald [24] have suggested that a locally covariant field should satisfy standard commutation relations, covariant conservation $\nabla^a T_{ab}(x)(\phi) = 0$ and suitable analyticity conditions in order to be a meaningful energy momentum tensor. For a free scalar field this implies that the most general energy momentum tensor is of the form

$$T_{ab}(x)(\phi) = T_{ab}^0(x)(\phi) + \alpha_1 g_{ab}(x) + \alpha_2 G_{ab}(x) + \alpha_3 I_{ab}(x) + \alpha_4 J_{ab}(x) \, , \qquad (6.2)$$

where $G_{ab}$ is the Einstein curvature tensor whereas $I_{ab}$ and $J_{ab}$ are local curvature tensors which are obtained as functional derivatives with respect to the metric of the action functionals $\int \sqrt{-g} R^2 d\,\mathrm{vol}(x)$ and $\int \sqrt{-g} R_{ab} R^{ab} d\,\mathrm{vol}(x)$ respectively. Moreover, a possible "model" $T_{ab}^0$ is the functional

$$T_{ab}^0(x)(\phi) = T_{ab}^{\mathrm{class}}(x)(\phi) + \lim_{x \to y} \left( D_{ab} - \frac{1}{3} g_{ab} P_x \right) w_N^h(x, y) \qquad N \geq 1 \qquad (6.3)$$

where $T_{ab}^{\mathrm{class}}$ is the classical energy momentum tensor of the scalar field, $D_{ab}$ is a second order bi-differential operator defined by $\lim_{x \to y} D_{ab} w(x, y) = \langle w, \frac{\delta^2}{\delta \phi^2} \rangle T_{ab}^{\mathrm{class}}(x)(\phi)$ and the modification term $-\frac{1}{3} g_{ab} P_x$ is necessary in order to have a covariantly conserved $T_{ab}^0$ [35]. The four parameters $\alpha_i$ are free parameters which can not be determined intrinsically within QFT on curved spacetimes, but only by measurements or within a more fundamental theoretical framework.

An alternative "model" $T_{ab}^0$ can be obtained by taking the functional derivative with respect to the inverse metric of the "one-loop effective Lagrangean"

$$\mathcal{L}^0(\phi)(x) = \mathcal{L}^{\mathrm{class}}(\phi)(x) + \left\langle w_N^{h_{dws}}, \frac{\delta^2}{\delta \phi^2} \right\rangle \mathcal{L}^{\mathrm{class}}(\phi)(x) \qquad N \geq 1.$$

Here $w_N^{h_{dws}}$ is the regular part of the deWitt-Schwinger Hadamard solution $h_{dws}$ which is a formal series in $\sigma(x, y)$ with purely geometric coefficients [19].

## 6.3 The Standard Cosmological Model in Quantum Field Theory on Curved Spacetimes

In the standard cosmological model the universe is modelled by a Friedmann–Lemaître–Robertson–Walker (FLRW) spacetime $(M, g)$ with manifold $M = I \times \mathbb{R}^3 \subset \mathbb{R}^4$ and metric $g = dt \otimes dt - a^2(t) dx^i \otimes dx_i$. We consider the case where the spatial slices are diffeomorphic to $\mathbb{R}^3$ for simplicity and because this is favoured by observations. Here $t$ is cosmological time, whereas the scale factor $a(t)$ is a smooth non-negative function whose logarithmic $t$-derivative is the Hubble rate $H$, which is assumed to be strictly positive in what follows. Further convenient time variables are the conformal time $\tau$, the scale factor $a$ and the redshift $z := a_0/a - 1$, where $a_0 = 1$ is the scale factor of today. These time variables are related by $dt = a d\tau = \frac{da}{aH} = -\frac{dz}{(1+z)H}$.

Given the high symmetry of $(M, g)$ and the Einstein equation $G_{ab} = 8\pi T_{ab}$, the energy momentum tensor $T_{ab}$ must be of perfect fluid form and thus determined by the energy density $\rho = (\partial_t)^a (\partial_t)^b T_{ab}$ and pressure $p$, which are related by the equation of state $p = p(\rho)$. Moreover, the Einstein equation is equivalent to the (first) Friedmann equation

$$H^2 = \frac{8\pi G}{3} \rho$$

and a conservation equation. According to the standard model of cosmology—the $\Lambda$CDM-model—our universe contains matter, radiation, and Dark Energy, modelled macroscopically as perfect fluids with equation of state $p = w\rho$, $w = 0, \frac{1}{3}, -1$ for matter, radiation and Dark Energy (assuming that the latter is just due to a cosmological constant) respectively. Consequently, the Friedmann equation can be conveniently rewritten as

$$\frac{H^2}{H_0^2} = \frac{\rho_{\Lambda\text{CDM}}}{\rho_0} = \Omega_\Lambda + \frac{\Omega_m}{a^3} + \frac{\Omega_r}{a^4}, \quad \rho_0 = \frac{3H_0^2}{8\pi G}, \tag{6.4}$$

where $H_0$ is the present Hubble rate—the Hubble constant—and the constants $\Omega_\Lambda$, $\Omega_m$, $\Omega_r$ denote the present fractions of the energy density due to Dark Energy, matter and radiation respectively. Observations indicate approximately

$$\Omega_m = 0.3, \quad \Omega_r = 10^{-4}, \quad \Omega_\Lambda = 1 - \Omega_m - \Omega_r \tag{6.5}$$

see [1] for the latest exact values from the Planck collaboration. In the context of cosmology the terms "matter" and "radiation" subsume all matter-energy with the respective macroscopic equation of state such that e.g. "radiation" does not encompass only electromagnetic radiation, but also the three left-handed neutrinos present in standard model of particle physics (SM) and possibly so-called Dark Radiation, and "matter" subsumes both the baryonic matter which is in principle well-understood in the SM and Dark Matter. Here, Dark Matter and Dark Radiation both quantify contributions to the macroscopic matter and radiation energy densities which exceed the ones expected from the knowledge of the SM and are believed to originate either from fields not present in the SM or from other sources, i.e. modifications of classical General Relativity.

Notwithstanding, at least the contributions to the macroscopic matter and radiation energy densities which are in principle well-understood originate microscopically from excitations of quantum fields, thence it should be possible to derive those from first principles within QFT on curved spacetimes. Such an analysis of the standard cosmological model within QFT on curved spacetimes has been performed by one of us in [18] and we shall review it in what follows.

A comprehensive analysis from this perspective could proceed as follows. One considers the full standard model of particle physics plus potential other fields and interactions as a perturbative interacting QFT on curved spacetime. One then aims to find a pair $(\omega, g)$, where $\omega$ is a Hadamard state on the algebra of this field model and $g$ is a metric on the manifold $M = I \times \mathbb{R}^3 \subset \mathbb{R}^4$ of FLRW type, such that (a) $(\omega, g)$ is a solution of the semiclassical Einstein equation

$$G_{ab} = 8\pi G \omega(T_{ab})$$

where $T_{ab}$ is the energy momentum tensor of the field model and (b) (6.4) are (6.5) are satisfied up to suitably small corrections. Unfortunately such an analysis is quite involved, but we can consider a number of simplifications. First, we disregard all field interactions. This is a legitimate approximation if we consider the cosmological evolution only after the primordial synthesis of light nuclei—the so-called Big Bang Nucleosynthesis (BBN)—as field interactions are usually assumed to be irrelevant for the large-scale properties of the quantum state after this era. In the standard cosmological model, this enters by assuming that the each component of the perfect fluid in (6.4) satisfies an individual conservation equation. As a further simplification, we disregard the spin of the quantum fields and model all massive fields, i.e. "matter", by a single massive scalar field, and all massless fields, i.e. "radiation", by a single massless scalar field, where both fields are considered to be conformally coupled to the scalar curvature ($\xi = \frac{1}{6}$). This is done for ease of presentation as computations with higher spin fields are in principle straightforward, see for instance [10, 11]; the conformal coupling $\xi = \frac{1}{6}$ is chosen because it simplifies computations and because the massless Dirac equation and the Maxwell equation are invariant under conformal isometries. Finally, provided one is able to assign a state $\omega$ to a FLRW metric $g$ in a coherent way, $\omega$ is in general a non-trivial functional of $g$ and thus obtaining an explicit solution of the semiclassical Einstein equation is at best difficult. In a recent yet unpublished work, Pinamonti and Siemssen have proven by a fixed point argument that the semiclassical Einstein equation can be uniquely solved for a linear scalar field model and a large class of initial conditions on a Cauchy surface, but for a quantitative analysis one needs to know the solution explicitly. We thus solve the semiclassical Einstein equation in the following approximate sense. We assume that the FLRW spacetime is given and determined by (6.4) and (6.5). On this spacetime we seek to find a pair of quantum states $\omega^m$ and $\omega^0$ for the massive and massless scalar field such that the sum of the energy densities in this states satisfies

$$\frac{\omega^0(\rho) + \omega^m(\rho)}{\rho_0} = \Omega_\Lambda + \frac{\Omega_m}{a^3} + \frac{\Omega_r}{a^4} = \frac{\rho_{\Lambda CDM}}{\rho_0} \qquad (6.6)$$

and (6.5) up to suitably small corrections in the time interval of interest $z \in [0, 10^9]$, where $z = 0$ marks the present and $z = 10^9$ is the redshift at which BBN took place.

In order to follow this program, it is useful to have at ones disposal a map which assigns a state $\omega$ to a FLRW metric $g$ *in a given coordinate system*; indeed this is necessary in order for the semiclassical Friedmann equation $3H^2 = 8\pi G\omega(\rho)$ to be well-defined in the first place. Such a construction is provided by the so-called states of low energy introduced by Olbermann [36]. These states minimize the energy density integrated in (cosmological) time with a sampling test function $f$ and are pure, Gaussian, isotropic and homogeneous states of Hadamard type. Their two-point Wightman function is (barring an $\epsilon$-prescription) of the form

$$\omega(x, y) = \frac{1}{8\pi^3 a(\tau_x) a(\tau_y)} \int_{\mathbb{R}^3} d\vec{k} \, \overline{\chi_k(\tau_x)} \chi_k(\tau_y) e^{i\vec{k}(\vec{x}-\vec{y})},$$

where the modes $\chi_k$ satisfy the ordinary differential equation

$$\left(\partial_\tau^2 + k^2 + m^2 a^2 + \left(\xi - \frac{1}{6}\right)Ra^2\right)\chi_k(\tau) = 0 \tag{6.7}$$

and the normalisation condition

$$\chi_k \partial_\tau \overline{\chi_k} - \overline{\chi_k}\partial_\tau \chi_k = i . \tag{6.8}$$

Here, $k = |\vec{k}|$ and $\bar{\phantom{x}}$ denotes complex conjugation. The modes $\chi_k$, which determine the state, are obtained by choosing arbitrary but fixed reference modes. The Bogoliubov coefficients in this mode basis are suitable functionals of the reference modes and the sampling function $f$. Olbermann has proven the Hadamard property of these states only for the case $\xi = 0$, but one can show that they are at least sufficiently regular in order to compute the energy density also in the case $\xi = \frac{1}{6}$. If $\xi = \frac{1}{6}$ and $m = 0$, then the Hadamard property follows from the fact that these states are related to the Minkowski vacuum state by a conformal isometry. In the following, we set $\xi = \frac{1}{6}$. A further assignment of a state to a FLRW spacetime in a given coordinate system is given by the so-called adiabatic states of order 0 introduced in [38] and further developed in [26, 34]. These are defined by the modes which satisfy (6.7) and the initial conditions $\chi_k(\tau)|_{\tau=\tau_0} = \tilde{\chi}_k(\tau)|_{\tau=\tau_0}$, $\partial_\tau \chi_k(\tau)|_{\tau=\tau_0} = \partial_\tau \tilde{\chi}_k(\tau)|_{\tau=\tau_0}$, where

$$\tilde{\chi}_k(\tau) = \frac{1}{\sqrt{2W(k,\tau)}}\exp\left(-i\int_{\tau_0}^{\tau} W(k,\tau')d\tau'\right), \quad W(k,\tau) = \sqrt{k^2 + m^2 a^2}. \tag{6.9}$$

The functions $\tilde{\chi}_k(\tau)$ solve (6.8) exactly but (6.7) only approximately with error terms quantified by $\frac{Hm}{W^2}$ and $\frac{\partial_\tau Hm}{W^3}$. A detailed discussion of the error terms can be found in [37].

In the $\Lambda$CDM model, the radiation contribution $\frac{\Omega_r}{a^4}$ to the energy density is mostly of thermal nature, while the matter contribution $\frac{\Omega_m}{a^3}$ is mostly due to Dark Matter, which in some scenarios is believed to be of thermal origin as well. Motivated by this we look for states which satisfy (6.6) and (6.5) among suitable "thermal excitations" of states of low energy. A fully satisfactory generalisation of the concept of thermal equilibrium to general curved spacetimes or even FLRW ones does not exist so far. Probably the most elaborated idea is the so-called local thermal equilibrium approach, see e.g. [45, 50] for a review. Here we take a more pragmatic approach and consider the states introduced in [11]. Given a pure, Gaussian, isotropic and homogeneous Hadamard state $\omega$ specified by modes $\chi_k$, one can construct a family of Gaussian Hadamard states

$\omega_{\beta,a_F}$ by defining the two-point Wightman function (up to an $\epsilon$-prescription) as

$$\omega(x, y) = \frac{1}{8\pi^3 a(\tau_x)a(\tau_y)} \int_{\mathbb{R}^3} d\vec{k}\, e^{i\vec{k}(\vec{x}-\vec{y})} \left( \frac{\chi_k(\tau_x)\overline{\chi_k(\tau_y)}}{1 - e^{-\beta k_0}} + \frac{\overline{\chi_k(\tau_x)}\chi_k(\tau_y)}{e^{\beta k_0} - 1} \right),$$

(6.10)

with $k_0 := \sqrt{k^2 + m^2 a_F^2}$. If $\chi_k$ are the modes of a state of low energy, these states match the almost equilibrium states introduced by Küskü in [33] up to the form of $k_0$. The Hadamard property of the states defined by (6.10) follows from results of [40]. In the massless case, these states are independent of $a_F$ and satisfy the conformal KMS condition with respect to the conformal Killing vector $\partial_\tau$. In the massive case, they are considered to describe approximately the quantum state of a field which has been in thermal equilibrium in the distant past, and has "frozen out" of equilibrium at the time $a = a_F$. This corresponds to the phenomenological picture behind Dark Matter of thermal origin in the standard literature see e.g. [32].

Given this choice of quantum states we are left with the cumbersome task to compute the energy density in these states. To this avail, we can rewrite the singular part $h_N^{sing}(x, y)$ of a Hadamard solution in terms of a Fourier integral in order to match the mode expansion of the states at hand, see [12, 14, 40, 43]. In this way we obtain a Fourier integral expression for the regular part $w_N^h(x, y)$ of the relevant two-point Wightman function. The energy density is obtained by applying to this regular object a second order bi-differential operator and then taking the limit $x \to y$, cf. (6.3). This is well defined and independent of $N$ if $N \geq 1$. As a result, we obtain the energy density as a convergent integral over $k$. In the massless case, this integral can be computed analytically. In the massive case however, both the integrand and the integral have been computed in [18] partly numerically and partly using analytical approximations. The reasons are manifold. To name a few, the mode equation (6.7) can not be solved analytically on FLRW spacetimes of the form (6.4) if $m > 0$. Moreover, even a numerical solution fails to be feasible for $m \gg H_0$—which is the realistic case as $H_0 \simeq 10^{-33}$ eV—because the modes oscillate heavily. To overcome the latter problem the approximate adiabatic modes $\tilde{\chi}_k(\tau)$, cf. (6.9), have been used as reference modes for the computation of the modes of the state of low energy, as they approximate the exact adiabatic modes of order zero particularly well exactly in the regime $m \gg H$.

Altogether the following results can be obtained. To discuss these, we rewrite the total energy density of the massless and massive conformally coupled scalar fields in the respective generalised thermal states (6.10) defined with respect to states of low energy as follows

$$\frac{\omega^0(\rho) + \omega^m(\rho)}{\rho_0} = \frac{\rho_{gvac}^m + \rho_{gvac}^0 + \rho_{gth}^m + \rho_{gth}^0}{\rho_0} + \gamma \frac{H^4}{H_0^4} + \Omega_\Lambda + \delta \frac{H^2}{H_0^2} + \epsilon \frac{J_{00}}{H_0^4}.$$

(6.11)

$\Omega_\Lambda$, $\delta$ and $\epsilon$ parametrise the freedom in the definition of the energy density as per (6.2). The number of free parameters in this equation has been reduced to three, because $I_{ab}$ and $J_{ab}$ are proportional in FLRW spacetimes. We take the point of view that $\delta$, which effectively renormalises Newton's constant, is not a free parameter because Newton's constant has been measured already. In order to do this, we have to fix a value for the inverse length scale $\mu$ in the singular part of a Hadamard solution $h_N^{\text{sing}}(x, y)$, we do this by confining $1/\mu$ to be a scale in the range in which the strength of gravity has been measured. Because of the smallness of the Planck length, the actual value of $1/\mu$ in this range does not matter as changing $1/\mu$ in this interval gives a negligible contribution to the energy density. One could also take a more conservative point of view and consider $\delta$ to be a free parameter, in this case comparison with cosmological data, e.g. from Big Bang Nucleosynthesis, would presumably constrain $\delta$ to be very small once $1/\mu$ is in the discussed range.

On this occasion, we would like to highlight the point of view on the so-called cosmological constant problem taken here, as well as in most works on QFT on curved spacetimes in the algebraic approach and e.g. the review [4]. It is often said that QFT *predicts* a value for the cosmological constant $\Lambda$ and thus for $\Omega_\Lambda$ which is way too large in comparison to the one measured. This conclusion is reached by computing one or several contributions to the vacuum energy in Minkowski spacetime $\Lambda_{\text{vac}}$ and finding them all to be too large, such that, at best, a fine-tuned subtraction in terms of a negative bare cosmological constant $\Lambda_{\text{bare}}$ is necessary in order to obtain the small value $\Lambda_{\text{vac}} + \Lambda_{\text{bare}}$ we observe. Here, we assume the point of view that it is not possible to provide an *absolute* definition of energy density within QFT on curved spacetimes, and thus neither $\Lambda_{\text{vac}}$ nor $\Lambda_{\text{bare}}$ have any physical meaning by themselves; only $\Lambda_{\text{vac}} + \Lambda_{\text{bare}}$ is physical and measurable and any cancellation which happens in this sum is purely mathematical. The fact that the magnitude of $\Lambda_{\text{vac}}$ depends on the way it is computed, e.g. the loop or perturbation order, cf. e.g. [44], is considered to be unnatural following the usual intuition from QFT on flat spacetime. However, it seems more convincing to us to accept that $\Lambda_{\text{vac}}$ and $\Lambda_{\text{bare}}$ have no relevance on their own, which does not lead to any contradiction between theory and observations, rather than the opposite. In the recent work [21] it is argued that a partial and unambiguous relevance can be attributed to $\Lambda_{\text{vac}}$ by demanding $\Lambda_{\text{bare}}$ to be analytic in all coupling constants and masses of the theory; taking this point of view, one could give the contribution to $\Lambda_{\text{vac}}$ which is non-analytic in these constants an unambiguous meaning. Indeed the authors of [21] compute a non-perturbative and hence non-analytic contribution to $\Lambda_{\text{vac}}$, which turns out to be small. In the view of this, one could reformulate the above statement and say that contributions to $\Lambda_{\text{vac}}$ and $\Lambda_{\text{bare}}$ which are analytic in masses and coupling constants have no physical relevance on their own.

The term in (6.11) proportional to $\gamma$, which is not present in the $\Lambda$CDM-model, appears due to the so-called trace anomaly, which is a genuine quantum and state-independent contribution to the quantum energy momentum tensor, see e.g. [51]. This term is fixed by the field content, $\gamma \simeq 10^{-122}$ for two scalar fields. As

$H < H_0 z^2$ in the $\Lambda$CDM-model for large redshifts, this term can be safely neglected for $z < 10^9$.

The first terms in (6.11) denote the genuinely quantum state dependent contributions to the energy densities of the two quantum fields. We have split these contributions into parts which are already present for infinite inverse temperature parameter $\beta$ in the generalised thermal states, and thus could be considered as contributions due to the states of low energy as generalised vacuum states ($\rho^m_{gvac}$, $\rho^0_{gvac}$), and into the remaining terms, which could be interpreted as purely thermal contributions ($\rho^m_{gth}$, $\rho^0_{gth}$). One can show that, up to the freedom parametrised by $\Omega_\Lambda$, $\delta$ and $\epsilon$, $\rho^0_{gvac} = 0$ for arbitrary sampling functions $f$, whereas $\rho^m_{gth}/\rho_{\Lambda CDM} \ll 1$ for small masses $m \simeq H_0$ and large masses $m \gg H_0$ if the sampling function $f$ defining the state of low energy has sufficiently large support in time. This generalises results obtained by Degner on de Sitter spacetime [12] and indicates that states of low energy with broad sampling functions are reasonable generalised vacuum states on FLRW spacetimes (Fig. 6.1).

As for the thermal contributions, one finds in the massless case

$$\rho^0_{gth} = \frac{\Omega_r}{a^4} \quad \text{with} \quad \Omega_r = \frac{\pi^2}{30\beta^4}.$$

Up to degree of freedom factors, this gives the $\Lambda$CDM value $\Omega_r \simeq 10^{-4}$ if the temperature parameter $1/\beta$ is in the range of the Cosmic Microwave Background temperature $1/\beta \simeq 2.7\,\mathrm{K}$. In the massive case, one can take typical values of $\beta$, $a_F$ and $m$ from Chapter 5.2 in [32] computed by means of effective Boltzmann

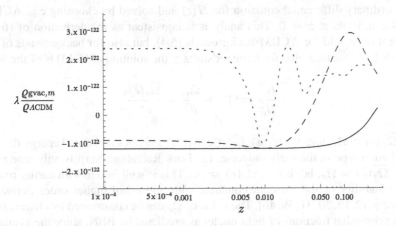

**Fig. 6.1** $\lambda \rho^m_{gvac}/\rho_{\Lambda CDM}$ for $z < 1$ for various values of $m$ (rescaled for ease of presentation). The *dotted line* corresponds to $m = 100 H_0$ and $\lambda = 10^{-2}$, the *dashed line* to $m = 10 H_0$ and $\lambda = 1$ and the *solid line* to $m = H_0$ and $\lambda = 10^2$. One sees nicely how the energy density is minimal in the support of the sampling function at around $z = 10^{-2}$

equations. A popular candidate for Dark Matter is a weakly interacting massive particle (WIMP), e.g. a heavy neutrino, for which [32] computes

$$x_F = \beta a_F m \simeq 15 + 3\log(m/\text{GeV}) \qquad a_F \simeq 10^{-12}(m/\text{GeV})^{-1} .$$

Using this one finds for large $m$

$$\rho_{\text{gth}}^m \simeq \frac{1}{(2\pi)^{3/2}} \frac{m}{\beta^3 a^3} x_F^{\frac{3}{2}} e^{-x_F} ,$$

and thus $\Omega_m \simeq 0.3$ for $m \simeq 100\,\text{GeV}$.

At this stage, we have already seen that there exist states for the field model under consideration for which the energy density in the time interval $z \in [0, 10^9]$ is of the form

$$\frac{\omega^0(\rho) + \omega^m(\rho)}{\rho_0} = \Omega_\Lambda + \frac{\Omega_m}{a^3} + \frac{\Omega_r}{a^4} + \epsilon\frac{J_{00}}{H_0^4} \qquad (6.12)$$

with $\Lambda$CDM values for $\Omega_m$, $\Omega_r$ and $\Omega_\Lambda$. This is the desired result up to the term $\epsilon\frac{J_{00}}{H_0^4}$ which is not present in the $\Lambda$CDM model, but quantified by the free parameter $\epsilon$. To analyse the influence of this term, we solve the equation

$$\frac{H^2}{H_0^2} = \Omega_\Lambda + \frac{\Omega_m}{a^3} + \frac{\Omega_r}{a^4} + \epsilon\frac{J_{00}}{H_0^4} . \qquad (6.13)$$

As $J_{00}$ contains second derivatives of $H$, this equation can be rewritten as a second order ordinary differential equation for $H(z)$ and solved by choosing e.g. $\Lambda$CDM initial conditions at $z = 0$. This analysis is consistent as the derivation of (6.12) does not only hold for $\Lambda$CDM-backgrounds (6.4), but also for backgrounds of the form (6.13). One finds that for large redshifts $z$, the solution of (6.13) is of the form

$$\frac{H^2}{H_0^2} = \Omega_\Lambda + \frac{\Omega_m}{a^3} + \frac{\widetilde{\Omega_r}(\epsilon)}{a^4}$$

with $\widetilde{\Omega_r}(\epsilon) \geq \Omega_r$, thus the term $\epsilon J_{00}$ effectively generates additional energy density of radiation type in the early universe, i.e. Dark Radiation. Surprisingly, one finds $\lim_{\epsilon\downarrow 0} \widetilde{\Omega_r}(\epsilon) = \Omega_r$, but $\lim_{\epsilon\uparrow 0} \widetilde{\Omega_r}(\epsilon) = \infty$. This is well in line with earlier results on the stability of the Einstein equation with additional higher order derivative terms, e.g. [2, 15, 20, 31, 39, 46]. The value of $\widetilde{\Omega_r}$ can be constrained by observations of the primordial fractions of light nuclei as predicted by BBN, since the synthesis of these nuclei depends sensitively on the Hubble rate at $z \simeq 10^9$. It turns out that $\widetilde{\Omega_r}(\epsilon)$ is in conflict with observations for $\epsilon < 0$, but that the BBN data are compatible with $0 \leq \epsilon < 2 \times 10^{-15}$ if all Dark Radiation is attributed to the origin discussed here.

The value of $\epsilon$ can be constrained also by other means. On the one hand, a further bound on $\epsilon$ can be obtained by analysing the effects of higher derivative contributions to the gravitational Lagrangean in the context of Inflation. In fact, an early inflationary model proposed by Starobinsky in [47] is based on an $\epsilon J_{00}$ contribution to the energy density. Confronting this inflationary model with current Cosmic Microwave Background data yields $\epsilon \simeq 10^{-113}$ [27]. Thus, if Inflation occurred due to the $\epsilon J_{00}$ contribution to the energy density, then $\epsilon$ is too small for generating a considerable amount of Dark Radiation. However, if Inflation has a different origin or did not occur at all, then one obtains the lower bound $\epsilon > 10^{-113}$. Finally, an upper bound on $\epsilon$ can be obtained by considering the Newtonian limit of the semiclassical Einstein equation. In this limit, the higher order derivative terms $I_{ab}$ and $J_{ab}$ in (6.2) generate two Yukawa corrections to the Newtonian potential of a point mass of opposite sign [48]. Assuming that these corrections don't cancel on the relevant length scales, one can obtain bounds on the strength and typical length scale of these Yukawa terms from torsion-balance experiments [28] and consequently the upper bound $\epsilon < 10^{-60}$ [9]. Again, this upper bound would imply that $\epsilon$ is too small for generating a considerable amount of Dark Radiation. However, it is still possible that the aforementioned Yukawa corrections cancel each other on the length scales relevant for the experiments described in [28], such that $\epsilon$ could be as large as our upper bound, which in this case would give a real bound on one and hence both Yukawa corrections. Moreover, the bounds inferred from [28] and from the analysis reviewed here stem from phenomena on completely different length scales. As a rough estimate we note that the diameter of our observable universe, which today is about $6/H_0 \simeq 10^{27}$ m, was at e.g. $z = 10^9$ still $10^{18}$ m and thus much larger than the submillimeter scales relevant for the torsion-balance experiments. Thus it could be that effects we have not considered yet, e.g. state-dependent effects which are due to the small-scale structure of the quantum states we have fixed only on cosmological scales so far, affect the comparison between the two different sources of input for the determination of $\epsilon$.

We conclude that a more fundamental understanding of the standard cosmological model appears to be possible within QFT on curved spacetimes. In this framework one even finds a new free parameter not present in the standard model. This parameter can potentially account for Dark Radiation, the existence and nature of which are currently topics of active research.

# References

1. Ade, P.A.R., et al. [Planck Collaboration]: Planck 2013 results. XVI. Cosmological parameters (2013) [arXiv:1303.5076 [astro-ph.CO]]
2. Anderson, P.: Effects of quantum fields on singularities and particle horizons in the early universe. Phys. Rev. D **28**, 271 (1983)
3. Benini, M., Dappiaggi, C., Hack, T. P.: Quantum field theory on curved backgrounds—a primer (2013) [arXiv:1306.0527 [gr-qc]]

4. Bianchi, E., Rovelli, C.: Why all these prejudices against a constant? (2010) [arXiv:1002.3966 [astro-ph.CO]]
5. Brunetti, R., Fredenhagen, K.: Microlocal analysis and interacting quantum field theories: renormalization on physical backgrounds. Commun. Math. Phys. **208**, 623 (2000) [arXiv:math-ph/9903028]
6. Brunetti, R., Fredenhagen, K., Kohler, M.: The microlocal spectrum condition and wick polynomials of free fields on curved space-times. Commun. Math. Phys. **180**, 633 (1996) [arXiv:gr-qc/9510056]
7. Brunetti, R., Fredenhagen, K., Rejzner, K.: Quantum gravity from the point of view of locally covariant quantum field theory (2013) [arXiv:1306.1058 [math-ph]]
8. Brunetti, R., Fredenhagen, K., Verch, R.: The generally covariant locality principle: a new paradigm for local quantum field theory. Commun. Math. Phys. **237**, 31 (2003) [arXiv:math-ph/0112041]
9. Calmet, X., Hsu, S.D.H., Reeb, D.: Quantum gravity at a TeV and the renormalization of Newton's constant. Phys. Rev. D **77**, 125015 (2008) [arXiv:0803.1836 [hep-th]]
10. Dappiaggi, C., Hack, T.-P., Pinamonti, N.: The extended algebra of observables for Dirac fields and the trace anomaly of their stress-energy tensor. Rev. Math. Phys. **21**, 1241 (2009) [arXiv:0904.0612 [math-ph]]
11. Dappiaggi, C., Hack, T.-P., Pinamonti, N.: Approximate KMS states for scalar and spinor fields in Friedmann-Robertson-Walker spacetimes. Ann. Henri Poincare **12**, 1449 (2011) [arXiv:1009.5179 [gr-qc]]
12. Degner, A.: Properties of States of Low Energy on Cosmological Spacetimes. Ph.D. thesis, University of Hamburg (2013). http://www.desy.de/uni-th/theses/Diss_Degner.pdf
13. Dimock, J.: Algebras of local observables on a manifold. Commun. Math. Phys. **77**, 219 (1980)
14. Eltzner, B., Gottschalk, H.: Dynamical backreaction in Robertson-Walker spacetime. Rev. Math. Phys. **23**, 531 (2011) [arXiv:1003.3630 [math-ph]]
15. Flanagan, E.E., Wald, R.M.: Does backreaction enforce the averaged null energy condition in semiclassical gravity? Phys. Rev. D **54**, 6233 (1996) [arXiv:gr-qc/9602052]
16. Fredenhagen, K., Rejzner, K.: Batalin-Vilkovisky formalism in perturbative algebraic quantum field theory. Commun. Math. Phys. **317**, 697 (2013) [arXiv:1110.5232 [math-ph]]
17. Haag, R.: Local Quantum Physics: Fields, Particles, Algebras. Texts and Monographs in Physics, 356 p. Springer, Berlin (1992)
18. Hack, T.-P.: The Lambda CDM-model in quantum field theory on curved spacetime and dark radiation (2013) [arXiv:1306.3074 [gr-qc]]
19. Hack, T.-P., Moretti, V.: On the stress-energy tensor of quantum fields in curved spacetimes—comparison of different regularization schemes and symmetry of the Hadamard/Seeley-DeWitt coefficients. J. Phys. A **45**, 374019 (2012) [arXiv:1202.5107 [gr-qc]]
20. Hänsel, M.: Stability of the Semiclassical Einstein Equations in FRW Spacetime. Master thesis, Universität Leipzig (2011)
21. Holland, J., Hollands, S.: A small cosmological constant due to non-perturbative quantum effects (2013) [arXiv:1305.5191 [gr-qc]]
22. Hollands, S., Wald, R.M.: Local Wick polynomials and time ordered products of quantum fields in curved space-time. Commun. Math. Phys. **223**, 289 (2001) [arXiv:gr-qc/0103074]
23. Hollands, S., Wald, R.M.: Existence of local covariant time ordered products of quantum fields in curved space-time. Commun. Math. Phys. **231**, 309 (2002) [arXiv:gr-qc/0111108]
24. Hollands, S., Wald, R.M.: Conservation of the stress tensor in interacting quantum field theory in curved spacetimes. Rev. Math. Phys. **17**, 227 (2005) [arXiv:gr-qc/0404074]
25. Hörmander, L.: The Analysis of Linear Partial Differential Operators I–IV. Springer, Berlin (1983)
26. Junker, W., Schrohe, E.: Adiabatic vacuum states on general spacetime manifolds: definition, construction, and physical properties. Ann. Poincare Phys. Theor. **3**, 1113 (2002) [arXiv:math-ph/0109010]
27. Kaneda, S., Ketov, S.V., Watanabe, N.: Fourth-order gravity as the inflationary model revisited. Mod. Phys. Lett. A **25**, 2753 (2010) [arXiv:1001.5118 [hep-th]]

28. Kapner, D.J., Cook, T.S., Adelberger, E.G., Gundlach, J.H., Heckel, B.R., Hoyle, C.D., Swanson, H.E.: Tests of the gravitational inverse-square law below the dark-energy length scale. Phys. Rev. Lett. **98**, 021101 (2007) [arXiv:hep-ph/0611184]
29. Kay, B.S.: Generally covariant perturbation theory: linear spin 0 quantum fields in external gravitational and scalar fields 2. Commun. Math. Phys. **71**, 29 (1980)
30. Kay, B.S., Wald, R.M.: Theorems on the uniqueness and thermal properties of stationary, nonsingular, quasifree states on space-times with a Bifurcate Killing Horizon. Phys. Rep. **207**, 49 (1991)
31. Koksma, J.F.: Dynamics driven by the trace anomaly in FLRW universes (2009) [arXiv:0911.2997 [gr-qc]]
32. Kolb, E.W., Turner, M.S.: The early universe. Front. Phys. **69**, 1 (1990)
33. Küskü, M.: A class of almost equilibrium states in Robertson-Walker spacetimes. DESY-THESIS-2008-020 (2009) [arXiv:0901.1440[hep-th]]
34. Lüders, C., Roberts, J.E.: Local quasiequivalence and adiabatic vacuum states. Commun. Math. Phys. **134**, 29–63 (1990)
35. Moretti, V.: Comments on the stress-energy tensor operator in curved spacetime. Commun. Math. Phys. **232**, 189 (2003) [arXiv:gr-qc/0109048]
36. Olbermann, H.: States of low energy on Robertson-Walker spacetimes. Class. Quant. Grav. **24**, 5011 (2007) [arXiv:0704.2986 [gr-qc]]
37. Olver, F.W.J.: Asymptotics and Special Functions. Academic, New York/London (1974)
38. Parker, L.: Quantized fields and particle creation in expanding universes 1. Phys. Rev. **183**, 1057 (1969)
39. Parker, L., Simon, J.Z.: Einstein equation with quantum corrections reduced to second order. Phys. Rev. D **47**, 1339 (1993) [arXiv:gr-qc/9211002]
40. Pinamonti, N.: On the initial conditions and solutions of the semiclassical Einstein equations in a cosmological scenario. Commun. Math. Phys. **305**, 563 (2011) [arXiv:1001.0864 [gr-qc]]
41. Radzikowski, M.J.: Micro-local approach to the Hadamard condition in quantum field theory on curved space-time. Commun. Math. Phys. **179**, 529 (1996)
42. Radzikowski, M.J.: A local to global singularity theorem for quantum field theory on curved space-time. Commun. Math. Phys. **180**, 1 (1996)
43. Schlemmer, J.: Ph.D. thesis, Universität Leipzig (2010)
44. Sola, J.: Cosmological constant and vacuum energy: old and new ideas (2013) [arXiv:1306.1527 [gr-qc]]
45. Solveen, C.: Local thermal equilibrium and KMS states in curved spacetime. Class. Quant. Grav. **29**, 245015 (2012) [arXiv:1211.0431 [gr-qc]]
46. Starobinsky, A.A.: A new type of isotropic cosmological models without singularity. Phys. Lett. **B91**, 99 (1980)
47. Starobinsky, A.A.: The perturbation spectrum evolving from a nonsingular initially De-Sitter cosmology and the microwave background anisotropy. Sov. Astron. Lett. **9**, 302 (1983)
48. Stelle, K.S.: Classical gravity with higher derivatives. Gen. Rel. Gravit. **9**, 353 (1978)
49. Verch, R.: Local definiteness, primarity and quasiequivalence of quasifree Hadamard quantum states in curved space-time. Commun. Math. Phys. **160**, 507 (1994)
50. Verch, R.: Local covariance, renormalization ambiguity, and local thermal equilibrium in cosmology. In: Finster, F., et al. (eds.) Quantum Field Theory and Gravity, p. 229. Birkhäuser, Basel (2012) [arXiv:1105.6249 [math-ph]]
51. Wald, R.M.: Trace anomaly of a conformally invariant quantum field in curved space-time. Phys. Rev. D **17**, 1477 (1978)
52. Wald, R.M.: Quantum Field Theory in Curved Space-Time and Black Hole Thermodynamics, 205 pp. Chicago University Press, Chicago (1994)

# Chapter 7
# Quantum Probability Theory and the Foundations of Quantum Mechanics

**Jürg Fröhlich and Baptiste Schubnel**

## 7.1 A Glimpse of Quantum Probability Theory and of a Quantum Theory of Experiments

By and large, people are better at coining expressions than at filling them with interesting, concrete contents. Thus, it may not be very surprising that there are many professional probabilists who may have heard the expression but do not appear to be aware of the need to develop *"quantum probability theory"* into a thriving, rich, useful field featured at meetings and conferences on probability theory. Although our aim, in this essay, is not to contribute new results on quantum probability theory, we hope to be able to let the reader feel the enormous potential and richness of this field. What we intend to do, in the following, is to contribute some novel points of view to the *"foundations of quantum mechanics"*, using mathematical tools from "quantum probability theory" (such as the theory of operator algebras).

The "foundations of quantum mechanics" represent a notoriously thorny and enigmatic subject. Asking 25 grown up physicists to present their views on the foundations of quantum mechanics, one can expect to get the following spectrum of reactions[1]: Two will refuse to talk—alluding to the slogan "shut up and calculate"— two will say that the problems encountered in this subject are so difficult that it

---

[1]This story is purely fictional, but quite plausible.

J. Fröhlich (✉)
Institut für Theoretische Physik, HIT K42.3, ETH Zürich, 8093 Zürich, Switzerland
e-mail: juerg@phys.ethz.ch

B. Schubnel
Departement Mathematik, ETH Zürich, 8092 Zürich, Switzerland
e-mail: baptiste.schubnel@math.ethz.ch

© Springer-Verlag Berlin Heidelberg 2015
P. Blanchard, J. Fröhlich (eds.), *The Message of Quantum Science*, Lecture Notes in Physics 899, DOI 10.1007/978-3-662-46422-9_7

might take another 100 years before they will be solved; five will claim that the "Copenhagen Interpretation", [75], has settled all problems, but they arc unable to say, in clear terms, what they mean; three will refer us to Bell's book [9] (but admit they have not understood it completely); two confess to be "Bohmians" [25] (but do not claim to have had an encounter with Bohmian trajectories); two claim that all problems disappear in the Dirac–Feynman path-integral formalism [23, 24, 30]; another two believe in "many worlds" [28] but make their income in our's, and two advocate "consistent histories" [41]; two swear on QBism [36], (but have never seen "les demoiselles d'Avignon"); two are convinced that the collapse of the wave function [38]—spontaneous or not—is fundamental; and one thinks that one must appeal to quantum gravity to arrive at a coherent picture, [60].

*Almost all of them are convinced that theirs is the only sane point of view.*[2] Many workers in the field have lost the ability to do technically demanding work or never had it. Many of them are knowingly or unknowingly envisaging an extension of quantum mechanics—but do not know how it will look like. But some claim that "quantum mechanics cannot be extended" [18], (perhaps unaware of the notorious danger of "no-go theorems"). See also [66, 72]

At least fifteen of the views those 25 physicists present logically contradict one another. Most colleagues are convinced that somewhat advanced mathematical methods are superfluous in addressing the problems related to the foundations of quantum mechanics, and they turn off when they hear an expression such as "$C^*$-algebra" or "type-III factor". Well, it might just turn out that they are wrong! What appears certain is that the situation is somewhat desperate, and this may explain why people tend to become quite emotional when they discuss the foundations of quantum mechanics; (see, e.g., [74]).

When the senior author had to start teaching quantum mechanics to students, many years ago, he followed the slogan "shut up and calculate"—until he decided that the situation described above, namely the fact that we do not really understand, in a coherent and conceptual way, what that most successful theory of physics called "quantum mechanics" tells us about Nature, represents an intellectual scandal.

Our essay will, of course, not remove this scandal. But we hope that, with some of our writings, (see also [32, 34]), we may be able to contribute some kind of intellectual "screw driver" useful in helping to unscrew[3] the enigmas at the root of the scandal, before very long. We won't attempt to extend or "complete" quantum mechanics (although we bear people no grudge who try to do so, and we wish them well). We are convinced that starting from simple, intuitive, general principles ("information loss" and "entanglement generation") and then elucidating

---

[2]And that Heisenberg's 1925 paper [46] cannot be understood.

[3]"dévisser les problèmes" (in reference to A. Grothendieck).

the *mathematical structure* inherent in quantum mechanics will lead to a better understanding of its deep message. (Of course, we realize that our hope is lost on people who are convinced that the mysteries surrounding the interpretation of quantum mechanics can be unravelled without any use of somewhat advanced mathematical concepts.)

Just to be clear about one point: We are not claiming to present any "revolutionary" new ideas; and we do not claim or expect to get much credit for our attempts.

But, by all means, let's get started! Quantum mechanics is "quantum", and it is intrinsically "probabilistic" [11, 27]. We should therefore expect that it is intimately connected to quantum probability theory, hence to "non-commutative measure theory", etc. However, in the end, *"quantum mechanics is quantum mechanics and everything else is everything else!"*[4]

## 7.1.1 Might Quantum Probability Theory be a Subfield of (Classical) Probability Theory?

And—if not—what's different about it? These questions are related to one concerning the existence of *hidden variables*. The first convincing results on hidden variables and on "Bell non-locality" were brought forward by Kochen and Specker [51] and (independently) by Bell [7–9]. These matters are so well known, by now, that we do not repeat them here. The upshot is that, loosely speaking, quantum probability theory cannot be imbedded in classical probability theory (except in the case of a two-level system).

The deeper problems of quantum mechanics can probably only be understood if we admit a notion of *time*, introduce time-evolution, proceed to consider *repeated measurements*, i.e., time-ordered sequences of observations or measurements resulting in a time-ordered sequence of events, and understand *in which way information gets lost for ever*, in the course of time evolution. (We believe that this will lead to an acceptable "ontology" of quantum mechanics [2, 25]) not involving any fundamental role of the "observer".)

In both worlds, the classical and the quantum world, physical quantities or (potential) properties are represented by self-adjoint operators, $a = a^*$, and possible events by spectral projections, $\Pi$, or certain products thereof (POVM's; see Appendix 7.1.A to Sects. 7.4 and 7.5.4). A successful measurement or observation of a physical quantity or property represented by an operator $a = a^*$ results in

---

[4] *"The one thing to say about art is that it is one thing. Art is art-as-art and everything else is everything else."* Ad Reinhardt, [63].

*one* of several possible events, $\Pi_1, \ldots, \Pi_k$ (spectral projections of $a$), with the properties that

$$(i) \quad \Pi_\alpha^2 = \Pi_\alpha = \Pi_\alpha^*, \; \alpha = 1, \ldots, k,$$

$$(ii) \quad \Pi_\alpha \Pi_\beta = \delta_{\alpha\beta} \Pi_\alpha,$$

$$(iii) \quad \sum_{\alpha=1}^{k} \Pi_\alpha = \mathbb{1}. \tag{7.1}$$

Suppose we carry out a sequence of *mutually "independent"* measurements or observations of physical quantities, $a_1, \ldots, a_n$, ordered in time, i.e., $a_1$ before $a_2$ before $a_3 \ldots$ before $a_n$ ($a_1 \prec a_2 \prec \ldots \prec a_n$). A physical theory should enable us to predict the probabilities for all possible "histories",

$$h_1^n(\underline{\alpha}) = \{\Pi_{\alpha_1}^{(1)}, \ldots, \Pi_{\alpha_n}^{(n)}\},$$

of events, where $\Pi_1^{(i)}, \ldots, \Pi_{k_i}^{(i)}$ are the possible events resulting from a successful measurement of $a_i$, $i = 1, \ldots, n$. On the basis of what prior knowledge? Well, we must know the time evolution of physical quantities and the "state", $\omega$, of the system, $S$, we observe. That means that, given a state $\omega$, there should exist a functional, $\text{Prob}_\omega$, that associates with each history $\{\Pi_{\alpha_1}^{(1)}, \ldots, \Pi_{\alpha_n}^{(n)}\}$—but for what *family* of histories, i.e., for which properties $a_1, \ldots, a_n$?—a probability

$$0 \leq \mu_\omega(\alpha_1, \ldots, \alpha_n) \equiv \text{Prob}_\omega\{\Pi_{\alpha_1}^{(1)}, \ldots, \Pi_{\alpha_n}^{(n)}\} \leq 1. \tag{7.2}$$

By property (iii) in Eq. (7.1),

$$\sum_{\alpha_1, \ldots, \alpha_n} \mu_\omega(\alpha_1, \ldots, \alpha_n) = 1, \tag{7.3}$$

because $\text{Prob}_\omega$ is normalized such that $\text{Prob}_\omega\{\mathbb{1}, \mathbb{1}, \ldots\} = 1$. In a classical theory, the projections $\{\Pi_{\alpha_i}^{(i)}\}_{\alpha_i=1}^{k_i}$, $i = 1, \ldots, n$, are characteristic functions on a measure space, $M_S$, and a state, $\omega$, is a probability measure on $M_S$. It then follows from property (iii) that

$$\sum_{\alpha=1}^{k_i} \text{Prob}_\omega\{\Pi_{\alpha_1}^{(1)}, \ldots, \Pi_\alpha^{(i)}, \ldots, \Pi_{\alpha_n}^{(n)}\} = \text{Prob}_\omega\{\Pi_{\alpha_1}^{(1)}, \ldots, \Pi_{\alpha_{i-1}}^{(i-1)}, \Pi_{\alpha_{i+1}}^{(i+1)}, \ldots, \Pi_{\alpha_n}^{(n)}\},$$

$$\tag{7.4}$$

for arbitrary $i = 1, \ldots, n$.

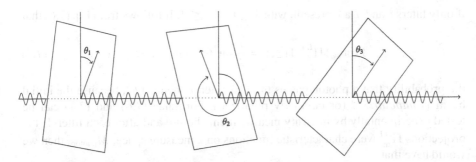

**Fig. 7.1** Beam of photons passing through polarization filters

If we consider a quantum mechanical system with finitely many degrees of freedom then the projections $\{\Pi^{(i)}_{\alpha_i}\}$ are orthogonal projections on a separable Hilbert space, $\mathcal{H}$, and, by Gleason's theorem [39], $\omega$ is given by a density matrix, $\rho_\omega$, on $\mathcal{H}$. Moreover, according to [50, 54, 64, 76],

$$\mathrm{Prob}_\omega\{\Pi^{(1)}_{\alpha_1}, \ldots, \Pi^{(n)}_{\alpha_n}\} = \mathrm{Tr}_\mathcal{H}\left(\Pi^{(n)}_{\alpha_n} \ldots \Pi^{(1)}_{\alpha_1} \rho_\omega \Pi^{(1)}_{\alpha_1} \ldots \Pi^{(n)}_{\alpha_n}\right). \tag{7.5}$$

The problem with Eq. (7.5) is that, most often, it represents physical and probability-theoretical nonsense. For example, it is usually left totally unclear what physical quantities or properties of $S$ will be measurable (i.e., which family of histories will become observable), given a time evolution $\tau_{t,s}$ and a state $\omega$. But such problems do not stop people from studying Eq. (7.5) again and again—and we are no exception. To address one of the key problems with Eq. (7.5), we study an example.

We consider a monochromatic beam of light, which, according to Einstein [26], consists of individual photons of fixed frequency. We then bring three filters into the beam that produce linearly polarized light. The direction of polarization is given by an angle $\theta$ that can be varied by rotating the filter around the axis defined by the beam; see Fig. 7.1.

With the filter $i$, we associate two possible events

$$\Pi^{(i)}_+ \leftrightarrow \text{a photon passes through filter } i$$

$$\Pi^{(i)}_- = \mathbb{1} - \mathbb{1}^{(i)}_+ \leftrightarrow \text{a photon does \textit{not} pass through filter } i.$$

Experimentally, one finds that, for any initially unpolarized beam of light, (meaning that the photons are all prepared in a state $\omega_0 \propto \frac{1}{2}\mathrm{Tr}_{\mathbb{C}^2}(\cdot)$),

$$\mathrm{Prob}_{\omega_0}\{\Pi^{(i)}_+, \Pi^{(j)}_+\} = \frac{1}{2}\cos^2(\theta_i - \theta_j), \ i < j, \tag{7.6}$$

if only filters $i$ and $j$ are present, with $1 \leq i < j \leq 3$. It follows from Eq. (7.6) that

$$\text{Prob}_{\omega_0}\{\Pi_+^{(i)}, \Pi_-^{(j)}\} = \frac{1}{2} \sin^2(\theta_i - \theta_j), \; i < j, \tag{7.7}$$

the probability that a photon passes the first filter, $i$, being $1/2$, because the initial beam is *unpolarized* (or circularly polarized). Formulae (7.6) and (7.7) can be tested experimentally by intensity measurements before and after each filter. If the projections $\Pi_\pm^{(i)}$ were characteristic functions on a measure space, $M_{\text{photon}}$, then we would have that

$$\text{Prob}_{\omega_0}\{\Pi_+^{(1)}, \Pi_-^{(3)}\} \leq \text{Prob}_{\omega_0}\{\Pi_+^{(1)}, \Pi_-^{(2)}\} + \text{Prob}_{\omega_0}\{\Pi_+^{(2)}, \Pi_-^{(3)}\}. \tag{7.8}$$

For,

$$\text{Prob}_{\omega_0}\{\Pi_+^{(1)}, \Pi_-^{(3)}\} = \text{Prob}_{\omega_0}\{\Pi_+^{(1)}, \Pi_-^{(2)}, \Pi_-^{(3)}\} + \text{Prob}_{\omega_0}\{\Pi_+^{(1)}, \Pi_+^{(2)}, \Pi_-^{(3)}\}$$
$$\leq \text{ right side of Eq. (7.8)}, \tag{7.9}$$

where Eq. (7.9) follows from the sum rule (7.4), and the upper bound (7.8) from the trivial inequality $0 \leq \Pi_\pm^{(i)} \leq 1$. Plugging expression (7.7) into (7.8). we conclude that

$$\frac{1}{2} \sin^2(\theta_1 - \theta_3) \leq \frac{1}{2} \sin^2(\theta_1 - \theta_2) + \frac{1}{2} \sin^2(\theta_2 - \theta_3). \tag{7.10}$$

Setting $\theta_1 = 0$, $\theta_2 = \pi/6$ and $\theta_3 = \pi/3$, Eq. (7.10) would imply that $3/8 \leq 1/8 + 1/8$, which is obviously wrong! What is going on? It turns out that the sum rule (7.9) is violated. The reason is that the projections $\Pi_\pm^{(2)}$ and $\Pi_\pm^{(3)}$ *do not commute*. This fact is closely related to *non-vanishing interference* between $\Pi_+^{(2)}$ and $\Pi_-^{(2)}$ analogous to the interference encountered in the *double-slit experiment*. Interference between $\Pi_+^{(2)}$ and $\Pi_-^{(2)}$ is measured by

$$\mathcal{I}(\Pi_+^{(2)}, \Pi_-^{(2)} \mid \Pi_\alpha^{(1)}, \Pi_\beta^{(3)}) := \text{Tr}_{\mathcal{H}}(\Pi_\beta^{(3)} \Pi_+^{(2)} \Pi_\alpha^{(1)} \rho_{\omega_0} \Pi_\alpha^{(1)} \Pi_-^{(2)} \Pi_\beta^{(3)}). \tag{7.11}$$

Choosing $\alpha = +$ and $\beta = -$ (for example), we find a non-vanishing interference term, which explains why the sum rule (7.9) is violated. What is the message? The first filter, 1, may be interpreted as *"preparing"* the photons in the beam hitting the filter 2 to be linearly polarized as prescribed by the angle $\theta_1$. In our experimental set-up there is *no* instrument measuring whether a photon has passed filter 2, or not. The *only* measurement is made after filter 3, where either a photon triggers a Geiger counter to click, or there is no photon triggering the Geiger counter. Let us denote the probability for the first event (Geiger counter clicks) by $p_+$, the second by $p_-$. The histories contributing to $p_-$ are

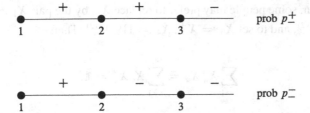

with $p_- = p_\pm^+ + p_-^-$. These two histories show interference. Given that a photon has passed filter 1, expressions (7.6) and (7.7) appear to imply that

$$p_\pm^+ = \cos^2(\theta_1 - \theta_2)\sin^2(\theta_2 - \theta_3)$$
$$p_-^- = \sin^2(\theta_1 - \theta_2). \tag{7.12}$$

The unique history contributing to $p_+$ appears to be

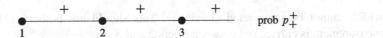

with

$$p_+^+ = \cos^2(\theta_1 - \theta_2)\cos^2(\theta_2 - \theta_3),$$

and, indeed,

$$p_+^+ + p_\pm^+ + p_-^- = 1.$$

These findings can be accounted for by associating with the event "+" the operator

$$X_+ = \Pi_+^{(3)}\Pi_+^{(2)}$$

and with the event "−" the operators

$$X_\pm^+ = \Pi_-^{(3)}\Pi_+^{(2)} \quad \text{and} \quad X_-^- = \Pi_-^{(2)}.$$

Then,

$$X_+^*X_+ + (X_\pm^+)^*X_\pm^+ + (X_-^-)^*X_-^- = \mathbb{1}.$$

It should however be noted that

$$X_+X_+^* + X_\pm^+(X_\pm^+)^* + X_-^-(X_-^-)^* \neq \mathbb{1}.$$

For this reason, some people may prefer to replace $X_+$ by the pair $X_1 := \Pi_|^{(3)} \Pi_+^{(2)}$, $X_2 := \Pi_+^{(3)} \Pi_-^{(2)}$, and to set $X_3 := X_-^+$, $X_4 := \Pi_-^{(3)} \Pi_-^{(2)}$. Then,

$$\sum_{\alpha=1}^{4} X_\alpha^* X_\alpha = \sum_{\alpha=1}^{4} X_\alpha X_\alpha^* = \mathbb{1}. \tag{7.13}$$

The family $(X_1, X_2, X_3, X_4)$ is called (the "square root" of) a *positive operator-valued measure* (POVM); (see [61], and Sects. 7.4.3 and 7.5.4). Note that

$$\mathrm{Tr}_{\mathcal{H}}(X_2 \rho_{\omega_0} X_2^*) = \mathrm{Prob}_{\omega_0}\{\Pi_-^{(2)}, \Pi_+^{(3)}\}$$

corresponds to the "virtual history"

which cannot be interpreted classically. This should not bother us, because no measurement is carried out between filters 2 and 3.

There is a more drastic way to present these findings: Consider N filters in series, the $j$th filter being rotated through an angle $j\pi/2N$. The probability for an initially vertically polarized photon ($\theta_0 = 0$) to be transmitted through all the filters is then given by

$$p_+ = \mathrm{Prob}_{\omega_{\theta_0}=0}\{\Pi_+^{(1)}, \ldots, \Pi_+^{(N)}\} = \left(\cos\left(\frac{\pi}{2N}\right)\right)^{2N} \xrightarrow[N\to\infty]{} 1. \tag{7.14}$$

If however, all filters, except for the $N$th one, are removed, then

$$p_+' := \mathrm{Prob}_{\omega_{\theta_0}=0}\{\Pi_+^{(N)}\} = \cos^2\left(\frac{\pi}{2}\right) = 0. \tag{7.15}$$

If $\Pi_+^{(1)}, \ldots, \Pi_+^{(N)}$ were "classical events", i.e., non-negative random variables, then one would have that $p_+ \le p_+'$. (See [9, 55] for closely related arguments.)

Actually, the discussion presented above, although often repeated, is somewhat misleading. The only measurement takes place *after* the last filter and is supposed to determine whether a photon has passed all the filters, or not. The corresponding physical quantity corresponds to the operators $\Pi_\pm^{(N)}$, where $N$ is the label of the last filter, and the measurement consists in verifying whether a Geiger counter placed after the last filter has clicked, or not. The filters have *nothing* to do with measurements, but determine (or, at least, affect) the form of the *time evolution* of the photons. The use of POVM's in discussing experiments like the ones above is not justified at a fundamental, conceptual level. It merely substitutes for a more precise understanding of time-evolution that involves including the filters in a quantum-mechanical description. It appears that, often, POVM's are used to cover up a lack of understanding of the time-evolution of large quantum systems. The role they play in a quantum theory of experiments is briefly described in Sect. 7.5.4.

A more compelling way of convincing oneself that quantum probability cannot be imbedded in classical probability theory than the one sketched above consists in studying *correlation matrices* of families of (non-commuting) possible events in two *independent* systems. One then finds that the numerical range of possible values of the matrix elements of such correlation matrices is strictly larger in quantum probability theory than in classical probability theory, as discovered by Bell [9, 71]. See [51] for an alternative approach.

### 7.1.2 The Quantum Theory of Experiments

We return to considering a system, $S$, and suppose that $n$ consecutive measurements have been carried out *successfully*, with the $i$th measurement described by spectral projections $\Pi_\alpha^{(i)} = (\Pi_\alpha^{(i)})^*$, $\alpha = 1, \ldots, k_i$, of a physical quantity $a_i = a_i^*$, with

$$\Pi_\alpha^{(i)} \Pi_\beta^{(i)} = \delta_{\alpha\beta} \Pi_\alpha^{(i)}, \qquad \sum_{\alpha=1}^{k_i} \Pi_\alpha^{(i)} = \mathbb{1}, \tag{7.16}$$

for all $i$. (We could also use POVM's, instead of projections, but let's not!) The probability of a history $\{\Pi_{\alpha_1}^{(1)}, \ldots, \Pi_{\alpha_n}^{(n)}\}$ in a state $\omega$ of $S$ given by a density matrix $\rho_\omega$ is then given by formula (7.5), above. The measurements can be considered to be successful only if the *sum rules* (7.4) *are very nearly satisfied, for all i*. Whether this is true, or not, can be determined by studying the interference between different histories. Given a state $\omega$, we define $N \times N$ matrices, $P^\omega = (P^\omega_{\underline{\alpha},\underline{\alpha}'})$, $N = k_1 \cdots k_n$, by

$$P^\omega_{\underline{\alpha},\underline{\alpha}'} := \omega \left( \Pi_{\alpha_1}^{(1)} \ldots \Pi_{\alpha_n}^{(n)} \Pi_{\alpha_n'}^{(n)} \ldots \Pi_{\alpha_1'}^{(1)} \right) = \mathrm{Tr} \left( \Pi_{\alpha_n'}^{(n)} \ldots \Pi_{\alpha_1'}^{(1)} \rho_\omega \Pi_{\alpha_1}^{(1)} \ldots \Pi_{\alpha_n}^{(n)} \right), \tag{7.17}$$

where $\omega(a)$ is the expectation of the operator $a$ in the state $\omega$. Measurements of the quantities $a_1, \ldots, a_n$ can be considered to be successful *only* if $P^\omega$ is approximately *diagonal*, i.e.,

$$|P^\omega_{\underline{\alpha},\underline{\alpha}'}| \ll \frac{1}{2} \left( P^\omega_{\underline{\alpha},\underline{\alpha}} + P^\omega_{\underline{\alpha}',\underline{\alpha}'} \right), \tag{7.18}$$

which is customarily called "*decoherence*"; see, e.g., [10, 37, 47, 49]. All this is discussed in much detail in Sects. 7.4.3 and 7.5. In particular, we will show that decoherence is a consequence of "*entanglement generation*" between the system $S$ and its environment $E$ and of "*information loss*", meaning that the original state of $S \vee E$ *cannot* be fully reconstructed from the results of arbitrary measurements carried out after some time $T$, long after the interactions between $S$ and $E$ have set in; see Sect. 7.5, and [17, 31]. In local relativistic quantum theory with massless particles (photons), the kind of information loss alluded to here is a general consequence of Huyghens' principle [14] and of "Einstein causality". It appears already in classical field theory. In local relativistic quantum theory it becomes manifest in the circumstance that the algebra of operators representing physical quantities measurable by a localized observer *after* some time $T$ does *not* admit any *pure* states. See [17].

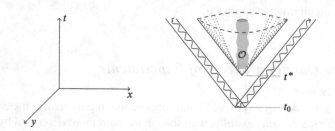

The event at time $t_0 < t^*$ involving photons can never be observed by the observer $\mathcal{O}$

The *key problem* in a quantum theory of experiments (or measurements/observations) is, however, to find out *which physical quantities will be measured* (i.e., what potential properties of a system will become *"empirical"* properties, or what families of histories of events can be expected to be observed) in the course of time, given the choice of a system, $S$, coupled to an environment, $E$, of a specific time evolution of $S \vee E$, and of a fixed state, $\omega$, of $S \vee E$. This is sometimes referred to as the problem of eliminating the mysterious role of the "observer" from quantum mechanics (making *many worlds* superfluous), and of determining the *"primitive ontology"* of quantum mechanics, [2]. This problem will be reckoned with in Sects. 7.5.3 and 7.5.4.

One customarily distinguishes between *"direct* (or von Neumann) *measurements"* and (indirect, or) *"non-demolition measurements"* carried out on a physical system $S$. It may be assumed that it is clear what is meant by a direct measurement. A non-demolition measurement is carried out by having a sequence of "probes" $(E_k)$ interact with the system $S$, one after another, with the purpose of measuring a physical quantity, $a = a^*$, of $S$ with (for simplicity) finite point spectrum, $\mathrm{spec}(a) = \{\alpha_1, \ldots, \alpha_n\}$. If $S$ is in an eigenstate, $| \alpha_i \rangle$, of $a$ corresponding to the eigenvalue $\alpha_i$ right before it starts to interact with the $k$th probe, $E_k$, the time-evolution of the composed system, $S \vee E_k$, is assumed to leave $|\alpha_i\rangle$ invariant but changes the state of $E_k$ in a manner that depends *non-trivially* on $\alpha_i$, for each $i = 1, \ldots, n$. This leads to entanglement between $S$ and $E_k$, $k = 1, 2, 3, \ldots$ If, for simplicity, it is assumed that the probes $E_1, E_2, E_3, \ldots$ are all *independent* of one another and that $E_k$ interacts with $S$ strictly *after* $E_{k-1}$ and strictly *before* $E_{k+1}$, then the state of $S$ *decohers* exponentially rapidly with respect to the basis $|\alpha_1\rangle, \ldots, |\alpha_n\rangle$, as $k \to \infty$. More precisely, if $\rho^{(k)}$ denotes the state of $S$ after its interaction with $E_k$ and before its interaction with $E_{k+1}$, with

$$\rho^{(k)}_{\alpha_i, \alpha_j} := \langle \alpha_i | \rho^{(k)} | \alpha_j \rangle, \tag{7.19}$$

then

$$\rho^{(k)}_{\alpha_i, \alpha_j} \longrightarrow \delta_{\alpha_i \alpha_j} \rho_{\alpha_i, \alpha_i}, \tag{7.20}$$

exponentially rapidly. This is easily verified; (see Sect. 7.5.6). A more subtle result on decoherence involving *correlated* probes that lead to *memory effects* has been established in [21].

One might ask what happens if a *direct measurement* is carried out on *every* probe $E_k$ after it has interacted with $S$, $k = 1, 2, 3, \ldots$. (We assume, for simplicity, that all probes $E_k$ are identical, independent and identically prepared, and that they are all subject to the *same* direct measurement.) Then one can show that, under natural non-degeneracy conditions, the state, $\rho^{(k)}$, of $S$, after the passage of $k$ probes $E_1, \ldots, E_k$, *converges to an eigenstate of $a$*, i.e.,

$$\rho^{(k)} \longrightarrow |\alpha_i\rangle \langle \alpha_i|, \tag{7.21}$$

as $k \to \infty$, for some $i$, and the probability of approach of $\rho^{(k)}$ to $|\alpha_i\rangle\langle\alpha_i|$ is given by $\rho_{\alpha_i, \alpha_i}$. This important result has been derived by Bauer and Bernard in [6] as a corollary of the *Martingale Convergence Theorem*; (see [1, 5, 56] for earlier ideas in this direction). The convergence claimed in Eq. (7.21) is remarkable, because it says that, asymptotically as $k \to \infty$, a *pure state* (some eigenstate of $a$) is approached; i.e., a very long sequence of *indirect* (non-demolition) *measurements* carried out on $S$ always results in a *"fact"* (namely, the state of $S$ approaches an eigenvector of the quantity $a$ that one intends to measure). Somewhat related results ("approach to a groundstate") for more realistic models have been proven in [22, 33, 35].[5]

In order to control the rate of convergence in Eqs. (7.20) and (7.21), it is helpful to make use of various notions of *quantum entropy*; (see, e.g., [20, 62]).

Some details concerning (indirect) non-demolition measurements and some remarks concerning interesting applications are sketched in Sect. 7.5.6; (but see [1, 6, 34, 42, 57]).

---

[5]A result of the form of Eq. (7.21) was conjectured by J.F. in the 1990s. But the proof remained elusive.

### 7.1.3   Organization of the Paper

In Sect. 7.2, we introduce an abstract algebraic framework for the formulation of mathematical models of physical systems that is general enough to encompass classical and quantum mechanical models. We attempt to clarify what kind of predictions a model of a physical system ought to enable us to come up with. Furthermore, we summarize some important facts about operator algebras needed in subsequent sections.

In Sect. 7.3, we describe *classical* models of physical systems within our algebraic framework and explain in which sense, and why, they are "realistic" and "deterministic".

In Sect. 7.4, we study a general class of quantum-mechanical models of physical systems within our general framework. We explain what some of the key problems in a quantum theory of observations and measurements are.

The most important section of this essay is Sect. 7.5. We attempt to elucidate the roles played by entanglement between a system and its environment and of information loss in understanding "decoherence" and "dephasing", which are key mechanisms in a quantum theory of measurements and experiments; see also [9, 37, 47, 49]. In particular, we point out that the state of the composition of a system with its environment can usually not be reconstructed from measurements long after interactions between the system and its environment have set in; ("information loss"). We also discuss the problem of "time in quantum mechanics" and sketch an answer to the question when an experiment can be considered to have been completed successfully; ("when does a detector click?"). Put differently, the "primitive ontology" of quantum mechanics is developed in Sects. 7.5.3 and 7.5.4. Finally, in Sect. 7.5.6, we briefly develop the theory of indirect non-demolition measurements, following [6].

An outline of *relativistic quantum theory* and of the role of space-time in relativistic quantum theory has been sketched in lectures and will be presented elsewhere; (see also [4]).

The main weakness of this essay (which might be fatal) is that we do not (and cannot) discuss sufficiently many simple, convincing examples illustrating the power of the general ideas presented here. This would simply take too much space. But examples will be discussed in [33, 34].

## 7.2   Models of Physical Systems

In this section, we sketch a somewhat abstract algebraic framework suitable to formulate *mathematical models of physical systems*. Our framework is general enough to encompass *classical and quantum-mechanical* models.

Throughout most of this essay, we consider *non-relativistic* models of physical systems, so that, in principle, all "observers" have access to the *same* observational data. For this reason, reference to "observers" is superfluous in the framework to be exposed here. This is radically different in causal relativistic models.

In every model of a physical system, $S$, one specifies $S$ in terms of (all) its *"potential properties"*, i.e., in terms of *"physical quantities"* or *"observables"* characteristic of $S$; see, e.g., [50]. No matter whether we consider classical or quantum-mechanical systems, "physical quantities" are represented, mathematically, by *bounded, self-adjoint, linear operators*. Thus, a system $S$ is specified by a list

$$\mathcal{P}_S = \{a_i\}_{i \in I_S} \tag{7.22}$$

of physical quantities, $a_i = a_i^*$, characteristic of $S$ that can be observed or measured in experiments.

In classical physics, a physical quantity, $a$, is given by a real-valued (measurable or continuous) function on a topological space, $M_S$, which is the *"state space"* of $S$ (the phase space if $S$ is Hamiltonian). Quantum-mechanically, more general linear operators are encountered, and, as is well known, the operators in $\mathcal{P}_S = \{a_i\}_{i \in I_S}$ need *not* all commute with one another. It is natural to assume that if $a \in \mathcal{P}_S$ is a physical quantity of $S$ then so is any polynomial, $p(a)$, in $a$ with real coefficients. It is, however, not very plausible that arbitrary real-linear combinations and/or symmetrized products of distinct elements in $\mathcal{P}_S$ would belong to $\mathcal{P}_S$. But, in non-relativistic physics, it has turned out to be reasonable to view $\mathcal{P}_S$ as a *self-adjoint subset of an operator algebra*, $\mathcal{A}_S$, usually taken to be a $C^*-$ or a von Neumann algebra, in terms of which a model of $S$ can be formulated. Physicists tend to be scared when they hear expressions like 'C*-' or 'von Neumann algebra'. Well, they shouldn't!

## 7.2.1 Some Basic Notions from the Theory of Operator Algebras

In order to render this paper comprehensible to the non-expert, we summarize some basic definitions and notions from the theory of operator algebras; for further details see [69, 70], and [16, 43, 44] .

An algebra, $\mathcal{A}$, over the complex numbers is a complex vector space equipped with a multiplication: If $a$ and $b$ belong to $\mathcal{A}$, then

- $\lambda a + \mu b \in \mathcal{A}, \quad \lambda, \mu \in \mathbb{C},$
- $a \cdot b \in \mathcal{A},$

where "·" denotes multiplication in $\mathcal{A}$. One says that $\mathcal{A}$ is a *algebra iff there exists an anti-linear involution, *, on $\mathcal{A}$, i.e., * : $\mathcal{A} \to \mathcal{A}$, with $(a^*)^* = a$, for all $a \in \mathcal{A}$, such that

$$(\lambda a + \mu b)^* = \bar{\lambda} a^* + \bar{\mu} b^*,$$

where $\bar{\lambda}$ is the complex conjugate of $\lambda \in \mathbb{C}$, and

$$(a \cdot b)^* = b^* \cdot a^*.$$

The algebra $\mathcal{A}$ is a *normed algebra* (Banach algebra) if it comes with a norm $\|(\cdot)\|$ satisfying

•

$$\|(\cdot)\| : \mathcal{A} \to [0, \infty[$$

•

$$\|a\| = 0, \text{ for } a \in \mathcal{A} \implies a = 0 \tag{7.23}$$

• ($\mathcal{A}$ is *complete* in $\|(\cdot)\|$, i.e., every Cauchy sequence in $\mathcal{A}$ converges to an element of $\mathcal{A}$).

A Banach algebra, $\mathcal{A}$, is a $C^*$-*algebra* iff

$$\|a^* \cdot a\| = \|a \cdot a^*\| = \|a\|^2, \ \forall a \in \mathcal{A}. \tag{7.24}$$

We define the *centre*, $\mathcal{Z}_\mathcal{A}$, of $\mathcal{A}$ to be the subset of $\mathcal{A}$ given by

$$\mathcal{Z}_\mathcal{A} := \{a \in \mathcal{A} \mid a \cdot b = b \cdot a, \forall b \in \mathcal{A}\}. \tag{7.25}$$

A *state*, $\omega$, on a *algebra $\mathcal{A}$ with identity $\mathbb{1}$ is a linear functional $\omega : \mathcal{A} \to \mathbb{C}$ with the properties that

$$\omega(a^*) = \overline{\omega(a)}, \qquad \omega(a^*a) \geq 0, \tag{7.26}$$

for all $a \in \mathcal{A}$, and

$$\omega(\mathbb{1}) = 1. \tag{7.27}$$

A state $\omega$ is *pure* if it cannot be written as a convex combination of two or more distinct states.

A *representation*, $\pi$, of a $C^*$-algebra $\mathcal{A}$ on a complex Hilbert space, $\mathcal{H}$, is a *homomorphism from $\mathcal{A}$ to the algebra, $\mathcal{B}(\mathcal{H})$, of all bounded linear operators on $\mathcal{H}$; i.e., $\pi$ is linear, $\pi(a \cdot b) = \pi(a) \cdot \pi(b)$, $\pi(a^*) = (\pi(a))^*$, and $\|\pi(a)\| \leq \|a\|$, (where $\|A\|$ is the operator norm of a bounded linear operator $A$ on $\mathcal{H}$).

A *automorphism*, $\alpha$, of a $C^*$-algebra $\mathcal{A}$ is a linear isomorphism from $\mathcal{A}$ onto $\mathcal{A}$ with the properties

$$\alpha(a \cdot b) = \alpha(a) \cdot \alpha(b),$$
$$\alpha(a^*) = (\alpha(a))^*, \tag{7.28}$$

for all $a, b \in \mathcal{A}$.

With a $C^*$-algebra $\mathcal{A}$ and a state $\omega$ on $\mathcal{A}$ we can associate a Hilbert space, $\mathcal{H}_\omega$, a unit vector $\Omega \in \mathcal{H}_\omega$, and a representation, $\pi_\omega$, of $\mathcal{A}$ on $\mathcal{H}_\omega$ such that $\{\pi_\omega(a)\Omega \mid a \in \mathcal{A}\}$ is dense in $\mathcal{H}_\omega$ (i.e. $\Omega$ is *cyclic* for $\pi_\omega(\mathcal{A})$), and

$$\omega(a) = \langle \Omega, \pi_\omega(a)\Omega \rangle, \tag{7.29}$$

where $\langle \cdot, \cdot \rangle$ is the scalar product on $\mathcal{H}_\omega$. This results from the so-called *Gel'fand–Naimark–Segal (GNS) construction*.

A theorem due to Gel'fand and Naimark says that every $C^*$-algebra, $\mathcal{A}$, can be viewed as a norm-closed subalgebra of $\mathcal{B}(\mathcal{H})$ closed under $*$, for some Hilbert space $\mathcal{H}$.

Thus, consider a $C^*$-algebra $\mathcal{A} \subset \mathcal{B}(\mathcal{H})$, for some Hilbert space $\mathcal{H}$. We define the commuting algebra, or commutant, $\mathcal{A}'$, of $\mathcal{A}$ by

$$\mathcal{A}' := \{a \in \mathcal{B}(\mathcal{H}) \mid a \cdot b = b \cdot a, \forall b \in \mathcal{A}\}. \tag{7.30}$$

The double commutant of $\mathcal{A}$, $\mathcal{A}''$, is defined by

$$\mathcal{A}'' \equiv (\mathcal{A}')' = \{a \in \mathcal{B}(\mathcal{H}) \mid a \cdot b = b \cdot a, \forall b \in \mathcal{A}'\} \supseteq \mathcal{A}. \tag{7.31}$$

It turns out that $\mathcal{A}'$ and $\mathcal{A}''$ are closed in the so-called weak $*$ topology of $\mathcal{B}(\mathcal{H})$; i.e., if $\{a_i\}_{i \in I}$ is a sequence (net) of operators in $\mathcal{A}'$ (or in $\mathcal{A}''$), with

$$\langle \varphi, a_i \psi \rangle \to \langle \varphi, a\psi \rangle, \quad \text{as } i \to \infty,$$

for *all* $\varphi, \psi \in \mathcal{H}$, where $a \in \mathcal{B}(\mathcal{H})$, then $a \in \mathcal{A}'$ (or $a \in \mathcal{A}''$, respectively). *Subalgebras of $\mathcal{B}(\mathcal{H})$ that are closed in the weak $*$ topology and contain the identity are called *von Neumann algebras* (or $W^*$-algebras). By a famous theorem of von Neumann, a *algebra $\mathcal{A}$ of operators on a Hilbert space is a von Neumann algebra if and only if $\mathcal{A} = \mathcal{A}''$.

Thus, if $\mathcal{A}$ is a $C^*$-algebra contained in $\mathcal{B}(\mathcal{H})$, for some Hilbert space $\mathcal{H}$, then $\mathcal{A}'$ and $\mathcal{A}''$ are von Neumann algebras. A von Neumann algebra $\mathcal{M} \subseteq \mathcal{B}(\mathcal{H})$ is called a *factor* iff its centre, $\mathcal{Z}_\mathcal{M}$, consists of multiples of the identity operator $\mathbb{1}$.

A von Neumann factor $\mathcal{M}$ is said to be of type $I$ iff $\mathcal{M}$ is isomorphic to $\mathcal{B}(\mathcal{H}_0)$, for some Hilbert space $\mathcal{H}_0$. A general von Neumann algebra, $\mathcal{N}$, is said to be of type I iff $\mathcal{N}$ is a direct sum (or integral) over its centre, $\mathcal{Z}_\mathcal{N}$, of factors of type I. A $C^*$-algebra $\mathcal{A}$ is called a type-I $C^*$-algebra, iff, for every representation $\pi$, of $\mathcal{A}$ on a Hilbert space $\mathcal{H}$,

$$\pi(\mathcal{A}) := \{\pi(a) \mid a \in \mathcal{A}\}$$

has the property that $\pi(\mathcal{A})''$ is a von Neumann algebra of type I. (For mathematical properties of type-I $C^*$-algebra see [40], and for examples relevant to quantum physics see [15].)

We define

$$\mathcal{A}' \cap \mathcal{B} := \{b \in \mathcal{B} \mid b \cdot a = a \cdot b, \ \forall a \in \mathcal{A}\}, \tag{7.32}$$

the "relative commutant" of $\mathcal{A}$ in $\mathcal{B}$.

Given a set $\mathcal{P} = \{a_i\}_{i \in I}$ of operators in a $C^*$-algebra $\mathcal{B}$, we define $\langle \mathcal{P} \rangle$ to be the $C^*$-subalgebra of $\mathcal{B}$ generated by $\mathcal{P}$, i.e., the norm-closure of arbitrary finite complex-linear combinations of arbitrary finite products of elements in the set $\{a_i, a_i^*\}_{i \in I}$, where $*$ is the $*$ operation on $\mathcal{B}$.

A trace $\tau : \mathcal{M}_+ \to [0, \infty]$ on a von Neumann Algebra $\mathcal{M}$ is a function defined on the positive cone, $\mathcal{M}_+$, of positive elements of $\mathcal{M}$ (i.e., elements $x \in \mathcal{M}$ of the form $x = y^*y$, $y \in \mathcal{M}$) that satisfies the properties

$$(i) \quad \tau(x + y) = \tau(x) + \tau(y), \qquad x, y \in \mathcal{M}_+$$

$$(ii) \quad \tau(\lambda x) = \lambda \tau(x), \qquad \qquad \lambda \in \mathbb{R}_+, \ x \in \mathcal{M}_+$$

$$(iii) \quad \tau(x^*x) = \tau(xx^*), \qquad \quad x \in \mathcal{M}.$$

A trace $\tau$ is said to be *finite* if $\tau(\mathbb{1}) < +\infty$. It can then be uniquely extended by linearity to a *state* $\tau$ on $\mathcal{M}$. Conversely, any state $\tau$ on $\mathcal{M}$ enjoying the property

$$\tau(a \cdot b) = \tau(b \cdot a), \qquad \forall a, b \in \mathcal{M}, \tag{7.33}$$

defines a finite trace on $\mathcal{M}$. We say that $\tau$ is faithful if $\tau(x) > 0$ for any non-zero element $x \in \mathcal{M}_+$. A trace $\tau$ is said to be *normal* if $\tau(\sup x_i) = \sup \tau(x_i)$ for every bounded net $(x_i)_{i \in I}$ of positive elements in $\mathcal{M}$, and *semifinite*, if, for any $x \in \mathcal{M}_+$, $x \neq 0$, there exists $y \in \mathcal{M}_+$, $0 < y \leq x$, such that $\tau(y) < \infty$. Traces play an important role in the classification of von Neumann algebras. It can be shown that a von Neumann algebra $\mathcal{M}$ is a direct sum (or direct integral) of factors of type $I_n$ and type $II_1$ if and only if it admits a faithful finite normal trace; see [69]. Similarly, $\mathcal{M}$ is a direct sum (or direct integral) of type I, type $II_1$ and type $II_\infty$ factors iff it admits a faithful semifinite normal trace. We use these results in Sect. 7.5 to characterize the centralizer of a state $\omega$.

For the time being, we do not have to know more about operator algebras than what has just been reviewed here. We can test our understanding of the notions introduced above on the example of direct sums of full finite-dimensional matrix algebras (block-diagonal matrices) and by doing some exercises, e.g., reproducing a proof of the GNS construction, or applying this material to group theory.

## 7.2.2   The Operator Algebras Used to Describe a Physical System

We have said that (a model of) a physical system, $S$, is specified by a list

$$\mathcal{P}_S = \{a_i\}_{i \in I_S}$$

of *physical quantities* or *potential properties*, $a_i = a_i^*$ ($i \in I_S$), characteristic of $S$ that can be observed or measured in experiments. (What is meant by this will hopefully become clear later, in Sects. 7.4 and 7.5.) We assume that $\mathcal{P}_S$ is a self-adjoint subset of a $C^*$-algebra. As explained in Sect. 7.2.1, we may then consider

$$\mathcal{A}_S := \langle \mathcal{P}_S \rangle, \tag{7.34}$$

the smallest $C^*$-algebra containing $\mathcal{P}_S$. The algebra $\mathcal{A}_S$ is called the *"algebra of observables"* defining $S$; (possibly a misnomer, because, a priori, only the elements of $\mathcal{P}_S$ correspond to observable physical quantities—but let's not worry about this). For physical systems with finitely many degrees of freedom, $\mathcal{A}_S$ is usually a type-I $C^*$-algebra.

We would like to have some natural notions of symmetries of a system $S$, including time evolution. Here we encounter, for the first but not the last time, the complication that $S$ is usually in contact with some *environment*, $E$, which may also include experimental equipment used to measure some observables of $S$. The environment is a physical system, too, and there usually are interactions between $S$ and $E$; in fact, only thanks to such interactions is it possible to retrieve information from $S$, i.e., measure a potential property $a_i$, $i \in I_S$, of $S$ in a certain interval of time. One typically chooses $E$ to be the *smallest system* with the property that the *composed system,* $S \vee E$, characterized by

$$\mathcal{P}_{S \vee E} = \{a, b \mid a \in \mathcal{P}_S, b \in \mathcal{P}_E\}, \tag{7.35}$$

can be viewed as a *"closed physical system"*.

What is a "closed physical system"? Let $\overline{S} := S \vee E$, and let $\mathcal{A}_{\overline{S}}$ denote the $C^*$-algebra generated by $\mathcal{P}_{S \vee E}$; i.e., $\mathcal{A}_{\overline{S}} = \langle \mathcal{P}_{S \vee E} \rangle$. We say that $\overline{S}$ is a *closed (physical) system* if the *time evolution* of physical quantities characteristic of $\overline{S}$ is given in terms of *\*automorphisms* of $\mathcal{A}_{\overline{S}}$; i.e., given two times, $s$ and $t$, $\tau_{t,s}$ is a *automorphism of $\mathcal{A}_{\overline{S}}$ that associates with every physical quantity in $\mathcal{A}_{\overline{S}}$ specified at time $s$ an operator in $\mathcal{A}_{\overline{S}}$ representing the *same* physical quantity at time $t$. We must require that

$$\tau_{t,s} \circ \tau_{s,u} = \tau_{t,u}, \tag{7.36}$$

for any triple of times $(t, s, u)$.

Given a physical system, $S$, we choose its environment $E$ such that, within a prescribed precision, $\overline{S} = S \vee E$ can be considered to be a closed physical system. "For all practical purposes" (FAPP, see [9]), i.e., within usually astounding precision, $\overline{S}$ is much ... much smaller than the entire universe; it does usually *not* include the experimentalist in the laboratory observing $S$ or the laptop of her theorist colleague next door, etc. To say that $\overline{S}$ is a *closed* physical system does, however, *not* exclude that $\overline{S}$ is entangled with another physical system, $S'$. Further discussion and examples of closed systems are presented in [29].

Given $S$ and $\overline{S} = S \vee E$, as above, we call $\mathcal{A}_{\overline{S}}$ the *"dynamical $C^*$-algebra"* of $S$.

Let $\mathcal{G}_S$ denote a group of *symmetries* of $S$. We will assume that every element $g \in \mathcal{G}_S$ can be represented by a *automorphism, $\sigma_g$, of $\mathcal{A}_{\overline{S}}$, with the property that

$$\sigma_{g_1} \circ \sigma_{g_2} = \sigma_{g_1 \circ g_2}, \tag{7.37}$$

i.e., $\sigma : \mathcal{G}_S \longrightarrow {}^*\mathrm{Aut}(\mathcal{A}_{\overline{S}})$ is a representation of $\mathcal{G}_S$ in the group, ${}^*\mathrm{Aut}(\mathcal{A}_{\overline{S}})$, of *automorphisms of $\mathcal{A}_{\overline{S}}$. We say that $\mathcal{G}_S$ is a group of *dynamical* symmetries of $S$ iff $\sigma_g$ and time evolution $\tau_{t,s}$ commute, for all $g \in \mathcal{G}_S$ and arbitrary pairs of times $(t, s)$.

By a *"state of a physical system"* $S$ we mean a state on the $C^*$-algebra $\mathcal{A}_{\overline{S}}$, in the sense of Eqs. (7.26) and (7.27) in Sect. 7.2.1. (This will turn out to be a misnomer when we deal with quantum systems. But the expression appears to be here to stay.) The set of all states of $S$ is denoted by $\mathcal{S}_{\overline{S}}$.

To summarize, *a (model of a) physical system, $S$,* is specified by the following data.

### Definition 2.1 (Algebraic Data Specifying a Model of a Physical System)

(I) A list of *physical quantities*, or *observables*, $\mathcal{P}_S = \{a_i = a_i^*\}_{i \in I_S}$, generating a $C^*$-algebra, $\mathcal{A}_S$, of "observables", that is contained in the $C^*$-algebra $\mathcal{A}_{\overline{S}}$ (the "dynamical $C^*$-algebra" of $S$) of a *closed* system, $\overline{S} = S \vee E$, containing $S$.

(II) The convex set, $\mathcal{S}_{\overline{S}}$, of *states of $S$*, interpreted as states on the $C^*$-algebra $\mathcal{A}_{\overline{S}}$.

(III) *Time translations of* $\overline{S}$, represented as *automorphisms $\{\tau_{t,s}\}_{t,s \in \mathbb{R}}$ on $\mathcal{A}_{\overline{S}}$ satisfying Eq. (7.36), and a group, $\mathcal{G}_S$, of *symmetries of $S$* represented by *automorphisms, $\{\sigma_g\}_{g \in \mathcal{G}_S}$, of $\mathcal{A}_{\overline{S}}$; (see Eq. (7.37)).

We should explain what is meant by *"time translations"*: For each time $t \in \mathbb{R}$, we have copies $\mathcal{P}_S(t)$ and $\mathcal{A}_S(t) = \langle \mathcal{P}_S(t) \rangle$ *isomorphic to $\mathcal{P}_S$ and $\mathcal{A}_S$, respectively, which are contained in $\mathcal{A}_{\overline{S}}$. If $a(s) \in \mathcal{P}_S(s)$ and $a(t) \in \mathcal{P}_S(t)$ are the operators in $\mathcal{A}_{\overline{S}}$ representing an arbitrary potential property, or observable, $a \in \mathcal{P}_S$, of $S$ at times $s$ and $t$, respectively, then

$$a(t) = \tau_{t,s}(a(s)), \tag{7.38}$$

with $\tau_{t,s} = \tau_{t,u} \circ \tau_{u,s}$, for arbitrary times $t$, $u$ and $s$ in $\mathbb{R}$.

We say that the system $\overline{S} = S \vee E$ is *autonomous* iff

$$\tau_{t,s} = \tau_{t-s} \tag{7.39}$$

where $\{\tau_t\}_{t \in \mathbb{R}}$ is a one-parameter group of *automorphisms of $\mathcal{A}_{\overline{S}}$.

We say that a system $S$ is a *subsystem* of a system $S'$ iff

$$\mathcal{P}_S \subset \mathcal{P}_{S'} \tag{7.40}$$

and

$$\mathcal{A}_{\overline{S}} \subseteq \mathcal{A}_{\overline{S'}}. \tag{7.41}$$

The composition, $S_1 \vee S_2$, of two systems, $S_1$ and $S_2$, can be defined by choosing

$$\mathcal{P}_{S_1 \vee S_2} := \mathcal{P}_{S_1} \cup \mathcal{P}_{S_2} \tag{7.42}$$

and $\mathcal{A}_{\overline{S_1 \vee S_2}}$ to contain the $C^*$-algebra generated by $\mathcal{A}_{\overline{S_1}}$ and $\mathcal{A}_{\overline{S_2}}$. (A more precise discussion would lead us into the theory of tensor categories.)

### 7.2.3   Potential Properties, Information Loss and Possible Events

Let $S$ be a physical system coupled to an environment $E$ and described, mathematically, by data

$$(\mathcal{P}_S, \mathcal{A}_{\overline{S}=S \vee E}, \{\tau_{t,s}\}_{t,s \in \mathbb{R}}, \mathcal{G}_S, \mathcal{S}_{\overline{S}}) \tag{7.43}$$

with properties as specified in points (I) through (III) of Definition 2.1, Sect. 7.2.2.

A "potential property" of $S$ is represented by an element $a \in \mathcal{P}_S$ or, more generally, by a self-adjoint operator $a = a^*$ in the algebra $\mathcal{A}_S$. An observation of a potential property, $a$, of $S$ at time $t$ will be described in terms of the operator $a(t) = \tau_{t,t_0}(a) \in \mathcal{A}_{\overline{S}}$, where $t_0$ is a fiducial time at which the state of $S$ is specified. Next, we have to clarify in which sense *information* is *lost*, as time increases. In *local, relativistic* quantum theory, a distinction between $S$ and $\overline{S}$ becomes superfluous, and one may usually identify $S$ with $\overline{S}$. Moreover, the finiteness of the speed of light, i.e., of the speed of propagation of arbitrary signals, and locality lead to an intrinsic notion of *information loss* [17, 31]—at least in theories with massless particles that satisfy Huyghens' Principle [14] and are allowed to escape to spatial $\infty$ (or fall into black holes). This is not so when one considers non-relativistic models of physical systems, with signals propagating arbitrarily fast ("Fernwirkung"). Nevertheless, one may argue that whenever properties of $S$ are observed successfully, thanks to interactions of $S$ with some environment/equipment $E$, then, as the price to pay, information is lost irretrievably: It disperses into the environment $E$, where it becomes inaccessible to experimental observation. Of course, this idea is plausible *only* if the cut between "system $S$" and "environment $E$", given a closed system $\overline{S}$, is made at the right place. To determine this cut, one must specify the list $\mathcal{P}_S$ of physical quantities characterising $S$ that are measurable in experiments, using $E$. Mathematically, the cut is determined by specifying the pair $(\mathcal{A}_S, \mathcal{A}_{\overline{S}})$ of algebras.

For the purpose of this essay, we adopt the point of view that the only properties of $\overline{S}$ that can potentially be observed, experimentally, are properties of $S$ represented by self-adjoint operators

$$a(t) = a^*(t), \quad \text{with } a \in \mathcal{P}_S, t \in \mathbb{R}. \tag{7.44}$$

In order to arrive at a mathematically precise concept of information loss (as time goes by), it is convenient to introduce the following algebras.

**Definition 2.2** The algebra, $\mathcal{E}_{\geq t}$, of potential properties observable after time $t$ is the $C^*$-subalgebra of $\mathcal{A}_{\overline{S}}$ generated by arbitrary finite linear combinations of arbitrary finite products

$$a_1(t_1) \ldots a_n(t_n), n = 1, 2, 3, \ldots,$$

where $t_i \geq t$ and $a_i \in \mathcal{A}_S$, $i = 1, \ldots, n$, (with $a(s)$ the operator in $\mathcal{A}_{\overline{S}}$ representing the operator $a \in \mathcal{A}_S$ at time $s$).

It follows from this definition that

$$\mathcal{E}_{\geq t} \subseteq \mathcal{E}_{\geq t'} \tag{7.45}$$

whenever $t > t'$, with $\mathcal{E}_{\geq t} \subseteq \mathcal{A}_{\overline{S}}$, for all $t \in \mathbb{R}$. We speak of *loss of information* iff

$$\mathcal{E}_{\geq t} \subsetneq \mathcal{E}_{\geq t'}, \tag{7.46}$$

for some times $t$ and $t'$, with $t > t'$. We define an algebra $\mathcal{E}_S$ by

$$\mathcal{E}_S := \overline{\bigvee_{t \in \mathbb{R}} \mathcal{E}_{\geq t}}^{\|\cdot\|} \tag{7.47}$$

It is one of the notorious problems in most approaches to a "quantum theory of experiments" that it is left unclear which self-adjoint operators in some very large algebra of operators correspond to potential properties of a quantum system that can actually be *measured* or observed. Most authors consider *far too many* operators as corresponding to potential properties of the system that are potentially measurable. As we will discuss in Sect. 7.5, it appears to be a general principle ("Duality between Observables and Indeterminates") that $\mathcal{E}_S \subsetneqq \mathcal{A}_{\overline{S}}$ and that the relative commutant of $\mathcal{E}_S$ inside $\mathcal{A}_{\overline{S}}$ contains a subalgebra isomorphic to $\mathcal{E}_S$. (Obviously, for classical systems—$\mathcal{A}_{\overline{S}}$ abelian, the commutant of $\mathcal{E}_S$ is all of $\mathcal{A}_{\overline{S}}$.)

Let $\omega \in \mathcal{S}_{\overline{S}}$ be a state of the system. Let $(\mathcal{H}_\omega, \pi_\omega, \Omega)$ denote the Hilbert space, the representation of $\mathcal{A}_{\overline{S}}$ on $\mathcal{H}_\omega$, and the cyclic vector in $\mathcal{H}_\omega$, respectively, associated to the pair $(\mathcal{A}_{\overline{S}}, \omega)$ by the GNS construction; see Sect. 7.2.1, Eq. (7.29). By $\mathcal{A}_{\overline{S}}^\omega$ we denote the von Neumann algebra corresponding to the weak closure of $\pi_\omega(\mathcal{A}_{\overline{S}})$ in the algebra, $\mathcal{B}(\mathcal{H}_\omega)$, of all bounded operators on $\mathcal{H}_\omega$.

**Definition 2.3** Given a physical system $S$, as in Definition 2.1, (I)–(III), above, and a state $\omega \in \mathcal{S}_{\overline{S}}$, a possible event in $S$ observable at time $t$ is a spectral projection,

$$P_{a(t)}(I), \tag{7.48}$$

of the operator $\pi_\omega(a(t)) \in \mathcal{A}_{\overline{S}}^\omega$ associated with a measurable subset $I \subseteq \mathrm{spec}\,\pi_\omega(a(t)) \subseteq \mathbb{R}$, where $a = a^* \in \mathcal{P}_S$ and $t \in \mathbb{R}$. (Here spec $A$ denotes the spectrum of a self-adjoint operator $A$ on $\mathcal{H}_\omega$.)

**Definition 2.4** The algebra, $\mathcal{E}_{\geq t}^\omega$, of all possible events observable at times $\geq t$, is the von Neumann algebra corresponding to the weak closure of $\pi_\omega(\mathcal{E}_{\geq t})$ in $B(\mathcal{H}_\omega)$. The von Neumann algebra $\mathcal{E}_S^\omega$ is defined similarly.

Note that if $\omega'$ is a state that is normal with respect to the state $\omega$ then $\mathcal{A}_{\overline{S}}^{\omega'} = \mathcal{A}_{\overline{S}}^\omega$, etc. The algebra $\mathcal{E}_{\geq t}^\omega$ contains the spectral projections $P_{a(s)}(I)$ describing possible events at times $s \geq t$; (see Eq. (7.48)). It is therefore justified to call $\mathcal{E}_{\geq t}^\omega$ the "*algebra of possible events observable at times $\geq t$* ". Loss of information may manifest itself in the property that the relative commutant

$$(\mathcal{E}_{\geq t}^\omega)' \cap \mathcal{E}_{\geq t'}^\omega \tag{7.49}$$

is *non-trivial*, for some $t > t'$.

We note that the algebra $\mathcal{E}_S$ carries an action of the group, $\mathbb{R}$, of time translations by $^*$automorphisms, $\{\overline{\tau}_t\}_{t \in \mathbb{R}}$, defined as follows: For $a_1(t_1) \ldots a_n(t_n) \in \bigvee_{t \in \mathbb{R}} \mathcal{E}_{\geq t}$, with $t_i \in \mathbb{R}, a_i \in \mathcal{A}_S, i = 1, \ldots, n$,

$$\overline{\tau}_t(a_1(t_1) \ldots a_n(t_n)) := a_1(t_1 + t) \ldots a_n(t_n + t). \tag{7.50}$$

The definition of $\overline{\tau}_t$ extends to all of $\mathcal{E}_S$ by linearity and continuity. One then has that

$$\overline{\tau}_t : \mathcal{E}_{\geq t'} \longrightarrow \mathcal{E}_{\geq t'+t} \subseteq \mathcal{E}_{\geq t'}, \tag{7.51}$$

for arbitrary $t \geq 0$.

Let $a \in \mathcal{P}_S$ be a potential property of $S$, and let $\omega$ be a state of $S$ (i.e., $\omega \in \mathcal{S}_{\overline{S}}$). Depending on the experimental equipment available to observe $a$, i.e., depending on the choice of the time evolution of $\overline{S} = S \vee E$, and depending on the choice of a state $\omega \in \mathcal{S}_{\overline{S}}$, an observation of $a$ may have different alternative outcomes; in particular, the resolution in an observation of $a$ at some time $t_*$ will depend on the choice of $(E, \{\tau_{t,s}\}_{t,s \in \mathbb{R}}, \omega)$. These alternative outcomes correspond to spectral projections $P_{a(t_*)}(I_\alpha), \alpha = 1, \ldots, k$, where $I_\alpha \cap I_\beta = \emptyset$, for $\alpha \neq \beta$, and $\cup_{\alpha=1}^k I_\alpha \supseteq$ spec $\pi_\omega(a(t_*))$. Then

$$P_{a(t_*)}(I_\alpha) P_{a(t_*)}(I_\beta) = \delta_{\alpha\beta} P_{a(t_*)}(I_\alpha), \tag{7.52}$$

and

$$\sum_{\alpha=1}^k P_{a(t_*)}(I_\alpha) = \mathbb{1}, \tag{7.53}$$

for an arbitrary $t_*$.

Traditionally, one says that the *purpose of a model of a physical system, $S$,* is to enable us to make predictions of the following kind: Suppose we are interested in testing some potential properties (or, put differently, measure some physical quantities) $a_1, \ldots, a_n$ characteristic of $S$ during intervals of time $\Delta_1 \prec \Delta_2 \prec \ldots \prec \Delta_n$, where

$$\Delta \prec \Delta' \text{ iff, } \forall t \in \Delta, \ \forall t' \in \Delta' : t \leq t'. \tag{7.54}$$

We assume that $S$ is in a state $\omega \in \mathcal{S}_{\overline{S}}$. Then a model of $S$ ought to tell us whether $a_1, \ldots, a_n$ will actually be measurable (i.e., are *"empirical"* properties) and predict the *probability* (frequency) that, in a test or measurement of $a_i$ at some time $t_i \in \Delta_i$, the event corresponding to the spectral projection $P_{a_i(t_i)}(I_{\alpha_i}^i)$, $\alpha_i = 1, \ldots, k_i$, is observed, (i.e., property $a_i(t_i)$ has a value in the interval $I_{\alpha_i}^i$), for all $i = 1, \ldots, n$, given the state $\omega \in \mathcal{S}_{\overline{S}}$; (the properties of the projections $P_{a_i(t_i)}(I_{\alpha_i}^i)$ are as in Eqs. (7.52), (7.53)).

We simplify our notation by setting

$$\Pi_{\alpha_i}^{(i)} \equiv \Pi_{\alpha_i}^{(i)}(t_i) := P_{a_i(t_i)}(I_{\alpha_i}^i), \tag{7.55}$$

with $t_i \in \Delta_i$, $a_i \in \mathcal{P}_S$, $i = 1, \ldots, n$, $\Delta_1 \prec \Delta_2 \prec \ldots \prec \Delta_n$. The time-ordered sequence

$$h_1^n(\underline{\alpha}) := \{\Pi_{\alpha_1}^{(1)}, \ldots, \Pi_{\alpha_n}^{(n)}\} \tag{7.56}$$

of possible events $\Pi_{\alpha_i}^{(i)}$ (as in Eq. (7.55)) is conventionally called a *"history"*. Given such a history, we define operators

$$H_k^n(\underline{\alpha}) := \Pi_{\alpha_n}^{(n)} \ldots \Pi_{\alpha_{k+1}}^{(k+1)} \Pi_{\alpha_k}^{(k)}, \tag{7.57}$$

with $\Pi_{\alpha_i}^{(i)}$ as in Eq. (7.55).

**Postulate 2.5** (see [59, 64, 76]) *Given a model of a physical system S, as specified in points (I)–(III) of Definition 2.1, Sect. 7.2.2, the probability of a history* $h_1^n(\underline{\alpha}) = \{\Pi_{\alpha_1}^{(1)}, \ldots, \Pi_{\alpha_n}^{(n)}\}$ *in a state* $\omega \in \mathcal{S}_{\overline{S}}$ *is predicted to be given by*

$$\mathrm{Prob}_\omega \, h_1^n(\underline{\alpha}) \equiv \mathrm{Prob}_\omega\{\Pi_{\alpha_1}^{(1)}, \ldots, \Pi_{\alpha_n}^{(n)}\} := \omega\left((H_1^n(\underline{\alpha}))^* H_1^n(\underline{\alpha})\right), \tag{7.58}$$

*with* $H_1^n(\underline{\alpha})$ *as in Eq. (7.57).* (It is assumed here that $a_1, \ldots, a_n$ are measurable, for the given time-evolution and state of the system; see Sect. 7.5.)

Much discussion in the remainder of this essay is devoted to finding out under what conditions formula (7.58), is meaningful, and—if it is—what it tells us about $S$. To give away our secrets, Postulate 2.5 is perfectly meaningful for *classical* models of physical systems, as discussed in Sect. 7.3, and it is most often meaningless for *quantum-mechanical* models. While FMPP ("for many practical purposes"), formula (7.58) is useful in quantum mechanics, conceptually it is misleading and often nonsensical! It does, however, pass some tests indicating that it defines a probability:

(1) $\mathrm{Prob}_\omega$ satisfies

$$0 \le \mathrm{Prob}_\omega\{\Pi_{\alpha_1}^{(1)}, \ldots, \Pi_{\alpha_n}^{(n)}\} \le 1, \tag{7.59}$$

for every state $\omega \in \mathcal{S}_{\overline{S}}$ and an arbitrary history $\{\Pi_{\alpha_1}^{(1)}, \ldots, \Pi_{\alpha_n}^{(n)}\}$.

(2)

$$\sum_{\alpha_i = 1, \ldots, k_i \, (i=1, \ldots, n)} \mathrm{Prob}_\omega\{\Pi_{\alpha_1}^{(1)}, \ldots, \Pi_{\alpha_n}^{(n)}\} = 1, \tag{7.60}$$

for arbitrary operators $a_1, \ldots, a_n$ and time intervals $\Delta_1 \prec \ldots \prec \Delta_n$, (with $\Pi_{\alpha_i}^{(i)}$ as in Eq. (7.55)).

Properties (1) and (2) show that $\mathrm{Prob}_\omega$ is a probability functional.

(3) As observed in [48, 59] and references given there, formula (7.58) represents the *"only possible"* definition of a probability functional on the lattice of possible events.

As already mentioned, formula (7.58) is perfectly adequate for an analysis of the predictions of *classical* models of physical systems. *Quantum-mechanically*, however, given

$$(\mathcal{A}_{\overline{S}}, \{\tau_{t,s}\}_{t,s \in \mathbb{R}}, \omega \in \mathcal{S}_{\overline{S}}),$$

one encounters plenty of sequences of potential properties,

$$\{a_1(t_1), \ldots, a_n(t_n)\},$$

with $a_i \in \mathcal{P}_S$, $t_i \in \Delta_i$, $i = 1, \ldots, n$, $\Delta_1 \prec \ldots \prec \Delta_n$, which turn out to be *incompatible* with one another. The question then arises which one among such sequences of potential properties of $S$ actually corresponds to a sequence of *empirical properties* of $S$ observed in the course of time; (assuming that there is only *one* rather than "many worlds"). Formula (7.58) does not tell us much about the answer to this question; but the idea of *loss of information*, as expressed in Eqs. (7.46) and (7.49), along with the phenomenon of *entanglement, does*! This is discussed in Sects. 7.5.3 and 7.5.4.

## 7.3 Classical ("Realistic") Models of Physical Systems

We start this section by recalling the usual distinction between *classical, realistic models* (abbreviated as "R-models") and *quantum-mechanical-models* (abbreviated as "Q-models") of physical systems: An *R-model* of a system $S$ is fully characterized by the property that its "dynamical $C^*$-algebra" $\mathcal{A}_{\overline{S}}$ (see Sect. 7.2.2) is *abelian* (commutative). Hence $\mathcal{A}_S$ is abelian, too.

A *Q-model* of a system $S$ differs from an R-model only in that the algebra $\mathcal{A}_S$ (and hence $\mathcal{A}_{\overline{S}}$) is *non commutative*. Apart from this crucial difference, the algebraic data defining an R- or a Q-model are as specified in points (I)–(III) of Definition 2.1, Sect. 7.2.2.

### 7.3.1  General Features of Classical Models

We recall a well-known theorem due to I.M. Gel'fand. Let $\mathcal{B}$ be an abelian $C^*$-algebra. The *spectrum*, $M$, of $\mathcal{B}$ is the space of all non-zero *homomorphisms from $\mathcal{B}$ into $\mathbb{C}$ (the "characters" of $\mathcal{B}$); $M$ is a locally compact topological (Hausdorff) space. If $\mathcal{B}$ contains an identity, $\mathbb{1}$, then $M$ is compact.

**Theorem 3.1 (Gel'fand)** *If $B$ is an abelian $C^*$-algebra then it is $^*$isomorphic to the $C^*$-algebra, $C_0(M)$, of continuous functions on $M$ vanishing at $\infty$, i.e.,*

$$B \simeq C_0(M). \tag{7.61}$$

*Furthermore, every state, $\omega$, on $B$ is given by a unique (Borel) probability measure, $d\mu_\omega$, on $M$ (and conversely).*

Every pure state is given by a Dirac $\delta$-function, $\delta_x$, on $M$, for some $x \in M$; i.e., the space of pure states can be identified with $M$, (which is why $M$ is called "*state space*"). Thus, the set of pure states of $B$ cannot be endowed with a linear or affine structure.

If $B_0 \subset B$ is a subalgebra of $B$ then any pure state of $B$ is also a pure state of $B_0$. If $B = \mathcal{A}_{\overline{S}}$ is the dynamical $C^*$-algebra of a realistic (classical) model of a physical system, $S$, we call $M =: M_S$ the state space of $S$. It is homeomorphic to the space of pure states of $\overline{S}$ and does not have a linear structure, i.e. there is *no superposition principle* for pure states. If $S = S_1 \vee S_2$ is the composition of two subsystems, $S_1$ and $S_2$, these systems are, of course, classical, too, and we have that any pure state of $S$ is also a pure state of $S_1$ and of $S_2$; i.e., there is *no* interesting notion of *entanglement*.

## 7.3.2  Symmetries and Time Evolution in Classical Models

According to point (III) of Definition 2.1 in Sect. 7.2.2, symmetries and time evolution of a system $S$ are given by $^*$automorphisms of its dynamical $C^*$-algebra $\mathcal{A}_{\overline{S}}$. If $B$ is an abelian $C^*$-algebra and $M$ denotes its spectrum then any $^*$automorphism, $\alpha$, of $B$ corresponds to a *homeomorphism*, $\phi_\alpha$, of $M$: If $a$ is an arbitrary element of $B$, thus given by a bounded continuous function (also denoted by $a$) on $M$, then

$$\alpha(a)(\xi) =: a(\phi_\alpha^{-1}(\xi)), \qquad \xi \in M. \tag{7.62}$$

Conversely, any homeomorphism, $\phi$, from $M$ to $M$ determines a $^*$automorphism, $\alpha_\phi$, by

$$\alpha_\phi(a)(\xi) := a(\phi^{-1}(\xi)), \qquad \xi \in M. \tag{7.63}$$

If $\{\alpha_{t,s}\}_{t,s \in \mathbb{R}}$ is a groupoid of $^*$automorphisms of $B$, with $\alpha_{t,s} \circ \alpha_{s,u} = \alpha_{t,u}$, then there exists a groupoid of homeomorphisms, $\{\phi_{t,s}\}_{t,s \in \mathbb{R}}$, of $M$, with $\phi_{t,s} \circ \phi_{s,u} = \phi_{t,u}$, such that

$$\alpha_{t,s}(a)(\xi) = a(\phi_{s,t}(\xi)), \qquad \xi \in M, \tag{7.64}$$

where $\phi_{s,t} = \phi_{t,s}^{-1}$.

Let us suppose that there is a subalgebra $\overset{\circ}{\mathcal{B}} \subset \mathcal{B}$ that is *norm-dense* in $\mathcal{B}$ such that $\alpha_{t,s}(a)$ is continuously differentiable in $t$ (and in $s$), for arbitrary $a \in \overset{\circ}{\mathcal{B}}$. We define

$$\delta_s(a) = \frac{d}{dt}\alpha_{t,s}(a)_{|t=s}, \qquad a \in \overset{\circ}{\mathcal{B}}. \tag{7.65}$$

Then $\delta_s$ is a *\*derivation* defined on $\overset{\circ}{\mathcal{B}}$. An operator $\delta : \mathrm{Dom}_\delta \to \mathcal{B}$ is a *\*derivation of $\mathcal{B}$ iff $\mathrm{Dom}_\delta \subseteq \mathcal{B}$ is norm-dense in $\mathcal{B}$, $\delta$ is linear, $\delta(a^*) = (\delta(a))^*$, and

$$\delta(a \cdot b) = \delta(a) \cdot b + a \cdot \delta(b) \qquad (\textit{Leibniz rule}), \tag{7.66}$$

for arbitrary $a, b \in \mathrm{Dom}_\delta$. If $\mathcal{B}$ is *abelian* then a *\*derivation $\delta$ of $\mathcal{B}$ corresponds to a *vector field* $X$ on $M$, (assuming that $M$ admits some vector fields):

$$\delta(a)(\xi) = (Xa)(\xi), \tag{7.67}$$

where $a$ corresponds to an arbitrary continuously differentiable function on $M$. If $\delta_s$ satisfies Eq. (7.65) then, for $a \in \overset{\circ}{\mathcal{B}} \subseteq \mathrm{Dom}_{\delta_s}$,

$$\frac{d}{dt}\alpha_{t,s}(a)_{|t=s} = \delta_s(a) = X_s a, \tag{7.68}$$

where, for each $s \in \mathbb{R}$, $X_s$ is a vector field on $M$. Equation (7.68) can be rewritten as

$$\frac{d}{dt}\phi_{t,s}(\xi) = -X_t(\phi_{t,s}(\xi)), \qquad \xi \in M. \tag{7.69}$$

Hence, at least formally, the homeomorphisms $\phi_{t,s}$ can be constructed from a family of vector fields $\{X_s\}_{s \in \mathbb{R}}$ by integrating the ordinary differential equations (7.69). These observations can be made precise if the spectrum $M$ of $\mathcal{B}$ admits a tangent bundle, $TM$, and the vector fields $X_s$ are globally Lipschitz and continuous in $s$, for all $s \in \mathbb{R}$. If $X_s \equiv X$ is independent of $s$ then $\phi_{t,s} = \phi_{t-s}$ is a one-parameter group of homeomorphisms of $M$, (and conversely).

All these remarks can be applied to a classical (model of a) physical system, $S$, with an abelian dynamical $C^*$-algebra $\mathcal{A}_{\overline{S}}$. One may then interpret the parameters $t, s \in \mathbb{R}$ of a groupoid $\{\tau_{t,s}\}_{t,s \in \mathbb{R}}$ of $^*$ automorphisms of $\mathcal{A}_{\overline{S}}$ as *times*; and we say that $S$ is *autonomous* iff $\tau_{t,s} = \tau_{t-s}$ belongs to a one-parameter group of *\*automorphisms of $\mathcal{A}_{\overline{S}}$, or if the vector field $X$ on $M_S = \mathrm{spec}\mathcal{A}_{\overline{S}}$ generating $\tau_t$ is *time-independent*. It is straightforward to describe general symmetries of $S$ in terms of groups of homeomorphisms of $M_S$.

### 7.3.3 Probabilities of Histories, Realism and Determinism

A physical quantity or property of a classical physical system $S$ is given by a continuous function, $a$, on $M_S$. We denote the family of all properties of $S$ specified at a fiducial time $t_0$ by $\mathcal{P}_S = \{a_i\}_{i \in I_S}$. A *possible event* in $S$ at a time $t$ corresponds to the characteristic function, $\chi_{\Omega_i^I(t)}$, of an open subset, $\Omega_i^I(t)$, of $M_S$ given by

$$\xi \in \Omega_i^I(t) \Leftrightarrow a_i(t)(\xi) \in I, \tag{7.70}$$

where $a_i \in \mathcal{P}_S$, $a_i(t) = \tau_{t,t_0}(a_i)$, and $I$ is an open subset of $\mathbb{R}$; (see Definition 2.3 in Sect. 7.2.3).

Let $\phi_{t,s}$ denote the homeomorphism of $M_S$ corresponding to $\tau_{t,s}$. Setting $\Omega_i^I := \phi_{t_0,t}(\Omega_i^I(t))$, we have that

$$\xi \in \Omega_i^I(t) \Leftrightarrow a_i(t)(\xi) \in I \Leftrightarrow \tau_{t,t_0}(a_i)(\xi) \in I$$
$$\Leftrightarrow a_i(\phi_{t_0,t}(\xi)) \in I \Leftrightarrow \eta := \phi_{t_0,t}(\xi) \in \Omega_i^I.$$

We choose $n$ properties, $a_1, \ldots, a_n$, of $S$ to be measured at times $t_1 \leq t_2 \leq \ldots \leq t_n$, with the measured value of $a_i$ contained in the interval $I_i$, $i = 1, \ldots, n$. We let $\Omega_i(t_i)$ be the open subset of $M_S$ given by

$$\xi \in \Omega_i(t_i) \Leftrightarrow a_i(t_i)(\xi) \in I_i, \tag{7.71}$$

$i = 1, \ldots, n$, and $\Omega_i = \phi_{t_0,t_i}(\Omega_i(t_i))$.

Let $\mu$ be a state of $S$, i.e., a probability measure on $M_S$. Every theoretical prediction concerning $S$ is the prediction of the *probability of a history*, $\{\xi_{t_i} := \phi_{t_0,t_i}(\xi) \in \Omega_i\}_{i=1}^n$:

$$\text{Prob}_\mu\{\chi_{\Omega_1(t_1)}, \ldots, \chi_{\Omega_n(t_n)}\} := \int_{M_S} d\mu(\xi) \prod_{i=1}^n \chi_{\Omega_i(t_i)}(\xi)$$

$$= \int_{M_S} d\mu(\xi) \prod_{i=1}^n \chi_{\Omega_i}(\phi_{t_0,t_i}(\xi)). \tag{7.72}$$

If $\mu$ is a pure state, i.e., $\mu = \delta_{\xi_0}$, for some $\xi_0 \in M_S$ then

$$\text{Prob}_{\delta_{\xi_0}}\{\chi_{\Omega_1(t_1)}, \ldots, \chi_{\Omega_n(t_n)}\} = \prod_{i=1}^n \chi_{\Omega_i(t_i)}(\xi_0) = \prod_{i=1}^n \chi_{\Omega_i}(\phi_{t_0,t_i}(\xi_0)), \tag{7.73}$$

i.e., the possible values of $\text{Prob}_{\delta_{\xi_0}}$ are 0 and 1, for any $\xi_0 \in M_S$ and *all* histories. If $\xi_t := \phi_{t_0,t}(\xi_0)$ is the trajectory of states with initial condition $\xi_0$ at time $t_0$ then

$$\text{Prob}_{\delta_{\xi_0}}\{\chi_{\Omega_1(t_1)}, \ldots, \chi_{\Omega_n(t_n)}\} = 1 \iff \xi_{t_i} \in \Omega_i, \tag{7.74}$$

for all $i = 1, \ldots, n$; otherwise, $\text{Prob}_{\delta_{\xi_0}}$ vanishes. If $\xi_0 \notin \Omega_i$ then the event $\{\phi_{t_0,t}(\xi) \in \Omega_i\}$ is first observed at time $t = \underline{t}_i$, where

$$\underline{t}_i := \inf \{t \mid \xi_{0,t} = \phi_{t_0,t}(\xi_0) \in \Omega_i\}, \tag{7.75}$$

and it is last seen at time $\bar{t}_i$, where

$$\bar{t}_i := \sup \{t \mid \xi_{0,t} = \phi_{t_0,t}(\xi_0) \in \Omega_i\}. \tag{7.76}$$

These features of classical physical systems, in particular the "0-1 laws" in Eq. (7.74), are characteristic of *realism* and *determinism*: Given that we know the state, $\xi_0$, of a system $S$ at some time $t_0$, we know its state, $\xi_t = \phi_{t_0,t}(\xi_0)$, and the value, $a_i(\xi_t)$, of an arbitrary property, $a_i \in \mathcal{P}_S$, of $S$, at an *arbitrary (earlier or later) time* $t$.

**Remark 3.2**  (i) A straightforward extension of Eq. (7.72) is the basis for a definition of the dynamical (Kolmogorov–Sinai) entropy of the state $\mu$; see [52, 65].

(ii) A special class of classical systems are *Hamiltonian systems*, $S$, for which $M_S$ is a symplectic manifold, and the homeomorphisms $\phi_{t,s}$ are symplectomorphisms.

## 7.4  Physical Systems in Quantum Mechanics

As indicated in the last section, the only feature distinguishing a quantum-mechanical model of a physical system $S$ (a $Q$-model) from a classical model (an $R$-model) is that, in a $Q$-model, $\mathcal{A}_S$ and hence $\mathcal{A}_{\overline{S}}$ are *non-commutative* algebras. This has profound consequences! In this section, we recall some of the better known ones among them; in particular those that concern problems with the Schwinger–Wigner formula; see Postulate 2.5, Eq. (7.58).

### 7.4.1  Complementary Possible Events Do Not Necessarily Exclude One Another

Let us recall the main task we are confronted with: We have to clarify what the mathematical data (see Definition 2.1, Sect. 7.2.2)

$$(\mathcal{P}_S, \mathcal{A}_{\overline{S}}, \{\tau_{t,s}\}_{t,s \in \mathbb{R}}, \omega \in \mathcal{S}_{\overline{S}}) \tag{7.77}$$

tell us about the "behaviour" of the system $S$, as time goes by; in particular about the empirical properties displayed by $S$ and the events happening in $S$. This task will be shouldered for quantum-mechanical models in Sect. 7.5; it has been dealt with for classical models in the last section, (see also [32]). To set the stage for the analysis of Sect. 7.5, it is useful to return to formulae (7.52), (7.53), (7.57) and, in particular, formula (7.58) for the probability of histories; see Sect. 7.2.3. Thus, we consider $n$ possible events associated with physical quantities/potential properties, $a_i \in \mathcal{P}_S$, of $S$ measured at times $t_i \in \Delta_i \subset \mathbb{R}$, $i = 1, \ldots, n$, with $\Delta_1 \prec \Delta_2 \prec \ldots \prec \Delta_n$. Given a state $\omega$ on $\mathcal{A}_{\overline{S}}$, possible events are represented by spectral projections, $\Pi_{\alpha_i}^{(i)} \in \mathcal{A}_{\overline{S}}^{\omega}$, of the operators $a_i(t_i) \in \mathcal{A}_{\overline{S}}$. The projections $\Pi_{\alpha_i}^{(i)}$ are given by

$$\Pi_{\alpha_i}^{(i)} \equiv \Pi_{\alpha_i}^{(i)}(t_i) := P_{a_i(t_i)}(I_{\alpha_i}^i), \tag{7.78}$$

$\alpha_i = 1, \ldots, k_i$, $i = 1, \ldots, n$, where $I_{\alpha_i}^i$ are disjoint measurable subsets of $\mathbb{R}$ with $\cup_{\alpha_i=1}^{k_i} I_{\alpha_i}^i \supseteq \operatorname{spec} \pi_\omega(a_i(t_i))$. It follows that

$$\sum_{\alpha_i=1}^{k_i} \Pi_{\alpha_i}^{(i)} = \mathbb{1}, \tag{7.79}$$

for all $i$. As in Eq. (7.57), we set

$$H_k^n(\underline{\alpha}) := \Pi_{\alpha_n}^{(n)} \cdots \Pi_{\alpha_k}^{(k)}, \quad 1 \leq k \leq n. \tag{7.80}$$

A stretch, $h_l^k(\underline{\alpha})$, of a history $h_1^n(\underline{\alpha})$ is defined by

$$h_l^k(\underline{\alpha}) := \{\Pi_{\alpha_l}^{(l)}, \ldots, \Pi_{\alpha_k}^{(k)}\}, \quad 1 \leq l \leq k \leq n, \tag{7.81}$$

with $h^n := h_1^n(\underline{\alpha})$. Furthermore, we set

$$h_{\check{k}}^n := \{\Pi_{\alpha_1}^{(1)}, \ldots, \Pi_{\alpha_{k-1}}^{(k-1)}, \Pi_{\alpha_{k+1}}^{(k+1)}, \ldots, \Pi_{\alpha_n}^{(n)}\}. \tag{7.82}$$

In the Schwinger–Wigner formula (7.58), the *probability of a history*, $h^n$, of $S$, given a state $\omega$, has been defined by

$$\operatorname{Prob}_\omega\{\mathbb{1}_{\alpha_1}^{(1)}, \ldots, \Pi_{\alpha_n}^{(n)}\} := \omega\left((H_1^n(\underline{\alpha}))^* H_1^n(\underline{\alpha})\right) = \omega(\Pi_{\alpha_1}^{(1)} \Pi_{\alpha_2}^{(2)} \cdots \Pi_{\alpha_n}^{(n)} \cdots \Pi_{\alpha_2}^{(2)} \Pi_{\alpha_1}^{(1)}), \tag{7.83}$$

with properties (1)–(3), (see Eqs. (7.59) and (7.60)).

Here we wish to point out some *fundamental problems* with formula (7.83) in quantum mechanics. Suppose that the complementary possible events

$\Pi_1^{(i)}, \ldots, \Pi_{k_i}^{(i)}$ were *mutually exclusive*, given that $\Pi_{\alpha_1}^{(1)}, \ldots, \Pi_{\alpha_{i-1}}^{(i-1)} \Pi_{\alpha_{i+1}}^{(i+1)}, \ldots, \Pi_{\alpha_n}^{(n)}$ are observed, for some $i < n$, then we would imagine that the *"sum rule"*

$$
\begin{aligned}
\sum_{\alpha_i=1}^{k_i} \mathrm{Prob}_\omega \, h_1^n(\underline{\alpha}) &= \sum_{\alpha_i=1}^{k_i} \mathrm{Prob}_\omega \{ \Pi_{\alpha_1}^{(1)}, \ldots, \Pi_{\alpha_i}^{(i)}, \ldots, \Pi_{\alpha_n}^{(n)} \} \\
&= \mathrm{Prob}_\omega \{ \Pi_{\alpha_1}^{(1)}, \ldots, \Pi_{\alpha_{i-1}}^{(i-1)}, \Pi_{\alpha_{i+1}}^{(i+1)}, \ldots, \Pi_{\alpha_n}^{(n)} \} \\
&= \mathrm{Prob}_\omega \, h_{\hat{\imath}}^n(\underline{\alpha})
\end{aligned}
\tag{7.84}
$$

holds; see Eq. (7.82). If $\Pi_{\alpha_i}^{(i)}$ commuted with the operator $H_{i+1}^n(\underline{\alpha})$, for all $\alpha_i$—as *is the case in every classical model*—then Eq. (7.84) would hold true. However, because of the non-commutative nature of $\mathcal{A}_{\overline{S}}$,

$$
[\Pi_{\alpha_i}^{(i)}, H_{i+1}^n(\underline{\alpha})] \neq 0,
\tag{7.85}
$$

in general. This leads to *non-vanishing interference terms*,

$$
\omega \left( (H_1^{i-1}(\underline{\alpha}))^* \Pi_{\alpha_i}^{(i)} (H_{i+1}^n(\underline{\alpha}))^* H_{i+1}^n(\underline{\alpha}) \Pi_{\beta_i}^{(i)} H_1^{i-1}(\underline{\alpha}) \right),
\tag{7.86}
$$

with $\alpha_i \neq \beta_i$. In the presence of non-vanishing interference terms the sum rule (7.84) is usually *violated*. This means that the complementary possible events $\Pi_1^{(i)}, \ldots, \Pi_{k_i}^{(i)}$, do, apparently, *not* mutually exclude one another, *given future events* $\Pi_{\alpha_{i+1}}^{(i+1)}, \ldots, \Pi_{\alpha_n}^{(n)}$ that cause interference. Put differently, a history $h^n$ does, in general, *not* result in the determination of a potential property $a_i$, of $S$ in the $i^{th}$ observation (or measurement), given the data in (7.77) (the time evolution $\{\tau_{t,s}\}_{t,s \in \mathbb{R}}$, and a state $\omega$). If the sum rule (7.84) is violated, then the operator $a_i(t_i)$ does not represent an *empirical property* of $S$, given *later* observations of physical quantities $a_{i+1}, \ldots, a_n$. Apparently, the operators $a \in \mathcal{P}_S$ do, in general, *not* represent properties of $S$ that exist *a priori*, but only *potential properties* of $S$ whose *empirical status* depends on the choice of the time evolution $\{\tau_{t,s}\}_{\tau,s \in \mathbb{R}}$ of $\overline{S} = S \vee E$ and of the state $\omega$. This will be made precise in Sect. 7.5.

### 7.4.2   The Problem with Conditional Probabilities

In Sect. 7.2.3, (7.59) and (7.60), we have seen that

$$
\mu_\omega(\underline{\alpha}) := \mathrm{Prob}_\omega \{ \Pi_{\alpha_1}^{(1)}, \ldots, \Pi_{\alpha_n}^{(n)} \}
\tag{7.87}
$$

is a *probability measure* on $\mathbb{Z}_{k_1} \times \ldots \times \mathbb{Z}_{k_n}$. Let us fix $\alpha_1, \ldots, \alpha_{i-1}, \alpha_{i+1}, \ldots, \alpha_n$, and ask what the *conditional probability*

$$\text{Prob}_{\mu_\omega}\{\Pi_{\alpha_i}^{(i)} \mid h_i^n(\underline{\alpha})\} \tag{7.88}$$

of the possible event $\Pi_{\alpha_i}^{(i)}$ is, given $\mu_\omega$ and $h_i^n$; (see Eq. (7.82)). Since (7.87) defines a probability measure, we may define

$$\text{Prob}_{\mu_\omega}\{\Pi_{\alpha_i}^{(i)} \mid h_i^n(\underline{\alpha})\} := \frac{\mu_\omega(\alpha_1, \ldots, \alpha_i, \ldots, \alpha_n)}{\sum_{\beta_i=1}^{k_i} \mu_\omega(\alpha_1, \ldots, \beta_i, \ldots, \alpha_n)}. \tag{7.89}$$

Unfortunately, there is a problem with this definition! Recall that $\Pi_{\beta_i}^{(i)}$ is a shorthand for the spectral projection $P_{a_i(t_i)}(I_{\beta_i}^i)$. We fix a subset $I_{\alpha_i}^i$, but introduce a *new* decomposition of spec $a_i$ into subsets

$$\tilde{I}_1^i := I_{\alpha_i}^i, \qquad \mathbb{R} \setminus I_{\alpha_i}^i = \cup_{\beta=2}^{m_i} \tilde{I}_\beta^i,$$

with $\tilde{I}_\beta^i \cap \tilde{I}_\gamma^i = \emptyset$, for $\beta \neq \gamma$, and define

$$\tilde{\Pi}_{\beta_i}^{(i)} := P_{a_i(t_i)}(\tilde{I}_{\beta_i}^i),$$

$\beta_i = 1, \ldots, m_i$. We define

$$\tilde{\mu}_\omega(\alpha_1, \ldots, \beta_i, \ldots, \alpha_n) := \text{Prob}_\omega\{\Pi_{\alpha_1}^{(1)}, \ldots, \tilde{\Pi}_{\beta_i}^{(i)}, \ldots, \Pi_{\alpha_n}^{(n)}\}.$$

Then

$$\tilde{\mu}_\omega(\alpha_1, \ldots, 1, \ldots, \alpha_n) = \mu_\omega(\alpha_1, \ldots, \alpha_i, \ldots, \alpha_n);$$

but, most often, the putative "conditional probabilities" are different,

$$\text{Prob}_{\tilde{\mu}_\omega}\{\Pi_{\alpha_i}^{(i)} \mid h_i^n(\underline{\alpha})\} \neq \text{Prob}_{\mu_\omega}\{\Pi_{\alpha_i}^{(i)} \mid h_i^n(\underline{\alpha})\}, \tag{7.90}$$

unless all possible interference terms *vanish*. Thus, in general, there is *no* meaningful notion of *"conditional probability"* in quantum mechanics.

It may be of interest to note that if the operators $a_i$ have pure-point spectrum with *only two* distinct eigenvalues then

$$\{\Pi_{\alpha_i}^{(i)}\}_{\alpha_i=1,2} = \{\tilde{\Pi}_{\beta_i}^{(i)}\}_{\beta_i=1,2},$$

and we have equality in Eq. (7.90). These findings may be viewed as a general version of the *Kochen–Specker theorem*, [51].

Let us recall a "test" for one of the possible events $\{\Pi_{\alpha_i}^{(i)}\}_{\alpha_i=1}^{k_i}$ to materialize in a measurement at time $t_i$ of the potential property of $S$ represented by the operator $a_i \in \mathcal{P}_S$; (see [32] and references given there). For this purpose, we introduce the matrix

$$P_{\underline{\alpha},\underline{\alpha}'}^{\omega} := \omega(\Pi_{\alpha_1}^{(1)} \ldots \Pi_{\alpha_n}^{(n)} \Pi_{\alpha_n'}^{(n)} \ldots \Pi_{\alpha_1'}^{(1)}), \tag{7.91}$$

with $\alpha_n = \alpha_n'$; see (7.17). Classically, $P^{\omega} = (P_{\underline{\alpha},\underline{\alpha}'}^{\omega})$ is always a diagonal matrix, because all the operators $\Pi_{\alpha_i}^{(i)}$ commute with one another and by Eq. (7.52). We say that a family of histories $\{h_1^n(\underline{\alpha})\}$ is *consistent* iff the commutators

$$[\Pi_{\alpha_i}^{(i)}, H_{i+1}^n(\underline{\alpha})]$$

*vanish*, for all $\alpha_i, \underline{\alpha}$ and $i = 1, \ldots, n$; (see [41]). If $\{h_1^n(\underline{\alpha})\}$ is consistent then $P_{\underline{\alpha},\underline{\alpha}'}^{\omega}$ is diagonal, and the sum rules (7.84) are valid for all $\underline{\alpha}$ and all $i = 1, \ldots, n$. We say that a family $\{h_1^n(\underline{\alpha})\}$ of histories is $\delta$-*consistent*$(0 \le \delta \le 1)$ iff

$$\|[\Pi_{\alpha_i}^{(i)}, H_{i+1}^n(\underline{\alpha})]\| \le 1 - \delta, \tag{7.92}$$

for all i.

A 1-consistent history is consistent. We define a diagonal matrix $\Delta^{\omega}$ by

$$\Delta_{\underline{\alpha},\underline{\alpha}'} := \begin{cases} P_{\underline{\alpha},\underline{\alpha}}^{\omega} & \text{if } \underline{\alpha} = \underline{\alpha}' \\ 0 & \text{else} \end{cases}$$

Clearly inequality (7.92) implies that

$$\|P^{\omega} - \Delta^{\omega}\| \le \text{const.}(1 - \delta). \tag{7.93}$$

This shows that, for a $\delta$-consistent family of histories, with $\delta \approx 1$, the sum rules (7.84) are very nearly satisfied, meaning that the events $\Pi_1^{(i)}, \ldots, \Pi_{k_i}^{(i)}$ mutually exclude one another FAPP ("for all practical purposes", [9]). In [32], we have called

$$e^{\omega} := 1 - \|P^{\omega} - \Delta^{\omega}\|$$

the "*evidence*" for $\Pi_1^{(i)}, \ldots, \Pi_{k_i}^{(i)}$ to mutually exclude one another, FAPP, $i = 1, \ldots, n$. Apparently, if $e^{\omega}$ is very close to 1, then everything might appear to be fine. Well, the appearance is deceptive, as we will explain below!

Dynamical mechanisms that imply that $\|P^{\omega} - \Delta^{\omega}\|$ becomes small, i.e., $e^{\omega}$ approaches 1, in suitable limiting regimes are known under the names of "*dephasing*" and "*decoherence*"; see [37, 47, 49, 75]. Understanding decoherence is clearly an important task. Here we summarize a few observations on those mechanisms; but see Sects. 7.5.3 and 7.5.4. (Some instructive examples will be discussed elsewhere.)

### 7.4.3 Dephasing/Decoherence

In our discussion of near (i.e., $\delta$-) consistency of families of histories, $h^n$, operators $Q_k^n(\underline{\alpha})$, defined by

$$Q_k^n(\underline{\alpha}) := (H_k^n(\underline{\alpha}))^* H_k^n(\underline{\alpha}) = \Pi_{\alpha_k}^{(k)}(t_k) \ldots \Pi_{\alpha_n}^{(n)}(t_n) \ldots \Pi_{\alpha_k}^{(k)}(t_k), \qquad (7.94)$$

$t_k < t_{k+1} < \ldots < t_n$, $1 \leq k \leq n$, play an important role. Inequality (7.92) implies that

$$\|[\Pi_{\alpha_i}^{(i)}, Q_{i+1}^n(\underline{\alpha})]\| \leq 2(1 - \delta) \ll 1 \qquad (7.95)$$

if $\delta$ is very close to 1. Condition (7.95) is slightly weaker than (7.92), so we will work with (7.95). If (7.95) holds, for all $i$ and all $\underline{\alpha}$, the sum rules (7.84) are satisfied, up to tiny errors, and the matrix $P^\omega$ is very nearly diagonal; so there is "decoherence". A (very stringent) sufficient condition for

$$[\Pi_{\alpha_i}^{(i)}, Q_{i+1}^n(\underline{\alpha})] = 0 \qquad (7.96)$$

to hold, for all $i$ and all $\underline{\alpha}$, i.e., for perfect decoherence to hold, is the following one: We observe that

$$Q_k^n(\underline{\alpha}) \in \mathcal{E}_{\geq t_k}^\omega, \quad \text{for all } \underline{\alpha}, \qquad (7.97)$$

where the von Neumann algebras $\mathcal{E}_{\geq t}^\omega$ of possible events observable at times $\geq t$ have been introduced in Definition 2.4, Sect. 7.2.3. If there is *loss of information*, in the sense of condition (7.49), more precisely if the relative commutants

$$(\mathcal{E}_{\geq t_{i+1}}^\omega)' \cap \mathcal{E}_{\geq \tilde{t}_i}^\omega, \qquad t_{i-1} < \tilde{t}_i \leq t_i, \qquad (7.98)$$

are *non-trivial*, for suitable choices of sequences of times $t_1 < t_2 < \ldots < t_n$, $\tilde{t}_1 < \tilde{t}_2 < \ldots < \tilde{t}_n$, and if the operator

$$a_i(t_i) \in (\mathcal{E}_{\geq t_{i+1}}^\omega)' \cap \mathcal{E}_{\geq \tilde{t}_i}^\omega, \qquad (7.99)$$

and hence $\Pi_{\alpha_i}^{(i)}$ belongs to $(\mathcal{E}_{\geq t_{i+1}}^\omega)' \cap \mathcal{E}_{\geq \tilde{t}_i}^\omega$, for all $\alpha_i = 1, \ldots, k_i$, with $t_{i-1} < \tilde{t}_i \leq t_i$, then

$$[\Pi_{\alpha_i}^{(i)}, Q_{i+1}^n(\underline{\alpha})] = 0, \qquad (7.100)$$

for all $\alpha_i$ and all $\underline{\alpha}$. If (7.99) and hence Eq. (7.100) hold, for all $i \leq n$, then there is perfect decoherence, and the histories $\{h_1^n(\underline{\alpha})\}$ form a consistent family.

The scenario for decoherence described here is encountered in *relativistic quantum field theories* with a massless particle (e.g., the photon), as can be inferred from results in [14, 17]. In *non-relativistic* quantum mechanics, the above scenario for decoherence remains plausible, provided one allows for small changes of the operators $a_i(t_i)$ into operators $\tilde{a}_i(t_i)$ that belong to $(\mathcal{E}^{\omega}_{\geq t_{i+1}})' \cap \mathcal{E}^{\omega}_{\geq \tilde{t}_i}$. In this connection the following result may be of interest.

**Theorem 4.1** *Let* $\Pi^{(1)}_{\alpha_1}, \ldots, \Pi^{(n)}_{\alpha_n}$ *be orthogonal projections, and let the operators* $Q^n_k(\underline{\alpha})$ *be defined as in Eq. (7.94). Suppose that*

$$\|[\Pi^{(i)}_{\alpha_i}, Q^n_{i+1}(\underline{\alpha})]\| < \epsilon, \tag{7.101}$$

*for all* $i = 1, \ldots, n-1$ *and all* $\underline{\alpha} = (\alpha_1, \ldots, \alpha_n)$, *with* $\epsilon$ *sufficiently small (depending on the total number,* $\sum_{i=1}^n k_i$, *of n-tuples* $\underline{\alpha}$, *with* $\alpha_i = 1, \ldots, k_i$). *Then there exist orthogonal projections* $\tilde{\Pi}^{(i)}_{\alpha_i}$, $\alpha_i = 1, \ldots, k_i$, $i = 1, \ldots, n$, *with*

$$\tilde{\Pi}^{(i)}_{\alpha_i} \tilde{\Pi}^{(i)}_{\beta_i} = \delta_{\alpha_i \beta_i} \tilde{\Pi}^{(i)}_{\alpha_i}, \qquad \sum_{\alpha_i=1}^{k_i} \tilde{\Pi}^{(i)}_{\alpha_i} = \mathbb{1}, \tag{7.102}$$

*such that*

$$\|\tilde{\Pi}^{(i)}_{\alpha_i} - \Pi^{(i)}_{\alpha_i}\| \leq C\epsilon, \tag{7.103}$$

*and*

$$[\tilde{\Pi}^{(i)}_{\alpha_i}, \tilde{Q}^n_{i+1}(\underline{\alpha})] = 0, \tag{7.104}$$

*for all* $\underline{\alpha}$ *and all* $i \leq n-1$. *The constant $C$ in Eq. (7.103) depends on* $\sum_{i=1}^n k_i$, *and* $\epsilon$ *must be chosen so small that* $C\epsilon < 1$; *(in which case* $\tilde{\Pi}^{(i)}_{\alpha_i}$ *and* $\Pi^{(i)}_{\alpha_i}$ *are unitarily equivalent).*

*Remark 4.2* The operators , $\tilde{Q}^n_k(\underline{\alpha})$ are defined as in Eq. (7.94), with $\Pi^{(i)}_{\alpha_i}(t_i) \equiv \Pi^{(i)}_{\alpha_i}$ replaced by $\tilde{\Pi}^{(i)}_{\alpha_i}$, for all $i$.

The proof of Theorem 4.1 can be inferred from Sect. 4.5 of [32], (Lemmata 7 and 8).

*Interpretation of Theorem 4.1.* Apparently, dephasing/decoherence in the form of inequalities (7.101) implies that if one reinterprets the measurements made at times $t_1 < t_2 < \ldots < t_n$ as observations of events $\tilde{\Pi}^{(1)}_{\alpha_1}, \ldots, \tilde{\Pi}^{(n)}_{\alpha_n}$ that differ slightly from the spectral projections $\Pi^{(1)}_{\alpha_1}, \ldots, \Pi^{(n)}_{\alpha_n}$ of potential properties $a_1, \ldots, a_n$ of $S$ then all interference terms (see (7.86), (7.91)) vanish, the matrix $P^\omega$ is diagonal, and the sum rules (7.84) hold. The family of histories $\{\tilde{\Pi}^{(1)}_{\alpha_1}, \ldots, \tilde{\Pi}^{(n)}_{\alpha_n}\}$ is consistent, and the complementary possible events $\tilde{\Pi}^{(i)}_1, \ldots, \tilde{\Pi}^{(i)}_{k_i}$ mutually exclude one another.

## Critique of the Concept of "Families of Consistent Histories"

(i) Given a measurement of a potential property $a_i \in \mathcal{P}_S$ of $S$ at some time $t_i$, the success of this measurement, as expressed in the decoherence of (absence of interference between) the events $\Pi_1^{(i)}, \ldots, \Pi_{k_i}^{(i)}$, apparently *not* only depends on the *past* but seems to depend on the *future*, namely on subsequent measurements of potential properties $a_{i+1}, \ldots, a_n$ at times $> t_i$. This is how conditions such as (7.92), (7.95) and (7.101) must be interpreted. The consistency of a family $\{h_1^i(\underline{\alpha})\}$ of stretches of histories (see Eq. (7.81) for the definition) can apparently only be assured if one also knows the family $\{h_{i+1}^n(\underline{\alpha})\}$ of stretches of histories in the *future* of $\{h_1^i(\underline{\alpha})\}$. This may be a deep aspect of quantum mechanics; but it is *more likely* an indication that there is something *wrong* with the concept of "consistent (families of) histories" and with a formulation of decoherence in the form of inequalities (7.101).

(ii) Accepting, temporarily, the idea of "consistent (families of) histories"—e.g., in the appealing form of conditions (7.99)—we encounter the following problem: Fixing the data

$$(\mathcal{P}_S, \mathcal{A}_{\overline{S}}, \{\tau_{t,s}\}_{t,s \in \mathbb{R}}, \omega \in \mathcal{S}_{\overline{S}}), \tag{7.105}$$

see (7.77), we may consider two (or more) families of potential properties of $S$,

$$\{a_1, \ldots, a_n\} \quad \text{and} \quad \{b_1, \ldots, b_m\}, \tag{7.106}$$

measured at times $t_1 < \ldots < t_n$ and $t_1' < \ldots < t_m'$, respectively, with $a_i \in \mathcal{P}_S$ and $b_j \in \mathcal{P}_S$, for all $i$ and $j$. Both families may give rise to families of consistent histories (e.g., if conditions (7.99) hold for the $a_i$'s and the $b_j$'s). Yet, there may *not* exist *any* family

$$\{c_1, \ldots, c_N\}, \ N \geq n + m,$$

of potential properties of $S$ ($c_j \in \mathcal{P}_S$, for all $j$) measured at times $T_1 < \ldots < T_N$, with

$$\{T_1, \ldots, T_N\} \supseteq \{t_1, \ldots, t_n\} \cup \{t_1', \ldots, t_m'\},$$

encompassing the two families in (7.106) and giving rise to a family of consistent histories. Since the data (7.105) are fixed, the confusing question arises *which one* of the two or more incompatible families of potential properties $\{a_1, \ldots, a_n\}$, $\{b_1, \ldots, b_m\}$, ... will actually be observed in the course of time, i.e., become *real*, (or, put differently, correspond to *empirical* properties). Some people suggest, following Everett [28], that there is a world for *every* family of potential properties of $S$ giving rise to a family of consistent histories to be observed. This is the *"many-worlds interpretation of quantum mechanics"*, which we find entirely unacceptable!

(iii) Unfortunately, the problem described in (ii) persists even in the decoherence scenario described in (7.96)–(7.100), above, because the von Neumann algebras

$$\mathcal{M}_i := (\mathcal{E}^\omega_{\geq t_{i+1}})' \cap \mathcal{E}^\omega_{\geq \tilde{t}_i} \qquad (t_{i-1} < \tilde{t}_i \leq t_i) \tag{7.107}$$

are usually *non-commutative*. If there are an $a_i$ and a $b_j$ from the sets of operators in (7.106) belonging to the *same* $\mathcal{M}_l$, and if

$$[a_i(t_i), b_j(t'_j)] \neq 0, \tag{7.108}$$

then the problem described in (ii) appears on the scene. It could be avoided if one assumed that $a_i(t_i)$ and $b_j(t'_j)$ must belong to the center, $\mathcal{Z}_{\mathcal{M}_l}$, of $\mathcal{M}_l$, because then the commutators on the left side in (7.108) would all vanish. The right version of something like this idea will be formulated in Sects. 7.5.3 and 7.5.4.

(iv) It has tacitly been assumed, so far, that the times at which quantum-mechanical measurements of potential properties of a system $S$ are carried out (we are talking of the times $t_i$ at which potential properties $a_i$ of $S$ are observed) can be *fixed precisely* (by an "observer"?). Obviously, this assumption is *nonsense* in quantum mechanics, (as opposed to classical physics); see Sect. 7.5.4.

In an appendix, the reader may find some remarks on positive operator-valued measures (POVM) [61] and their uses; (but see also the end of Sect. 7.5.4 and [33]).

## 7.4.A    Appendix to Sect. 7.4: Remarks on Positive Operator-Valued Measures (POVM)

It may and will happen sometimes that the commutators

$$[\Pi^{(i)}_{\alpha_i}, Q^n_{i+1}(\underline{\alpha})]$$

are *not* small in norm, and the matrix $P^\omega$ defined in Eq. (7.91) has "large" off-diagonal elements. Then some of the operators $a_i$ representing potential properties of $S$ are not measurable and do apparently *not* represent *empirical properties* of $S$, given the data

$$(\mathcal{P}_S, \mathcal{A}_{\overline{S}}, \{\tau_{t,s}\}_{t,s \in \mathbb{R}}, \omega \in \mathcal{S}_{\overline{S}}).$$

While this is a perfectly interesting piece of information, it raises the question whether formula (7.83) continues to contain interesting information, although the sum rule (7.84) may be strongly violated. A conventional answer to this question involves the notion of *"positive operator-valued measures"* (POVM): For $k^- < k^+$, we define

$$H_{k^-}^{k^+}(\underline{\alpha}) := \Pi_{\alpha_{k^+}}^{(k^+)} \Pi_{\alpha_{k^+-1}}^{(k^+-1)} \dots \Pi_{\alpha_{k^-+1}}^{(k^-+1)} \Pi_{\alpha_{k^-}}^{(k^-)}. \tag{7.109}$$

We observe that

$$\sum_{\underline{\alpha}} \left( H_{k^-}^{k^+}(\underline{\alpha}) \right)^* H_{k^-}^{k^+}(\underline{\alpha}) = \mathbb{1}, \tag{7.110}$$

(and

$$\sum_{\underline{\alpha}} H_{k^-}^{k^+}(\underline{\alpha}) \left( H_{k^-}^{k^+}(\underline{\alpha}) \right)^* = \mathbb{1}.)$$

Consider

$$\mathrm{Prob}_\omega \{ \Pi_{\alpha_1}^{(1)}, \dots, \Pi_{\alpha_n}^{(n)} \} \equiv \mathrm{Prob}_\omega \{ h_1^{k^--1}(\underline{\alpha}), h_{k^-}^{k^+}(\underline{\alpha}), h_{k^++1}^n(\underline{\alpha}) \}$$

$$:= \omega \left( (H_1^{k^--1})^* (H_{k^-}^{k^+})^* (H_{k^++1}^n)^* H_{k^++1}^n H_{k^-}^{k^+} H_1^{k^--1} \right) \tag{7.111}$$

We may say that $h_{k^-}^{k^+}(\underline{\alpha})$ represents a *single experiment* on the system $S$ if the sum rule (7.84) is *violated substantially*, for all $i = k^-, k^- + 1, \dots, k^+$, but

$$\sum_{\alpha_{k^-}, \dots, \alpha_{k^+}} \mathrm{Prob}_\omega \{ h_1^{k^--1}(\underline{\alpha}), h_{k^-}^{k^+}(\underline{\alpha}), h_{k^++1}^n(\underline{\alpha}) \} \approx \mathrm{Prob}_\omega \{ h_1^{k^--1}(\underline{\alpha}), h_{k^++1}^n(\underline{\alpha}) \}, \tag{7.112}$$

up to an error that is so small that it is below the experimental resolution. In view of Eq. (7.110), our discussion can be formalized as follows.

**Definition 4.3** The "square root" of a positive operator-valued measure (POVM) is a (finite) family of operators

$$\underline{X} - \{ X_\alpha \}_{\alpha=1}^N \tag{7.113}$$

with the property that

$$\sum_{\alpha=1}^N X_\alpha^* X_\alpha = \mathbb{1}. \tag{7.114}$$

The "positive operator-valued measure" is then given by the operators $\{ X_\alpha^* X_\alpha \}_{\alpha=1}^N$.

Given a time-ordered sequence of ("square roots" of) POVM's, $\underline{X}^{(1)}, \ldots, \underline{X}^{(n)}$, the probability of observing a "generalized history"

$$h_1^n(\underline{\alpha}) = \{X_{\alpha_1}^{(1)}, \ldots, X_{\alpha_n}^{(n)}\} \tag{7.115}$$

is given by

$$\text{Prob}_\omega\{X_{\alpha_1}^{(1)}, \ldots, X_{\alpha_n}^{(n)}\} := \omega\left((X_{\alpha_1}^{(1)})^* \ldots (X_{\alpha_n}^{(n)})^* X_{\alpha_n}^{(n)} \ldots X_{\alpha_1}^{(1)}\right). \tag{7.116}$$

The probabilities of such generalized histories have the desirable properties (7.59) and (7.60). We say that $\{\underline{X}^{(1)}, \ldots, \underline{X}^{(n)}\}$, with $\underline{X}^{(i)}$ (the square root of) a POVM, for all $i$, describes a time-ordered sequence of $n$ successful experiments, or observations, iff

$$\sum_{\alpha_i} \text{Prob}_\omega\{X_{\alpha_1}^{(1)}, \ldots, X_{\alpha_i}^{(i)}, \ldots, X_{\alpha_n}^{(n)}\} \approx \text{Prob}_\omega\{X_{\alpha_1}^{(1)}, \ldots, X_{\alpha_{i-1}}^{(i-1)}, X_{\alpha_{i+1}}^{(i+1)}, \ldots, X_{\alpha_n}^{(n)}\},$$

$$\tag{7.117}$$

up to a tiny error below the experimental resolution, for all $i = 1, \ldots, n$ and all $\underline{\alpha}$. An example of events described by POVM's is described in Sect. 7.5.4; (see also [33]).

All the concepts and notions introduced in Sect. 7.4 can be carried over to this generalized setup, after replacing $a_i$ by $\underline{X}^{(i)}$ and $\Pi_{\alpha_i}^{(i)} \equiv \Pi_{\alpha_i}^{(i)}(t_i)$ by $X_{\alpha_i}^{(i)} \equiv X_{\alpha_i}^{(i)}(\Delta_i)$ (or their adjoints), $i = 1, \ldots, n$, with $\Delta_1 \prec \ldots \prec \Delta_n$. Wherever possible, we will, however, consider self-adjoint operators and their spectral projections, instead of POVM's, throughout this essay; (but see Remark 5.8, Sect. 7.5.4).

## 7.5  Removing the Veil: Empirical Properties of Physical Systems in Quantum Mechanics

In a classical model of a physical system, $S$, properties of $S$ exist a priori. They are represented by real-valued continuous (or measurable) functions on the state space, $M_S$, of the system. In contrast, in a quantum-mechanical model of a physical system, the system can still be characterized by a list, $\mathcal{P}_S$, of *potential properties* (represented by self-adjoint operators); but these properties do *not* exist a priori. Whether they correspond to *empirical properties* of $S$, or not, depends on the choice of the environment $E$; (e.g., on the experiments that are made). The question then arises what the *empirical properties* are that will be observed in the course of time, given the time evolution $\{\tau_{t,s}\}_{t,s \in \mathbb{R}}$ of $\overline{S} = S \vee E$ and its state $\omega \in \mathcal{S}_{\overline{S}}$; (see Definition 2.1, Sect. 7.2.2). In (7.43), we have identified the fundamental data underlying a model of $S$,

$$(\mathcal{P}_S, \mathcal{A}_{\overline{S}}, \{\tau_{t,s}\}_{t,s \in \mathbb{R}} \subset {}^*\text{Aut}(\mathcal{A}_{\overline{S}}), \omega \in \mathcal{S}_{\overline{S}}), \tag{7.118}$$

see also (7.77) and Sect. 7.2.2. These data ought to *determine* which *empirical properties* S exhibits and what *family* of histories of events (but, of course, not *which* history) will be recorded in the course of time if $S$ is monitored/observed when coupled to a given environment/equipment $E$. We have seen in Sect. 7.4 that the answer to the question of what exactly the data in (7.118) determine is *not obvious*.

## 7.5.1   Information Loss and Entanglement

Let $a$ be a potential property of $S$ ($a = a^* \in \mathcal{P}_S$). We assume, for simplicity, that spec $a$ consists of finitely many eigenvalues, $\alpha_1, \alpha_2, \ldots, \alpha_k$. Let $\omega$ be the state of $\overline{S} = S \vee E$, and let us suppose that, thanks to an appropriate choice of $E$, the potential property $a$ is *observed* (i.e., becomes an empirical property of $S$) around some time $t$. According to almost everybody's understanding of quantum mechanics, the following claim appears to be reasonably plausible: After the observation of $a$ at a time $\approx t$, $S$ evolves *as if* its state where given by

$$\omega \longrightarrow \sum_{i=1}^{k} p_i \omega_i, \qquad (7.119)$$

where $p_i$ is the probability to observe the value $\alpha_i$ of $a$, and $\omega_i$ is a state with the property that if $a$ were observed in a system prepared in the state $\omega_i$ at time $\approx t$ then its value would be $\alpha_i$ with certainty. If no measurements are made before $a$ is observed then, according to Born [11],

$$p_i = \omega(\Pi_i(t)),$$

where $\Pi_i(t)$ is the spectral projection of the operator $a(t) = \tau_{t,t_0}(a)$ corresponding to the eigenvalue $\alpha_i$, (with $t$ the time of measurement of $a$). Note that the state in (7.119) is usually a *mixed state*, i.e., an incoherent superposition of the states $\omega_i$, *even* if $\omega$ is a *pure* state. It is perceived as one aspect of the *"measurement problem"* to understand how a pure state can evolve into a mixture. (Another aspect is to understand why the state of $S$ is given by $\omega_i$, right after the measurement of $a$, if $a$ is measured to have the value $\alpha_i$, for some $i = 1, \ldots, k$. This will be discussed in Sects. 7.5.4 and 7.5.6.)

In order to explain why the first aspect of the measurement problem does not represent a serious problem, we have to return to an analysis of two fundamental phenomena: $(\mathcal{L}o\mathcal{I})$ *Loss of information* into $E$; and $(\mathcal{E})$ *Entanglement* between $S$ and $E$.

In Definition 2.2 of Sect. 7.2.3, we have introduced *algebras*, $\mathcal{E}_{\geq t}$, *of potential properties of* $S$ *observable/measurable after time* $t$. These algebras are $C^*$-subalgebras of the algebra $\mathcal{A}_{\overline{S}}$. We have denoted by $\mathcal{E}_S$ the smallest $C^*$-algebra containing $\mathcal{E}_{\geq t}$, for all $t \in \mathbb{R}$; see Eq. (7.47). Clearly $\mathcal{E}_S \subset \mathcal{A}_{\overline{S}}$. As indicated in Sect. 7.2.3, it is the consequence of a **general principle**—"*Duality between Observables* and *Indeterminates*"—that $\mathcal{E}_S$ is *properly* contained in $\mathcal{A}_{\overline{S}}$ (and that the relative commutant of $\mathcal{E}_S$ inside $\mathcal{A}_{\overline{S}}$ contains a subalgebra isomorphic to $\mathcal{E}_S$). This principle will be discussed in the context of examples in a forthcoming communication.

The algebra $\mathcal{E}_S$ carries an action of the group $\mathbb{R}$ of time translations by *automorphisms $\{\overline{\tau}_t\}_{t \in \mathbb{R}}$, where $\overline{\tau}_t$ determines *morphisms

$$\overline{\tau}_t : \mathcal{E}_{\geq t'} \longrightarrow \mathcal{E}_{\geq t+t'} \subseteq \mathcal{E}_{\geq t'},$$

for all $t' \in \mathbb{R}$ and all $t \geq 0$; see (7.50) and (7.51).

Thus, in hindsight, the mathematical data enabling one to predict the behavior of a physical system $S$ in the course of time, given its state, can be chosen to consist of the *filtration* of algebras

$$\mathcal{A}_{\overline{S}} \supset \mathcal{E}_S \supseteq \mathcal{E}_{\geq t} \supseteq \mathcal{E}_{\geq t'} \supset \{\mathbb{C}\mathbb{1}\}, \qquad t' \geq t, \tag{7.120}$$

along with a specification of *morphisms* (*time translations*)

$$\overline{\tau}_t : \mathcal{E}_{\geq t'} \longrightarrow \mathcal{E}_{\geq t+t'} \subseteq \mathcal{E}_{\geq t'}, \tag{7.121}$$

for $t' \in \mathbb{R}$, $t \geq 0$, and of a *state* $\omega$,

$$\omega : \text{state on } \mathcal{A}_{\overline{S}}. \tag{7.122}$$

In Definition 2.4, Sect. 7.2.3, we have introduced the von Neumann algebras $\mathcal{E}_{\geq t}^{\omega}$, $t \in \mathbb{R}$, and $\mathcal{E}_S^{\omega}$. (We recall that if $\omega'$ is an arbitrary state on $\mathcal{E}_S$ *normal* with respect to $\omega$ then $\mathcal{E}_S^{\omega'} = \mathcal{E}_S^{\omega}$ and $\mathcal{E}_{\geq t}^{\omega'} = \mathcal{E}_{\geq t}^{\omega}$, for all $t$.)

*Loss of information* ($\mathcal{LoI}$) is the phenomenon that if successful measurements of potential properties of $S$ have been made between some times $t$ and $t' > t$ then $\mathcal{E}_{\geq t'}$ is *strictly contained* in $\mathcal{E}_{\geq t}$. Together with the phenomenon of *entanglement* ($\mathcal{E}$), this may entail that the *restriction of the state $\omega$ to the algebra $\mathcal{E}_{\geq t'}$ is a mixture* (i.e., an incoherent superposition of approximate eigenstates of some physical quantity, as in (7.119)), *even if $\omega$ is a pure* state of $\mathcal{E}_S$.

While ($\mathcal{LoI}$) is common to classical and quantum-mechanical models of physical systems, ($\mathcal{E}$) and (7.119) (with $p_i > 0$, for *two* or *more* choices of $i$) is specific to quantum-mechanical models. We have seen in Sect. 7.2.3 that, quantum-mechanically, ($\mathcal{LoI}$) may manifest itself in the property that some of the relative commutants,

$$(\mathcal{E}_{\geq t}^{\omega})' \cap \mathcal{E}_{\geq t'}^{\omega} \tag{7.123}$$

are non-trivial, for some $t > t'$; (see (7.49)). If $\mathcal{E}^{\omega}_{\geq t}$ is a factor (i.e., a von Neumann algebra with trivial center, as defined in Sect. 7.2.1, (7.25)) then (7.5.6) implies that

$$\mathcal{E}^{\omega}_{\geq t} \subsetneq \mathcal{E}^{\omega}_{\geq t'}. \tag{7.124}$$

## 7.5.2 Preliminaries Towards a Notion of "Empirical Properties" of Quantum Mechanical Systems

Let $a = a^* \in \mathcal{E}_S$ be an operator representing a potential property (or physical quantity) of $S$ (i.e., $a = \tau_{t',t_0}(c)$, $c \in \mathcal{P}_S$), and let $\omega$ denote the state of $S$. We assume that $a$ has a finite spectrum,

$$a = \sum_{i=1}^{k} \alpha_i \Pi_i, \qquad k < \infty, \tag{7.125}$$

where $\alpha_1, \ldots, \alpha_k$ are the eigenvalues of $a$ (now viewed as a self-adjoint operator in the von Neumann algebra $\mathcal{E}^{\omega}_S$), and $\Pi_i \equiv \Pi^{(i)}_{\alpha_i} \in \mathcal{E}^{\omega}_S$ is the spectral projection of $a$ corresponding to $\alpha_i$, $i = 1, \ldots, k$. How should we define *empirical properties* of $S$? To say that $a$ is an empirical property of $S$ at some time $t'$ earlier than $t$, i.e., that $a$ is measured (or observed) before time $t$, means that

$$\omega(b) \approx \sum_{i=1}^{k} \omega(\Pi_i b \Pi_i), \tag{7.126}$$

for all $b \in \mathcal{E}^{\omega}_{\geq t}$; i.e., $\omega_{|\mathcal{E}^{\omega}_{\geq t}}$ is close to an incoherent superposition (mixture) of eigenstates, $p_i^{-1}\omega(\Pi_i(\cdot)\Pi_i)$ ($p_i \neq 0$), of $a$, where $p_i = \omega(\Pi_i)$, (and $p_i > 0$, for at least one choice of $i$). A sufficient condition for Eq. (7.126) to hold is that

$$a \in (\mathcal{E}^{\omega}_{\geq t})' \cap \mathcal{E}^{\omega}_S. \tag{7.127}$$

If there existed a sequence of times, $t_1 < t_2 < \ldots < t_n$, and self-adjoint operators $a_1, \ldots, a_n$, with finite point spectra, as above, and

$$a_l \in (\mathcal{E}^{\omega}_{\geq t_l+1})' \cap \mathcal{E}^{\omega}_{\geq t_l},$$

$l = 1, \ldots, n - 1$, $a_n \in \mathcal{E}_{\geq t_n}$, then the family of histories

$$h_1^n(\underline{j}) = \{\Pi^{(1)}_{j_1}, \ldots, \Pi^{(n)}_{j_n}\},$$

where $\Pi_{ji}^{(l)}$ is the spectral projection of $a_l$ corresponding to the eigenvalue $\alpha_{ji}^{(l)}$ of $a_l, l = 1, \ldots, n$, is *consistent*; see (7.96)–(7.100), Sect. 7.4.3. For this observation to be interesting, the relative commutants $(\mathcal{E}_{\geq t_l+1}^{\omega})' \cap \mathcal{E}_{\geq t_l}^{\omega}$ would have to be non-trivial and if we wish to escape from critique (iii) at the end of Sect. 7.4.3 the algebras $(\mathcal{E}_{\geq t_l+1}^{\omega})' \cap \mathcal{E}_{\geq t_l}^{\omega}$ would have to be *abelian*, for all $l$. This does *not* look like a satisfactory or plausible assumption, and we have to continue our search for a good notion of "empirical properties"!

**Definition 5.1** (i) Given von Neumann algebras $\mathcal{M} \subseteq \mathcal{N}$, a state $\omega$ on $\mathcal{N}$ and an operator $a \in \mathcal{N}$, we define $\{a, \omega\}_{\mathcal{M}}$ to be the bounded linear functional on $\mathcal{M}$ defined by

$$\{a, \omega\}_{\mathcal{M}}(b) := \omega([a, b]), \qquad b \in \mathcal{M}. \tag{7.128}$$

(ii) The centralizer (or stabilizer), $\mathcal{C}_{\mathcal{M}}^{\omega}$, of $\omega$ is the subalgebra of $\mathcal{M}$ defined by

$$\mathcal{C}_{\mathcal{M}}^{\omega} := \{a \in \mathcal{M} \mid \{a, \omega\}_{\mathcal{M}} = 0\}. \tag{7.129}$$

It is easy to see that $\omega$ defines a *trace* on $\mathcal{C}_{\mathcal{M}}^{\omega}$. This means that $\mathcal{C}_{\mathcal{M}}^{\omega}$ is a direct sum (or integral) of finite-dimensional matrix algebras, type-$II_1$ factors and abelian algebras.

*Remark 5.2* Centralizers of states or weights on von Neumann algebras play an interesting role in the classification of von Neumann algebras, (in particular in the study of type-III factors); see [19, 45]. In an appendix to Sect. 7.5, we recall a few relevant results on centralizers.

Obviously, strict equality in Eq. (7.126) follows from the assumption that

$$\{a, \omega\}_{\mathcal{E}_{\geq t}^{\omega}} = 0, \qquad a \in \mathcal{E}_{\geq t}^{\omega}. \tag{7.130}$$

In other words, condition (7.130) implies that, as a state on the algebra $\mathcal{E}_{\geq t}^{\omega}$ of possible events in $S$ observable after time $t$, $\omega$ is an incoherent superposition of eigenstates of $a$, *even* if, as a state on $\mathcal{E}_S$, $\omega$ is *pure*. However, to convince oneself that $\omega$ is a mixture (incoherent superposition) it is often enough to assume that the norm of the linear functional $\{a, \omega\}_{\mathcal{E}_{\geq t}^{\omega}}$, with $a \in \mathcal{E}_{\geq t}^{\omega}$, is *small*. Let us suppose that $a$ is self-adjoint and that its spectrum consists of finitely many eigenvalues $\alpha_1 > \alpha_2 > \ldots > \alpha_k$. Then

$$a = \sum_{i=1}^{k} \alpha_i \Pi_i,$$

where $\Pi_1, \ldots, \Pi_k$ are the spectral projections of $a$ satisfying $\Pi_i = \Pi_i^*$, $\Pi_i \Pi_l = \delta_{il} \Pi_l$, for all $i, l = 1, \ldots, k$, and $\sum_{i=1}^{k} \Pi_i = \mathbb{1}$. The following result is easily proven.

**Lemma 5.3** *The following assertions are equivalent:*

(i) $|\{a, \omega\}_{\mathcal{E}_{\geq t}^\omega} (b)| < \epsilon \|b\|$, $\forall b \in \mathcal{E}_{\geq t}^\omega$

(ii) $|\omega(b) - \sum_{i=1}^{k} \omega(\Pi_i b \Pi_i)| \leq const. \, \epsilon \|b\|$, $\forall b \in \mathcal{E}_{\geq t}^\omega$.

In view of Lemma 5.3, one might be tempted to identify elements of the centralizer

$$\mathcal{C}_{\geq t}^\omega := \mathcal{C}_{\mathcal{E}_{\geq t}^\omega}^\omega \qquad (7.131)$$

with *empirical properties of $S$ observable at times $\geq t$*. Yet, this is not quite the right idea!

(1) A family of operators, $a_1, \ldots, a_n$, with

$$a_i \in \mathcal{C}_{\geq t_i}^\omega,$$

$i = 1, \ldots, n$, $t_1 < t_2 < \ldots < t_n$, does not necessarily give rise to a family of consistent histories. The reason is exceedingly simple: Let $\Pi_l^{(i-1)}$, $l = 1, \ldots, k_{i-1}$, be the spectral projections of $a_{i-1} \in \mathcal{C}_{\geq t_{i-1}}$. Let $\omega_l$ denote the state

$$\omega_l(b) = p_l^{-1} \omega(\Pi_l^{(i-1)} b \Pi_l^{(i-1)}),$$

where $p_l = \omega(\Pi_l^{(i-1)}) > 0$. Let us assume that $p_l > 0$ for at least *two* distinct values of $l$. The problem is that, in general, the assumption that $a_i \in \mathcal{C}_{\geq t_i}^\omega$ does *not* imply that $a_i \in \mathcal{C}_{\geq t_i}^{\omega_l}$, for all $l = 1, \ldots, k_{i-1}$ for which $p_l > 0$; this is the phenomenon of *"spontaneous symmetry breaking"*. This means that the "sum rule" (7.84), Sect. 7.4.1, may be violated at the $i$th slot, for some $1 < i < n$. Hence the family $a_1, \ldots, a_n$ may *not* give rise to a family of consistent histories.

(2) In general, the centralizers $\mathcal{C}_{\geq t}^\omega$ are *non-abelian* algebras. If the centralizers $\mathcal{C}_{\geq t}^\omega$ are non-commutative algebras then identifying empirical properties of $S$ observable at times $\geq t$ with elements of $\mathcal{C}_{\geq t}^\omega$ is subject to critique (ii), Sect. 7.4.3. Our task is then to find out *which elements of $\mathcal{C}_{\geq t}^\omega$ may correspond to empirical properties of $S$*. (The center of $\mathcal{C}_{\geq t}^\omega$ is denoted by $\mathcal{Z}_{\geq t}^\omega$. If $\mathcal{Z}_{\geq t}^\omega$ were known to contain operators representing potential properties of $S$ then these operators could be interpreted as *empirical properties* of $S$ observed at some times $> t$, and critique (ii) of Sect. 7.4.3 would not apply, anymore.)

## 7.5.3   So, What are "Empirical Properties" of a Quantum-Mechanical System?

Consider the data characterizing a physical system as specified in (7.120)–(7.122). Let $\mathcal{E}_{\geq t}$ be the algebra of physical quantities pertaining to a system $S$ that can be observed at times $\geq t$, and let $\mathcal{E}_S$ be the $C^*$-algebra obtained as the norm closure of $\bigvee_{t \in \mathbb{R}} \mathcal{E}_t$. Let $\omega$ be a state on $\mathcal{E}_S$. By $\mathcal{C}_{\geq t}^{\omega}$ we have denoted the centralizer of the state $\omega$ (viewed as a state on the von Neumann algebra $\mathcal{E}_{\geq t}^{\omega}$ corresponding to the weak closure of $\mathcal{E}_{\geq t}$ in the GNS representation associated with $(\mathcal{E}_S, \omega)$). We have seen, after definition (7.129), that $\omega_{|\mathcal{C}_{\geq t}^{\omega}}$ is a *trace* on $\mathcal{C}_{\geq t}^{\omega}$. This implies that

$$\mathcal{C}_{\geq t}^{\omega} = \int_{\Lambda}^{\oplus} \mathcal{C}_{\geq t, \lambda}^{\omega}, \tag{7.132}$$

where every algebra $\mathcal{C}_{\geq t, \lambda}^{\omega}$, $\lambda \in \Lambda \equiv \Lambda_{\omega}$, is either a finite-dimensional matrix algebra, $\approx \mathbb{M}_{n_\lambda}(\mathbb{C})$, of $n_\lambda \times n_\lambda$ matrices, with $1 \leq n_\lambda < \infty$, or a type-$II_1$ factor; (see [69], Theorem 8.21 in Chapter 4, and Theorem 2.4 in Chapter 5). If $\mathcal{C}_{\geq t, \lambda}^{\omega}$ is isomorphic to $\mathbb{M}_{n_\lambda}(\mathbb{C})$ then

$$\omega_{|\mathcal{C}_{\geq t, \lambda}^{\omega}} \propto \mathrm{tr}_{\mathbb{C}^{n_\lambda}}(\cdot). \tag{7.133}$$

Let us assume, temporarily, that $\Lambda$ is discrete, and

$$\mathcal{C}_{\geq t}^{\omega} = \bigoplus_{\lambda \in \Lambda} \mathcal{C}_{\geq t, \lambda}^{\omega}, \tag{7.134}$$

with

$$\mathcal{C}_{\geq t, \lambda}^{\omega} \simeq \mathbb{M}_{n_\lambda}(\mathbb{C}), \qquad n_\lambda < \infty, \tag{7.135}$$

for all $\lambda \in \Lambda$. Then $\mathcal{E}_{\geq t}^{\omega}$ is a von Neumann algebra of type I and

$$\omega_{|\mathcal{E}_{\geq t}^{\omega}} =: \rho_{\geq t}^{\omega}, \tag{7.136}$$

where $\rho_{\geq t}^{\omega}$ is a *density matrix*, so that

$$\rho_{\geq t}^{\omega} = \sum_{\lambda \in \Lambda} p_\lambda(t) \Pi_\lambda(t), \tag{7.137}$$

and the operators $\Pi_\lambda(t) \equiv \Pi_\lambda^{\omega}(t)$ are the eigenprojections of $\rho_{\geq t}^{\omega}$, with $\dim(\Pi_\lambda(t)) = n_\lambda < \infty$, the weights $p_\lambda(t) \equiv p_\lambda^{\omega}(t) \geq 0$ are the eigenvalues of $\rho_{\geq t}^{\omega}$, arranged in decreasing order, and

$$\mathrm{tr}(\rho_t^{\omega}) = \sum_{\lambda \in \Lambda} p_\lambda(t) \dim(\Pi_\lambda(t)) = 1.$$

Then $\mathcal{C}_{\geq t, \lambda}^{\omega} \simeq \mathbb{M}_{n_\lambda}(\mathbb{C})$ is the algebra of all bounded operators from the eigenspace Ran $\Pi_\lambda(t)$ to itself, and

$$\omega_{|\mathcal{C}_{\geq t, \lambda}^{\omega}} = p_\lambda(t) \mathrm{tr}(\Pi_\lambda(t)(\cdot)).$$

Any operator $a \in \mathcal{E}_{\geq t}^{\omega}$ commuting with all the projections $\Pi_\lambda(t)$, $\lambda \in \Lambda$, belongs to $\mathcal{C}_{\geq t}^{\omega}$, and any operator in the center $\mathcal{Z}_{\geq t}^{\omega}$ of $\mathcal{C}_{\geq t}^{\omega}$ is a function of the projections $\Pi_\lambda(t)$, $\lambda \in \Lambda_\omega$. In particular $\Pi_\lambda(t) \in \mathcal{Z}_{\geq t}^{\omega} \subset \mathcal{C}_{\geq t}^{\omega}$, for all $\lambda$, (and hence the eigenprojections of $\rho_{\geq t}^{\omega}$ might qualify as empirical properties of $S$).

Henceforth, we consider the special case specified in Eqs. (7.134)–(7.137); (but see Remark (1) of Sect. 7.5.5, and Appendix 7.5.A).

**Definition 5.4** Let $a = a^*$ be an operator in $\mathcal{E}_{\geq t}^{\omega}$. We define

$$\bar{a}^\lambda := \frac{1}{n_\lambda}\mathrm{tr}(\Pi_\lambda(t)a). \tag{7.138}$$

If $\lambda$ is such that $p_\lambda(t) > 0$ then

$$\bar{a}^\lambda := \frac{1}{p_\lambda(t)n_\lambda}\omega(\Pi_\lambda(t)a).$$

Note that $\bar{\mathbb{1}}^\lambda = 1$. We set

$$a^\omega := \sum_{\lambda \in \Lambda}\bar{a}^\lambda \Pi_\lambda(t) \in \mathcal{Z}_{\geq t}^{\omega} \subset \mathcal{C}_{\geq t}^{\omega} \tag{7.139}$$

and define the "variance of $a$ in $\omega$" by

$$\Delta_t^\omega a := \sqrt{\sum_{\lambda \in \Lambda}p_\lambda(t)\mathrm{Tr}(\Pi_\lambda(t)(a - \bar{a}^\lambda \cdot \mathbb{1})^2)} = \sqrt{\omega((a - a^\omega)^2)}. \tag{7.140}$$

We observe that if $\Delta_t^\omega a = 0$ then $a \in \mathcal{C}_{\geq t}^{\omega}$, and, on the range of $\rho_t^\omega$, $a|_{\mathrm{Ran}\rho_t^\omega} = a^\omega|_{\mathrm{Ran}\rho_t^\omega}$ is a function of $\rho_t^\omega$, i.e., $a|_{\mathrm{Ran}\rho_t^\omega} \in \mathcal{Z}_{\geq t}^{\omega}$. For a general element, $a$, of $\mathcal{E}_{\geq t}^{\omega}$,

$$|\{a, \omega\}_{\mathcal{E}_t^\omega}(b)| = |\omega([a, b])| = |\omega([a - a^\omega, b])|$$
$$\leq 2\sqrt{\omega((a - a^\omega)^2)\omega(b^*b)} \leq 2\Delta_t^\omega a \, \|b\|, \tag{7.141}$$

for arbitrary $b \in \mathcal{E}_{\geq t}^{\omega}$. Thus, if $\Delta_t^\omega a$ is small then $\|\{a, \omega\}_{\mathcal{E}_t^\omega}\|$ is small, too, and Lemma 5.3 then tells us that $\omega|_{\mathcal{E}_t^\omega}$ is close to an incoherent superposition of eigenstates of $a$.

Let $d\mu_\lambda(\alpha)$ denote the spectral measure of the operator $a = a^* \in \mathcal{E}_{\geq t}^{\omega}$ in the state

$$n_\lambda^{-1}\mathrm{tr}(\Pi_\lambda(t)(\cdot)).$$

Then

$$0 \leq \omega((a - a^\omega)^2) = \sum_{\lambda \in \Lambda} p_\lambda(t) n_\lambda \frac{1}{n_\lambda} \operatorname{tr}(\Pi_\lambda (a - \overline{a}^\lambda)^2)$$

$$= \sum_{\lambda \in \Lambda} p_\lambda(t) n_\lambda \int d\mu_\lambda(\alpha)(\alpha - \overline{a}^\lambda)^2.$$

Thus,

$$p_\lambda(t) n_\lambda \int d\mu_\lambda(\alpha)(\alpha - \overline{a}^\lambda)^2 \leq (\Delta_t^\omega a)^2,$$

for every $\lambda \in \Lambda$. We conclude that if, for some $\lambda \in \Lambda$,

$$\frac{1}{p_\lambda(t) n_\lambda} (\Delta_t^\omega a)^2 < \epsilon^2,$$

for some $\epsilon > 0$, then $a$ has spectrum at a distance less than $\epsilon$ from $\overline{a}^\lambda$. In particular, if $a$ has discrete spectrum then $a$ has at least one eigenvalue $\alpha_\lambda$, with

$$|\alpha_\lambda - \overline{a}^\lambda| < \epsilon. \tag{7.142}$$

Next, let $a \in \mathcal{P}_S$ be the operator representing some *potential property* of $S$. Then $a(t) := \tau_{t,t_0}(a) \in \mathcal{E}_{\geq t}^\omega$.

**Definition 5.5** We say that a potential property of $S$ represented by an operator $a \in \mathcal{P}_S$ is an *empirical property of $S$ at time $t$ within an uncertainty (of size) $\delta \geq 0$* iff

$$\Delta_t^\omega a(t) \leq \delta. \tag{7.143}$$

*Remark 5.6* If $\delta$ is below the resolution threshold of the equipment used to monitor $S$ then, FAPP, $a(t)$ indeed represents an empirical property of $S$ at time $t$, in the following sense:

(1) $\|\{a(t), \omega\}_{\mathcal{E}_{\geq t}^\omega}\|$ is so small that it cannot be distinguished from 0;
(2) $\omega(b) \approx \sum_i \omega(\Pi_i(t) b \Pi_i(t))$, for all $b \in \mathcal{E}_{\geq t}^\omega$, where $\Pi_1(t), \Pi_2(t), \ldots$ are the spectral projections of $a(t)$, (assuming $a = a^*$ has discrete spectrum; see Lemma 5.3 for a precise statement);
(3) on the range of the density matrix $\rho_{\geq t}^\omega$, $a(t)$ is "close" to the operator $a(t)^\omega \in \mathcal{Z}_{\geq t}^\omega$;
(4) $a$ has eigenvalues near the numbers $\overline{a(t)}^\lambda$, for all $\lambda \in \Lambda_\omega$ for which $(p_\lambda(t) n_\lambda)^{-1} \delta^2$ is small.

One may then argue that if $\Delta_t^\omega a(t)$ is very small, and if a measurement or observation of $a \in \mathcal{P}_S$ at a time $\approx t$ indicates that it has a value $\alpha \approx \overline{a(t)}^\lambda$ then one may use the state

$$\omega_\lambda := \frac{1}{n_\lambda} \text{tr}(\Pi_\lambda(t)(\cdot)) \qquad (7.144)$$

to predict the behavior of the system $S$ at times later than $t$. This idea, reminiscent of "state collapse", will be further discussed below.

Note that the maximal uncertainty $\delta$ admissible in statement (2) above depends on the spectrum of the operator $a$.

## 7.5.4 When Does an Observation or Measurement of a Physical Quantity Take Place?

Let $a = a^* \in \mathcal{P}_S$ represent a potential property of a quantum-mechanical system $S$, which is assumed to be prepared in a state $\omega$ on the algebra $\mathcal{E}_S$. We propose to analyze *whether* and *when* $a$ corresponds to an *empirical property* of $S$, in the sense that, given the time evolution $\{\tau_{t,s}\}_{t,s \in \mathbb{R}}$ of $S$ and the state $\omega$, $a$ is measurable (i.e., the value of $a$ can be measured or observed) at some finite time. Definition 5.5 and the discussion thereafter suggest to consider the variance $\Delta_t^\omega a(t)$ $(a(t) = \tau_{t,t_0}(a))$, of $a(t)$ as a *function of time* $t$. This function is non-negative and bounded. Let $\delta$ be some non-negative number below the resolution threshold of the equipment used to monitor $S$. Let $t_*$ be defined as the smallest time such that

$$\Delta_{t_*}^\omega a(t_*) \leq \delta. \qquad (7.145)$$

Then it is reasonable to say that $a$ is observed/measured—put differently, $a$ becomes an empirical property of $S$ within an uncertainty of size $\delta$—at a time $\gtrsim t_*$. If the equipment $E$ used to monitor $S$ is only sensitive to observing the eigenvalue $\alpha_i$ of $a$, i.e., to the possible event $\Pi_i$ (spectral projection of $a$ corresponding to the eigenvalue $\alpha_i$) then one may plausibly say that the possible event $\Pi_i$ is observed at a time $\gtrsim t_*$ iff

$$\Delta_{t_*}^\omega a(t_*) + 1 - \omega(\Pi_i(t_*))$$

is *very small*. In this case, we say that the equipment $E$ prepares the state of $S$ to lie in the range of the projection $\Pi_i(t) \approx \sum_{\lambda \in \Lambda_\omega^{(i)}} \Pi_\lambda(t)$, with $t \gtrsim t_*$, where $\Lambda_\omega^{(i)}$ is defined by the property that $|\alpha_i - \overline{a(t)}^\lambda| < \delta$, for all $\lambda \in \Lambda_\omega^{(i)}$. Thus, the function

$$T_{\omega,a}(t) := \Delta_t^\omega a(t) \qquad (7.146)$$

contains all important information on the time around which the potential property
$a$ of $S$ becomes an empirical property; and the function

$$T^i_{\omega,a}(t) := \Delta^\omega_t a(t) + 1 - \omega(\Pi_i(t)) \tag{7.147}$$

tells us when (around which time) a detector sensitive to the possible event $\Pi_i$
"clicks"; (see also [13, 73] for some ideas on this matter that will not be pursued
here).

Next, we analyze *repeated observations/measurements*, as in Sect. 7.4.1. It
suffices to consider only *two* subsequent measurements. Let $a = a^* \in \mathcal{P}_S$ represent
a potential property of $S$, and let $\delta \geq 0$ be a measure for the resolution of the
equipment $E$ used to monitor $S$ in a measurement of $a$.

**Definition 5.7** For $a = a^* \in \mathcal{P}_S$, $\delta \geq 0$, and a time $t_* > -\infty$, we define a subset
of states on $\mathcal{A}_{\overline{S}}$ (or on $\mathcal{E}_S \subset \mathcal{A}_{\overline{S}}$) by

$$S(a, \delta, t_*) := \{\omega \in S_{\overline{S}} \mid \inf_{t \geq t_*} \Delta^\omega_t a(t) < \delta\}, \tag{7.148}$$

where $\delta$ is so small that properties (1) through (4) in Remark 5.6, above, are valid.

Apparently, $S(a, \delta, t_*)$ is the set of states of $S$ with the property that, given the
time evolution $\{\tau_{t,s}\}_{t,s \in \mathbb{R}}$, the operator $a$ corresponds to an *empirical property* of $S$,
within an uncertainty of size $\delta$, that is measurable at some time after $t_*$.

Next, we consider two potential properties of $S$ represented by two self-adjoint
operators, $a_1$ and $a_2$, and we suppose that, first, $a_1$ and, afterwards, $a_2$, are measured.
For simplicity we suppose that the spectra of $a_1$ and $a_2$ consist of finitely many
eigenvalues $\alpha^{(i)}_j$, $j = 1, \ldots, k_i < \infty$, $i = 1, 2$. We assume that the state, $\omega$, of
$S$ before the measurement of $a_1$, belongs to $S(a_1, \delta_1, t_{1*})$, for a sufficiently small
number $\delta_1$ (below a threshold of resolution). Then $\Delta^\omega_{t_1} a_1(t_1) \leq \delta_1$, at some time
$t_1 \geq t_{1*}$. A successful measurement of $a_1$ around some time $t_1 \geq t_{1*}$ results in the
assignment of a value $\alpha^{(1)}_j \approx \overline{a_1(t_1)}^\lambda$, $\lambda \in \Lambda^{(j)}_\omega$, to the physical quantity represented
by $a_1$, where

$$\Lambda^{(j)}_\omega := \{\lambda \in \Lambda_\omega \mid |\overline{a_1(t_1)}^\lambda - \alpha^{(1)}_j| < \delta_1\}. \tag{7.149}$$

(For consistency, we assume that $\min_{j \neq l} |\alpha^{(1)}_j - \alpha^{(1)}_l| > 2\delta_1$.) The *probability* of this
measurement outcome is given by

$$P^{(1)}_j(t_1) = \sum_{\lambda \in \Lambda^{(j)}_\omega} \omega(\Pi^\omega_\lambda(t_1)) = \sum_{\lambda \in \Lambda^{(j)}_\omega} p^\omega_\lambda(t_1) n^\omega_\lambda = \omega(\Pi^{(1)}_j(t_1)) + \mathcal{O}(\delta_1),$$

$$\tag{7.150}$$

where $p_\lambda(t_1) \equiv p^\omega_\lambda(t_1)$, $n_\lambda \equiv n^\omega_\lambda = \dim \Pi^\omega_\lambda(t_1)$, and $\Pi_\lambda(t_1) \equiv \Pi^\omega_\lambda(t_1)$ are as
defined in Eqs. (7.136) and (7.137), (the superscript "$\omega$" is supposed to highlight the

dependence on the state $\omega$), and $\Pi_j^{(1)}(t_1)$ is the eigenprojection of the operator $a_1(t_1)$ corresponding to the eigenvalue $\alpha_j^{(1)}$. If $P_j^{(1)}(t_1)$ is very small one can ignore the possibility that, for a system $S$ prepared in the state $\omega$, an observation/measurement of $a_1$ will yield a value $\approx \alpha_j^{(1)}$.

Let $\omega_j$ denote the state

$$\omega_j(b) = \frac{\sum_{\lambda \in \Lambda_\omega^{(j)}} \omega(\Pi_\lambda^\omega(t_1) b \Pi_\lambda^\omega(t_1))}{P_j^{(1)}(t_1)} = \frac{\omega(\Pi_j^{(1)}(t_1) b \Pi_j^{(1)}(t_1))}{\omega(\Pi_j^{(1)}(t_1))} + \mathcal{O}(\delta_1),$$
(7.151)

for an arbitrary operator $b \in \mathcal{E}_{\geq t}^\omega$, with $t \geq t_1$; (recall that $\mathcal{E}_{\geq t}^\omega \subseteq \mathcal{E}_{\geq t_1}^\omega$, for $t \geq t_1$).

Let us suppose that, for all $j \in \{1, \ldots, k_1\}$ for which $P_j^{(1)}(t_1) > \delta_2 > 0$,

$$\omega_j \in \mathcal{S}(a_2, \delta_2, t_{2*}^{(j)}),$$
(7.152)

for some time $t_{2*}^{(j)} > t_1$. If $\delta_2$ is chosen small enough one may expect to be able to successfully measure the quantity represented by $a_2$ at a time $t_2 \geq t_{2*}^{(j)}$, assuming that, at a time $t_1 < t_{2*}^{(j)}$, $a_1$ was found to have a value $\approx \alpha_j^{(1)}$.

The *joint probability* to find a value $\approx \alpha_j^{(1)}$ in a measurement of $a_1$ around some time $t_1$ and, in a subsequent measurement around a time $t_2 > t_1$, a value $\approx \alpha_l^{(2)}$ of the quantity represented by $a_2$, (with $l \in \{1, \ldots, k_2\}$), is given by

$$\text{Prob}_\omega\{\Pi_j^{(1)}(t_1), \Pi_l^{(2)}(t_2)\} = P_j^{(1)}(t_1) \sum_{\lambda \in \Lambda_{\omega_j}^{(l)}} \omega_j(\Pi_\lambda^{\omega_j}(t_2))$$

$$= \omega(\Pi_j^{(1)}(t_1) \Pi_l^{(2)}(t_2) \Pi_j^{(1)}(t_1)) + \mathcal{O}(\delta_1 \vee \delta_2),$$
(7.153)

where $\Lambda_{\omega_j}^{(l)} = \{\lambda \in \Lambda_{\omega_j} \mid |\overline{a_2(t_2)}^\lambda - \alpha_l^{(2)}| < \delta_2\}$, and $\delta_1 \vee \delta_2 = \max\{\delta_1, \delta_2\}$.

The definitions of centralizers, $\mathcal{C}_{\geq t_1}^\omega$, etc., and of the variance $\Delta_t^\omega a(t)$ readily imply that

$$\sum_{j=1}^{k_1} \omega(\Pi_j^{(1)}(t_1) \Pi_l^{(2)}(t_2) b \Pi_l^{(2)}(t_2) \Pi_j^{(1)}(t_1)) = \omega(\Pi_l^{(?)}(t_2) b \Pi_l^{(2)}(t_2)) + \mathcal{O}(\delta_1),$$
(7.154)

and if $\omega_j \in \mathcal{S}(a_2, \delta_2, t_{2*}^{(j)})$ then

$$\sum_{l=1}^{k_2} \omega(\Pi_j^{(1)}(t_1) \Pi_l^{(2)}(t_2) b \Pi_l^{(2)}(t_2) \Pi_j^{(1)}(t_1)) = \omega(\Pi_j^{(1)}(t_1) b \Pi_j^{(1)}(t_1)) + \mathcal{O}(\delta_1 \vee \delta_2),$$
(7.155)

for an arbitrary operator $b \in \mathcal{E}^{\omega}_{\geq t}$, with $t > \max_j t_{2*}^{(j)}$. It is clear how to extend our discussion to an arbitrary chronological (time-ordered) sequence of measurements of quantities $a_1, \ldots, a_n$, $(a_i \in \mathcal{P}_S, \forall i)$. Moreover, the mathematical relationship between Eqs. (7.154) and (7.155), on one side, and $\delta$-consistent families of histories—see (7.92) and (7.93), Sect. 7.4.2—on the other side, is easy to unravel. We do not wish to discuss further details.

*Remark 5.8 (Remark on the Role of POVM's)* It may and *will* occasionally happen that, given that a quantity represented by an operator $a_1$ has been observed/measured, the quantity represented by the operator $a_2$ can be measured, subsequently, *only for certain*, but *not all*, outcomes of the measurement of $a_1$. More precisely, it may happen that, for some eigenvalues $\alpha_j^1$, $j \in G$, of $a_1$, $\omega_j \in \mathcal{S}(a_2, \delta_2, t_{2*}^{(j)})$, while, for $i \in B := \{1, \ldots, k_1\} \setminus G$,

$$\omega_i \notin \mathcal{S}(a_2, \delta_2, t_{2*}), \tag{7.156}$$

for any $t_{2*} < \infty$; ($\delta_1$ and $\delta_2$ being chosen appropriately, depending on the resolution of the corresponding measurements, as discussed above).

If $B \neq \emptyset$ then one must take the position that the observations of $a_1$ and $a_2$ represent *one single* measurement, which must be described using "*positive operator-valued measures*" (POVM's)—see Appendix 7.4.A, Eqs. (7.113) and (7.114):

$$X = \{X_{jl}, X_i \mid j \in G, l = 1, \ldots, k_2, i \in B\} \tag{7.157}$$

where, for $j \in G$,

$$X_{jl} = \sum_{\lambda_1 \in \Lambda_{\omega}^{(j)}} \sum_{\lambda_2 \in \Lambda_{\omega_j}^{(l)}} \Pi_{\lambda_2}^{\omega_j}(t_2^{(j)}) \Pi_{\lambda_1}^{\omega}(t_1) \approx \Pi_l^{(2)}(t_2^{(j)}) \Pi_j^{(1)}(t_1), \tag{7.158}$$

(up to a small perturbation of $\mathcal{O}(\delta_1 \vee \delta_2)$), while, for $i \in B$,

$$X_i = \sum_{\lambda_1 \in \Lambda_{\omega}^{(i)}} \Pi_{\lambda_1}^{\omega}(t_1) \approx \Pi_i^{(1)}(t_1), \tag{7.159}$$

where $t_1$ and $t_2^{(j)}$ are the times of measurement of $a_1$ and $a_2$, respectively. Then

$$\sum_{j \in G} \sum_{l=1}^{k_2} X_{jl}^* X_{jl} + \sum_{i \in B} X_i^* X_i = \mathbb{1}. \tag{7.160}$$

The use of POVM's will be discussed in more detail and in connection with concrete examples elsewhere. Here we just remark that simple examples showing

why one needs to introduce POVM's are encountered in the analysis of repeated
Stern–Gerlach measurements of atomic spins (followed by detectors sensitive to the
arrival of the atoms).

### 7.5.5 Generalizations and Summary

(1) In order to keep our exposition reasonably simple, we have made the simplify-
    ing assumptions (7.134) and (7.135). It is, however, not very hard to develop our
    ideas in full generality. For this purpose, we must return to formula (7.132): The
    space $\Lambda = \Lambda_\omega$ appearing in (7.132) is the *spectrum* of the *center*, $\mathcal{Z}^\omega_{\geq t}$, of the
    *centralizer*, $\mathcal{C}^\omega_{\geq t}$, of the state $\omega$, viewed as a state on the algebra $\mathcal{E}^\omega_{\geq t}$. The theory
    of "conditional expectations" [68] enables us (under fairly general hypotheses)
    to construct a conditional expectation $\hat{\epsilon}_{\geq t} : \mathcal{E}^\omega_{\geq t} \to \mathcal{Z}^\omega_{\geq t}$, which permits us to
    associate with every operator $a \in \mathcal{E}^\omega_{\geq t}$ an operator $a^\omega \in \mathcal{Z}^\omega_{\geq t}$. The map $a \mapsto a^\omega$
    is linear, and $(a^\omega)^\omega = a^\omega$. (In the special case where Eqs. (7.134) and (7.135)
    hold it is given by formula (7.139).) Having constructed $a^\omega$, we set

$$\Delta^\omega_t a := \sqrt{\omega((a - a^\omega)^2)}.$$

From this point on, we may follow the arguments from (7.141) onwards, and in
Sect. 7.5.4.

(2) In our approach to the "quantum theory of experiments/quantum measurement
    theory", the *"ontology"* underlying a quantum-mechanical model of a physical
    system $S$ is represented by

    (a) a set, $\mathcal{P}_S$, of physical quantities characterizing $S$;
    (b) a filtration of $C^*$-algebras

$$\mathcal{E}_S \supseteq \mathcal{E}_{\geq t} \supseteq \mathcal{E}_{\geq t'} \supset \{\mathbb{C} \cdot \mathbb{1}\}, \qquad t \leq t',$$

    and *morphisms

$$\bar{\tau}_t : \mathcal{E}_{\geq t'} \longrightarrow \mathcal{E}_{\geq t'+t} \subseteq \mathcal{E}_{\geq t'},$$

    for $t \geq 0$;
    (c) a state $\omega$ on $\mathcal{E}_S$;
    (d) the centralizers $\mathcal{C}^\omega_{\geq t}$ of $\omega_{|\mathcal{E}_{\geq t}}$ and their centers $\mathcal{Z}^\omega_{\geq t}$.

If $S$ is prepared in a state $\omega$ before one attempts to measure a physical quantity
represented by an operator $a \in \mathcal{P}_S$ then the measurement is successful around
some time $t$ if $a(t) = \tau_{t,t_0}(a)$ is "close" to an operator in $\mathcal{Z}^\omega_{\geq t}$, in the sense that
the variance, $\Delta^\omega_t a(t)$, of $a(t)$ in $\omega$ is small.

(3) Let us return to the special situation described in Eq. (7.134) through
    Eq. (7.137). Suppose that all the algebras $\mathcal{E}^{\omega}_{\geq t}$, $t \in \mathbb{R}$, are isomorphic to a
    fixed factor $\mathcal{E} \simeq B(\mathcal{H})$ of type $I_{\infty}$. Then our approach is "dual" to one
    where the density matrices $\{\rho^{\omega}_{\geq t}\}_{t \in \mathbb{R}}$ are interpreted as states on $\mathcal{E}$ and are
    considered to be the fundamental objects, and time evolution is described in
    terms of completely positive maps on the space of density matrices. With
    the idealization/approximation that time evolution is given by a groupoid of
    completely positive maps, this is the point of view popular among quantum
    information scientists; (see, e.g., [53]).

    The trajectories of density matrices $\{\rho^{\omega}_{\geq t}\}_{t \in \mathbb{R}}$ are then what replaces the
    trajectories $\{\xi_t = \phi_{t,t_0}(\xi_0)\}_{t \in \mathbb{R}}$ of a classical system (as discussed in Sect. 7.3).
    However, because of the phenomena of information loss and entanglement, the
    density matrices $\rho^{\omega}_{\geq t}$ tend to describe *mixed* states, *even* if the state $\omega$ is a pure
    state of the algebra $\mathcal{E}_S$, and hence only yield probabilistic predictions, while the
    states $\xi_t$ of a classical system are *pure*, for all $t$, provided the initial state is pure,
    and hence yield deterministic predictions.

(4) It is clearly important to extend our theory to local relativistic quantum theory
    (LRQT). In LRQT, the algebras $\mathcal{E}_{\geq t}$, $t \in \mathbb{R}$, are replaced by algebras, $\mathcal{E}_P$,
    of "observables" localized inside the *forward light cone* of a point $P$ (the
    momentary position of an observer) on a time-like curve in space-time, (the
    observer's world line). If the theory describes a massless photon and if $\omega$
    is a state normal to the vacuum then the von Neumann algebras $\mathcal{E}^{\omega}_P$ are all
    isomorphic to the hyperfinite factor of type $III_1$, as discussed in [17]. Hence
    the algebras $\mathcal{E}^{\omega}_P$ do *not* have *any pure states*, and the principle of Loss of
    Information ($\mathcal{LoI}$) is a fundamental feature of the theory. We will return to
    this topic elsewhere.

(5) It is clearly important to understand how quantum-mechanical systems can be
    prepared in *specific states* ("preparation of states"). This topic will be discussed
    in [34]; but see also (7.144) and the remark right above (7.146). Moreover, it is
    quite crucial to back up the general analysis presented in this essay with simple
    models of "information loss" and "decoherence/dephasing". This will be done
    in a forthcoming publication.

The last topic we briefly address in this essay is a theory of weak (non-
demolition) experiments, following [6]. This theory explains why in many experiments, the system ends up being in an eigenstate of the operator representing the
quantity that is measured, i.e., why *"facts"* emerge in non-demolition measurements.

## 7.5.6 Non-demolition Measurements

After having presented a long and rather abstract discussion of "direct (or von
Neumann) measurements", in Sects. 7.5.3–7.5.5, we wish to sketch the theory of

"indirect (non-demolition) measurements". The main results described here have recently appeared in [6]; see also [1]. The practical importance of these results comes from recent experiments; see, e.g., [42].

We consider a physical system $S$ (e.g., the quantized electromagnetic field in a cavity). We wish to measure a physical quantity represented by an operator $a = a^* \in \mathcal{A}_S$ (e.g., the photon number inside the cavity) with the help of "non-demolition measurements". For this purpose, we bring $S$ into contact with a sequence, $E_1, E_2, E_3, \ldots$, of identical "probes" (e.g., excited atoms passing through the cavity); the interaction of $E_k$ with $S$ is supposed to take place in the time interval $[k - 1, k]$ and is supposed to be turned off during all other times. Actually, after some direct measurement of a property $b_k = b_k^* \in \mathcal{A}_{E_k}$ at a time later than $k$—as described abstractly in Sect. 7.5.4—probe $E_k$ "gets lost for ever", in the sense that no further information about $E_k$ can be retrieved, anymore.

Let $\rho$ denote the initial state of $S$ and $\psi^{(k)} := \psi$ the initial state of probe $E_k$, (the same for all $k$). For simplicity, we assume that the spectrum of the operator $a$ representing the physical property of $S$ to be measured is finite pure-point spectrum. We denote the spectral projection corresponding to an eigenvalue $\alpha$ of $a$ by $\Pi_\alpha = \Pi_\alpha^*$. Then

$$\Pi_\alpha \Pi_\beta = \delta_{\alpha\beta} \Pi_\alpha, \qquad \sum_\alpha \Pi_\alpha = \mathbb{1}.$$

Next, we specify the time-evolution of the composed system $S \vee E_1 \vee E_2 \vee \ldots$: Up to time $k = 1, 2, 3, \ldots$, the time evolution of $E_j$ is assumed to be *trivial*, for all $j > k$. For the subsystem $S \vee E_1 \vee \ldots \vee E_k$ it is specified as follows: Let $A_{\alpha,\alpha'}$ be an arbitrary operator in $\mathcal{A}_S$ mapping Ran $\Pi_{\alpha'}$ to Ran $\Pi_\alpha$, with $\Pi_\beta A_{\alpha,\alpha'} \Pi_{\beta'} = \delta_{\alpha\beta}\delta_{\alpha'\beta'} A_{\alpha,\alpha'}$. Let $B_j$ be an operator in $\mathcal{A}_{E_j}$, $j \leq k$. Then the time-evolution of $A_{\alpha,\alpha'} \otimes B_1 \otimes \ldots \otimes B_k$ from time 0 to time $k$ in the Heisenberg picture is given by

$$\tau_{k,0}(A_{\alpha,\alpha'} \otimes B_1 \otimes \ldots \otimes B_k) := A_{\alpha,\alpha'} \otimes U_\alpha B_1 U_{\alpha'}^* \otimes \ldots \otimes U_\alpha B_k U_{\alpha'}^*,$$

where $U_\alpha$ is a unitary operator in $\mathcal{A}_{E_k} \simeq \mathcal{A}_E$, for all $\alpha \in \mathrm{spec}\, a$. Defining

$$U(i, i - 1) := \sum_\alpha \Pi_\alpha \otimes \mathbb{1} \otimes \ldots \otimes U_\alpha \otimes \mathbb{1} \otimes \ldots,$$

with $U_\alpha$ inserted in the $(i + 1)^{st}$ factor of the tensor product, we have that

$$\tau_{k,0}(A_{\alpha,\alpha'} \otimes B_1 \otimes \ldots \otimes B_j) = \prod_{i=k}^{1} U(i, i - 1)(A_{\alpha,\alpha'} \otimes B_1 \otimes \ldots \otimes B_j) \prod_{i=1}^{k} U(i, i - 1)^*$$

$$= \tau_{k,0}(A_{\alpha,\alpha'} \otimes B_1 \otimes \ldots \otimes B_k) \otimes B_{k+1} \otimes \ldots \otimes B_j,$$
$$(7.161)$$

for arbitrary $j \geq k$. This is a typical (albeit highly idealized) example of time-evolution in a non-demolition measurement. Let $\Psi := \rho \otimes \psi \otimes \psi \otimes \ldots$ denote the initial state of the composed system, $S \vee E_1 \vee E_2 \vee \ldots$. If we set

$$B_1 = B_2 = \ldots = B_{k_0} = \mathbb{1},$$

for some $k_0 < \infty$ then

$$
\begin{aligned}
&\Psi\left(\tau_{k,0}(A_{\alpha,\alpha'} \otimes \mathbb{1} \otimes \ldots \otimes \mathbb{1} \otimes B_{k_0+1} \otimes \ldots \otimes B_{k_0+l})\right) \\
&= \rho(A_{\alpha,\alpha'})\psi(U_\alpha U_{\alpha'}^*)^{k_0} \prod_{i=k_0+1}^{k} \psi(U_\alpha B_i U_{\alpha'}^*) \prod_{i=k+1}^{k_0+l} \psi(B_i),
\end{aligned}
\tag{7.162}
$$

for $k_0 \leq k \leq k_0 + l$. Because $U_\alpha$ is unitary, for all $\alpha \in \mathrm{spec}\, a$,

$$|\Psi(U_\alpha U_{\alpha'}^*)| \leq 1, \qquad \text{for all } \alpha, \alpha',$$

by the Cauchy–Schwarz inequality. We assume that

$$|\Psi(U_\alpha U_{\alpha'}^*)| \leq \mu < 1, \qquad \text{for } \alpha \neq \alpha'. \tag{7.163}$$

Then, for $\alpha \neq \alpha'$

$$|\Psi\left(\tau_{k,0}(A_{\alpha,\alpha'} \otimes \mathbb{1} \otimes \ldots \otimes \mathbb{1} \otimes B_{k_0+1} \otimes \ldots \otimes B_{k_0+l})\right)| \leq \mu^{k_0}, \tag{7.164}$$

which, by Eq. (7.163), tends to 0 exponentially fast, as $k_0 \to \infty$, for arbitrary $A_{\alpha,\alpha'}, B_{k_0+1}, \ldots, B_{k_0+l}$, with $\|A_{\alpha,\alpha'}\|, \|B_{k_0+1}\|, \ldots, \|B_{k_0+l}\|$ bounded by 1. This is *"decoherence"* over the spectrum of the operator $a$ representing the quantity to be measured:

$$\Psi_{|\mathcal{E}_{\geq k_0}} \longrightarrow \sum_\alpha \Psi(\Pi_\alpha(\cdot)\Pi_\alpha)_{|\mathcal{E}_{\geq k_0}}, \tag{7.165}$$

as $k_0 \to \infty$, where $\mathcal{E}_{\geq k_0}$ is the algebra introduced in Definition 2.2. Henceforth, we choose an initial state, $\rho$, for $S$ satisfying

$$\rho = \sum_\alpha \rho(\Pi_\alpha(\cdot)\Pi_\alpha) = \sum_\alpha p_\alpha \rho_\alpha,$$

where

$$p_\alpha = \rho(\Pi_\alpha), \quad \rho_\alpha = p_\alpha^{-1}\rho(\Pi_\alpha(\cdot)\Pi_\alpha). \tag{7.166}$$

We assume that (after many identical probes have interacted with $S$, so that decoherence over the spectrum of the observable $a$ has set in) a *direct measurement* of a physical quantity represented by an operator $b = b^* \in \mathcal{A}_E$ is carried out on *every probe* $E_k \simeq E$, *after* it has interacted with $S$. We assume that the spectrum of $b$ is pure-point, with eigenvalues denoted by $\xi$ and corresponding spectral projections written as $\pi_\xi$. Then $\pi_\xi = \pi_\xi^*$ and

$$\pi_\xi \pi_{\xi'} = \delta_{\xi\xi'} \pi_\xi, \qquad \sum_\xi \pi_\xi = \mathbb{1}. \qquad (7.167)$$

The probability, $\mu(\underline{\xi}_k | \alpha)$, of a history

$$\underline{\xi}_k := \{\pi_{\xi_1}, \ldots, \pi_{\xi_k}\} \qquad (7.168)$$

of possible outcomes of those direct measurements in the state $\Psi_\alpha$ defined by

$$\Psi_\alpha := \rho_\alpha \otimes \psi \otimes \psi \otimes \ldots,$$

with $\rho_\alpha$ as in Eq. (7.166), is given by

$$\mu(\underline{\xi}_k | \alpha) = \prod_{i=1}^k p(\xi_i | \alpha), \qquad (7.169)$$

where

$$p(\xi | \alpha) := \psi(U_\alpha \pi_\xi U_\alpha^*). \qquad (7.170)$$

Note that $\sum_\xi p(\xi | \alpha) = 1$, by Eq. (7.167) and the unitarity of $U_\alpha$. In the following, we identify $\pi_\xi$ with $\xi$ and use the notation $\underline{\xi}_k = (\underline{\xi}_{k-1}, \xi_k)$. In the initial state $\Psi$, the probability of the history $\underline{\xi}_k$ is then given by

$$\mu(\underline{\xi}_k) = \sum_\alpha p_\alpha \mu(\underline{\xi}_k | \alpha). \qquad (7.171)$$

Next, we calculate the probability, $p^{(k)}(\alpha | \underline{\xi}_k)$ of the possible event $\Pi_\alpha$, given that a history $\underline{\xi}_k$ is observed on the first $k$ probes, and given the initial state $\Psi$. By Eqs. (7.166) and (7.169)–(7.171),

$$p^{(k)}(\alpha | \underline{\xi}_k) = p_\alpha \frac{\mu(\underline{\xi}_k | \alpha)}{\mu(\underline{\xi}_k)}, \qquad (7.172)$$

(with $p_\alpha = \rho(\Pi_\alpha)$; see Eq. (7.166)). These probabilities have the following properties:

(i)

$$0 \le p^{(k)}(\alpha|\underline{\xi}_k) \le 1, \quad \text{and} \quad \sum_\alpha p^{(k)}(\alpha|\underline{\xi}_k) = 1.$$

(ii)

$$p^{(k)}(\alpha|\underline{\xi}_k) = p_\alpha \frac{\mu(\alpha|\underline{\xi}_{k-1})}{\mu(\underline{\xi}_k)} p(\xi_k|\alpha)$$

$$= p^{(k-1)}(\alpha|\underline{\xi}_{k-1}) \frac{\mu(\underline{\xi}_{k-1})}{\mu(\underline{\xi}_k)} p(\xi_k|\alpha) \qquad (7.173)$$

$$= p^{(k-1)}(\alpha|\underline{\xi}_{k-1}) \frac{p(\xi_k|\alpha)}{\sum_\beta p^{(k-1)}(\beta|\underline{\xi}_{k-1}) p(\xi_k|\beta)},$$

because, by Eqs. (7.171), (7.169) and (7.172),

$$\frac{\mu(\underline{\xi}_k)}{\mu(\underline{\xi}_{k-1})} = \sum_\beta p_\beta \frac{\mu(\underline{\xi}_{k-1}|\beta)}{\mu(\underline{\xi}_{k-1})} p(\xi_k|\beta)$$

$$= \sum_\beta p^{(k-1)}(\underline{\xi}_{k-1}|\beta) p(\xi_k|\beta). \qquad (7.174)$$

(iii) The expectation, $\mathbb{E}_k$, of $p^{(k)}(\alpha|\underline{\xi}_k)$, given $\alpha$ and $\underline{\xi}_{k-1}$, satisfies

$$\mathbb{E}_k p^{(k)}(\alpha|\underline{\xi}_k) := \sum_{\xi_k} p^{(k)}(\alpha|\underline{\xi}_{k-1}, \xi_k) \frac{\mu(\underline{\xi}_{k-1}, \xi_k)}{\sum_{\xi_k} \mu(\underline{\xi}_{k-1}, \xi_k)}$$

$$= \sum_{\xi_k} p^{(k)}(\alpha|\underline{\xi}_{k-1}, \xi_k) \frac{\mu(\underline{\xi}_k)}{\mu(\underline{\xi}_{k-1})}$$

$$= \sum_{\xi_k} p_\alpha \frac{\mu(\underline{\xi}_{k-1}|\alpha) p(\xi_k|\alpha)}{\mu(\underline{\xi}_k)} \frac{\mu(\underline{\xi}_k)}{\mu(\underline{\xi}_{k-1})} \qquad (7.175)$$

$$= \sum_{\xi_k} p^{(k-1)}(\alpha|\underline{\xi}_{k-1}) p(\xi_k|\alpha) = p^{(k-1)}(\alpha|\underline{\xi}_{k-1}),$$

(see below Eq. (7.170)).

Properties (i) and (iii) identify the random variables $\{p^{(k)}(\alpha|\underline{\xi}_k) \mid \alpha \in \text{spec } a\}$ as *bounded martingales*. The Martingale Convergence Theorem (see e.g., [58]) then implies that

$$p^{(k)}(\alpha|\underline{\xi}) \xrightarrow[k \to \infty]{} p^{(\infty)}(\alpha|\underline{\xi}),$$

where $\underline{\xi} = \underline{\xi}_\infty$, and $p^{(k)}(\alpha|\underline{\xi})$ does not depend on $\xi_{k+1}, \xi_{k+2}, \ldots$. Property (ii) then implies that, for every $\xi_\infty \in \text{spec } b$,

$$p^{(\infty)}(\alpha|\underline{\xi}) = p^{(\infty)}(\alpha|\underline{\xi}) \frac{p(\xi_\infty|\alpha)}{\sum_\beta p^{(\infty)}(\beta|\underline{\xi})p(\xi_\infty|\beta)}. \tag{7.176}$$

If for all $\alpha, \beta \in \text{spec } a$ with $\alpha \neq \beta$, there exists $\xi \in \text{spec } b$ such that $p(\xi,|\alpha) \neq p(\xi|\beta)$ then Equation (7.176) and

$$p^{(\infty)}(\alpha|\underline{\xi}) = \delta_{\alpha\alpha_0}, \tag{7.177}$$

for some $\alpha_0$ (depending on $\underline{\xi}$).

Thus, for almost every history $\underline{\xi} = \underline{\xi}_\infty$ of outcomes of "von Neumann measurements" of the probes $E_1, E_2, \ldots$, the state $\Psi \circ \tau_{k,0}$, conditioned on $\underline{\xi}_\infty$, converges on $\mathcal{A}_S$ to an *eigenstate* of the operator $a \in \mathcal{A}_S$ representing the physical quantity to be measured, as $k \to \infty$. The probability (with respect to the histories $\underline{\xi}_\infty$) of convergence to an eigenstate corresponding to the eigenvalue $\alpha$ of $a$ is given by $p_\alpha$; (see Eq. (7.166)). Stated differently, the range of values of the functions $p^{(\infty)}(\alpha|\cdot)$ on the space of histories consists of $\{0, 1\}$, and, for almost every history $\underline{\xi}_\infty$, $\sum_\alpha p^{(\infty)}(\alpha|\underline{\xi}_\infty) = 1$. These are the results that have been announced in Sect. 7.1.2; see (7.21).

It is not hard to see that the approach of the state of $S$ to an eigenstate of $a$ is exponential in the time $k$. This is a "large-deviation estimate" established in [6]. It involves use of a "dynamical relative entropy". The techniques sketched in this subsection have interesting applications to Mott's problem of *"particle tracks"* in quantum theory.

For a mathematical theory of *"preparation of states"* in quantum mechanics, see [33, 35]. Simple models of "information loss" and "decoherence" will be proposed and studied in a separate publication.

## 7.5.A Appendix to Sect. 7.5

The purpose of this appendix is to describe some mathematical structure useful to imbed the material in Sects. 7.5.3 and 7.5.4 into a more general context. In particular, we do not wish to assume that the algebras $\mathcal{E}^\omega_{\geq t}$ are type-I von Neumann algebras; (i.e., we do not start from Eqs. (7.134)–(7.136)). To begin with, we summarize some further basic facts concerning von Neumann algebras; (see also Sect. 7.2.1). Let $\mathcal{M}$ be a von Neumann algebra, and let $\omega$ be a normal state on $\mathcal{M}$. Then $(\pi_\omega, \mathcal{H}_\omega, \Omega)$ stands for the representation, $\pi_\omega$, of $\mathcal{M}$ on the Hilbert space $\mathcal{H}_\omega$, with $\Omega$ the cyclic unit vector in $\mathcal{H}_\omega$ (unique up to a phase) such that

$$\omega(a) = \langle \Omega, \pi_\omega(a)\Omega \rangle_{\mathcal{H}_\omega}. \tag{7.178}$$

This is the GNS construction applied to $(\mathcal{M}, \omega)$; see Eq. (7.29), Sect. 7.2.1. We say that $\omega$ is *separating* for $\mathcal{M}$ iff, for any $a \in \mathcal{M}$,

$$\omega(ba) = 0, \ \forall b \in \mathcal{M} \Longrightarrow a = 0; \tag{7.179}$$

or, equivalently, $\pi_\omega(a)\Omega = 0$ (in $\mathcal{H}_\omega$) implies that $a = 0$; (it is assumed that $\pi_\omega$ is *faithful*, and we will henceforth write $a$ for $\pi_\omega(a)$).

Given a separating state, $\omega$, on a von Neumann algebra $\mathcal{M}$, Tomita–Takesaki theory [12, 67] guarantees that there is a one-parameter unitary group $\{\Delta_\omega^{i\sigma}\}_{\sigma \in \mathbb{R}}$, where $\Delta_\omega > 0$ is a self-adjoint operator on $\mathcal{H}_\omega$ (the Tomita–Takesaki modular operator) and an anti-unitary involution, $J_\omega$, on $\mathcal{H}_\omega$, with the properties

$$\Delta_\omega^{i\sigma} a \Delta_\omega^{-i\sigma} \in \mathcal{M}, \quad J_\omega a J_\omega \in \mathcal{M}', \tag{7.180}$$

for all $a \in \mathcal{M}$ and for all $\sigma \in \mathbb{R}$, ($\mathcal{M}'$ is the commutant of $\mathcal{M}$),

$$\Delta_\omega^{i\sigma}\Omega = \Omega, \quad J_\omega\Omega = \Omega, \tag{7.181}$$

for all $\sigma$, and

$$\langle \Omega, ab\Omega \rangle_{\mathcal{H}_\omega} = \langle \Omega, b\Delta_\omega a\Omega \rangle_{\mathcal{H}_\omega}, \tag{7.182}$$

for arbitrary $a, b \in \mathcal{M}$; (KMS condition). If $\varphi$ is a linear functional on $\mathcal{M}$ we define

$$\|\varphi\| := \sup_{b \in \mathcal{M}} \frac{|\varphi(b)|}{\|b\|} \tag{7.183}$$

Eqs. (7.178) and (7.182) then show that if $\omega$ is separating for $\mathcal{M}$,

$$\|\{a, \omega]_{\mathcal{M}}\| < \epsilon \iff \|(\Delta_\omega a - a)\Omega\|_{\mathcal{H}_\omega} < \epsilon, \tag{7.184}$$

for any $a \in \mathcal{M}$; (recall that $\{a, \omega]_{\mathcal{M}}(b) = \omega([a, b])$, $b \in \mathcal{M}$—see Eq. (7.128), Sect. 7.5.2). In Eq. (7.129), we have defined the centralizer of $\omega$ to be the subalgebra of $\mathcal{M}$ given by

$$\mathcal{C}_{\mathcal{M}}^\omega := \{a \in \mathcal{M} \mid \{a, \omega]_{\mathcal{M}} = 0\}. \tag{7.185}$$

We recall that $\omega$ defines a *trace* on $\mathcal{C}_{\mathcal{M}}^\omega$. By (7.184),

$$C^\omega_M = \{a \in M \mid \Delta_\omega a\Omega = a\Omega\}, \qquad (7.186)$$

assuming that $\omega$ is separating for $M$. The following claim is easy to verify (using Liouville's theorem for analytic functions of one complex variable, and Eq. (7.186)): If $\omega$ is separating for $M$

$$\{a, \omega\}_M = 0 \iff \Delta_\omega^{i\sigma} a \Delta_\omega^{-i\sigma} = a, \quad \forall \sigma \in \mathbb{R}, \qquad (7.187)$$

for any $a \in M$; (see, e.g., [3]). The group, $\{\alpha_\sigma\}_{\sigma \in \mathbb{R}}$, of *automorphisms of $M$ defined by $\alpha_\sigma(a) = \Delta_\omega^{i\sigma} a \Delta_\omega^{-i\sigma}$ is called the Tomita–Takesaki modular automorphism group. The equivalence in (7.187) together with Eq. (7.185) show that if $\omega$ is separating for $M$ then the centralizer, $C^\omega_M$, is nothing but the subalgebra of $M$ of fixed points under the Tomita–Takesaki modular automorphism group. The following result is due to Takesaki, [68]: Let $N$ be a von Neumann subalgebra of $M$, and let $\omega$ be a faithful, normal, separating state on $M$. Then the following statements are equivalent:

(i)   $N$ is invariant under the modular automorphism group $\{\alpha_\sigma\}_{\sigma \in \mathbb{R}}$ associated with $(M, \omega)$.

(ii)  There exists a ($\sigma$-weakly) continuous projection, $\epsilon$, of norm 1 (a "conditional expectation") of $M$ onto $N$ such that

$$\omega(a) = \omega_{|N}(\epsilon(a)), \qquad (7.188)$$

for all $a \in M$.

**Remark 5.9** For $a, b$ in $N$ and $x \in M$, we have that

$$\left.\begin{array}{r} \epsilon(x^* x) \geq \epsilon(x)^* \epsilon(x) \geq 0, \\ \epsilon(axb) = a\epsilon(x)b. \end{array}\right\} \qquad (7.189)$$

As a corollary of Takesaki's result on conditional expectations, we have that if $\omega$ is separating for $M$ then

(a)   there is a conditional expectation, $\epsilon = \epsilon^\omega$, from $M$ onto the centralizer $C^\omega_M$ of $\omega$ satisfying (7.188); and

(b)   there is a conditional expectation, $\dot\epsilon^\omega$, from $M$ onto the center, $Z^\omega_M$, of $C^\omega_M$ satisfying (7.188).

**Definition 5.10** The variance of an operator $a \in M$ in the state $\omega$ is defined by

$$\Delta^\omega_M a := \sqrt{\omega((a - a^\omega))}, \qquad (7.190)$$

where $a^\omega := \dot\epsilon^\omega(a)$.

These general results can be applied to the considerations in Sects. 7.5.2–7.5.4, with the following identifications:

$$\mathcal{M} \to \mathcal{E}^{\omega}_{\geq t}, \quad \mathcal{C}^{\omega}_{\omega} \to \mathcal{C}^{\omega}_{\geq t}, \quad \mathcal{Z}^{\omega}_{\mathcal{M}} \to \mathcal{Z}^{\omega}_{\geq t}. \tag{7.191}$$

We then use the notations $\epsilon^{\omega} \to \epsilon^{\omega}_{\geq t}, \dot{\epsilon}^{\omega} \to \dot{\epsilon}^{\omega}_{\geq t}$ and $\Delta^{\omega}_{\mathcal{M}} a \to \Delta^{\omega}_t a$; (see Eq. (7.140), Sect. 7.5.3). Concerning the special case introduced in Eqs. (7.134)–(7.136), we remark that $\omega$ is separating for $\mathcal{E}^{\omega}_{\geq t}$ iff all eigenvalues of the density matrix $\rho^{\omega}_{\geq t}$ introduced in Eq. (7.136) are strictly positive (which is generically the case). As an exercise, the reader may enjoy deriving the explicit formulae for $\epsilon^{\omega}_{\geq t}$ and $\dot{\epsilon}^{\omega}_{\geq t}$; (see Eq. (7.139)). The material sketched here is important in relativistic quantum theory (LRQT).

**Acknowledgements** A rough first draft of this paper has been written during J.F.'s stay at the School of Mathematics of the Institute for Advanced Study (Princeton), 2012/2013. His stay has been supported by the 'Fund for Math' and the 'Monell Foundation'. He is deeply grateful to Thomas C. Spencer for his most generous hospitality. He acknowledges useful discussions with Ph. Blanchard, P. Deift, S. Kochen and S. Lomonaco. He thanks D.Bernard for drawing his attention to [6] and W. Faris for correspondence. He is grateful to D. Buchholz, D. Dürr, S. Goldstein, J. Yngvason and N. Zanghi for numerous friendly and instructive discussions, encouragement and for the privilege to occasionally disagree in mutual respect and friendship.

# References

1. Adler, S.L., Brody, D.C., Brun, T.A., Hughston, L.P.: Martingale models for quantum state reduction. J. Phys. A **34**(42), 8795 (2001)
2. Allori, V., Goldstein, S., Tumulka, R., Zanghì, N.: Predictions and primitive ontology in quantum foundations: a study of examples. Br. J. Philos. Sci. (2013)
3. Araki, H.: Multiple time analyticity of a quantum statistical state satisfying the KMS boundary condition. Publ. Res. I. Math. Sci. **4**(2), 361–371 (1968)
4. Bannier, U.: Intrinsic algebraic characterization of space-time structure. Int. J. Theor. Phys. **33**(9), 1797–1809 (1994)
5. Barchielli, A., Paganoni, A.: On the asymptotic behaviour of some stochastic differential equations for quantum states. Infinite Dimens. Anal. Quantum Probab. Relat. Top. **6**(02), 223–243 (2003)
6. Bauer, M., Bernard, D.: Convergence of repeated quantum non-demolition measurements and wave-function collapse. Phys. Rev. A **84**(4), 44103 (2011)
7. Bell, J.S.: On the Einstein-Podolsky-Rosen paradox. Physics **1**(3), 195–200 (1964)
8. Bell, J.S.: On the problem of hidden variables in quantum mechanics. Rev. Mod. Phys. **38**(3), 447–452 (1966)
9. Bell, J.S.: Speakable and Unspeakable in Quantum Mechanics: Collected Papers on Quantum Philosophy. Cambridge University Press, Cambridge (2004)
10. Blanchard, P., Olkiewicz, R.: Decoherence induced transition from quantum to classical dynamics. Rev. Math. Phys. **15**(3), 217–244 (2003)
11. Born, M.: Quantenmechanik der Stoßvorgänge. Z. Phys. **38**(11–12), 803–827 (1926)
12. Bratteli, O., Robinson, D.W.: Operator Algebras and Quantum Statistical Mechanics, vol. 1–2. Springer, New York (2003)
13. Brunetti, R., Fredenhagen, K.: When does a detector click? Phys. Rev. A **66**, 044101 (2001)
14. Buchholz, D.: Collision theory for massless bosons. Commun. Math. Phys. **52**(2), 147–173 (1977)
15. Buchholz, D., Grundling, H.: Lie algebras of derivations and resolvent algebras. Commun. Math. Phys. **320**(2), 455–467 (2012)

16. Buchholz, D., Grundling, H.: Quantum systems and resolvent algebras (2013). arXiv preprint arXiv:1306.0860
17. Buchholz, D., Roberts, J.E.: New light on infrared problems: sectors, statistics, symmetries and spectrum (2013). arXiv preprint arXiv:1304.2794
18. Colbeck, R., Renner, R.: Quantum theory cannot be extended. Bull. Am. Phys. Soc. **56**(1), 513 (2011)
19. Connes, A.: Une classification des facteurs de type III. Ann. Sci. École Norm. Sup. **6**(2), 133–252 (1973)
20. Connes, A., Narnhofer, H., Thirring, W.: Dynamical entropy of C*algebras and von Neumann algebras. Commun. Math. Phys. **112**(4), 691–719 (1987)
21. De Roeck, W., Fröhlich, J.: Diffusion of a massive quantum particle coupled to a quasi-free thermal medium. Commun. Math. Phys. **303**(3), 613–707 (2011)
22. De Roeck, W., Kupiainen, A.: Approach to ground state and time-independent photon bound for massless spin-boson models. Ann. Henri Poincaré **14**(2), 253–311 (2013)
23. Dirac, P.A.M.: The Lagrangian in quantum mechanics. Phys. Z. **3**(1), 64–72 (1933)
24. Dowker, F., Johnston, S., Sorkin, R.D.: Hilbert spaces from path integrals. J. Phys. A **43**(27), 275–302 (2010)
25. Dürr, D., Teufel, S.: Bohmian Mechanics. Springer, New York (2009)
26. Einstein, A.: Über einen die Erzeugung und Verwandlung des Lichtes betreffenden heuristischen Gesichtspunkt. Ann. Phys.-Berlin **322**(6), 132–148 (1905)
27. Einstein, A.: Zur Quantentheorie der Strahlung. Phys. Z. **18**, 121–128 (1917)
28. Everett, H.: "Relative state" formulation of quantum mechanics. Rev. Mod. Phys. **29**(3), 454 (1957)
29. Faupin, J., Fröhlich, J., Schubnel, B.: On the probabilistic nature of quantum mechanics and the notion of closed systems to appear in Commun. Math. Phys. (2014, submitted)
30. Feynman, R.P., Hibbs, A.R.: Quantum Mechanics and Path Integrals: Emended Edition. Dover, Mineola (2012)
31. Fröhlich, J.: Abschied von Determinismus und Realismus in der Physik des 20. Jahrhunderts. Akademie der Wissenschaften und der Literatur zu Mainz, Abhandlungen der Mathematisch-naturwissenschaftlichen Klasse **1**, 1–22 (2011)
32. Fröhlich, J., Schubnel, B.: Do we understand quantum mechanics—finally? In: Wolfgang Reiter et al.(eds.), *Erwin SchrödingerÛ50 years after*, Zürich : European Mathematical Society Publ., 2013, pages 37–84.
33. Fröhlich, J., Schubnel, B.: Paper in preparation
34. Fröhlich, J., Schubnel, B.: On the preparation of states in quantum mechanics. J. Math. Phys. (to appear)
35. Fröhlich, J., Griesemer, M., Schlein, B.: Asymptotic completeness for Rayleigh scattering. Ann. Henri Poincaré **3**(1), 107–170 (2002)
36. Fuchs, C.A.: Qbism, the perimeter of quantum Baycsianism (2010). arXiv preprint arXiv:1003.5209
37. Gell-Mann, M., Hartle, J.B.: Classical equations for quantum systems. Phys. Rev. D **47**(8), 3345–3382 (1993)
38. Ghirardi, G.C., Rimini, A., Weber, T.: Unified dynamics for microscopic and macroscopic systems. Phys. Rev. D **34**(2), 470 (1986)
39. Gleason, A.M.: Measures on the closed subspaces of a Hilbert space. J. Math. Mech. **6**(6), 885–893 (1957)
40. Glimm, J.: Type I C*-algebras. Ann. Math. **73**(3), 572–612 (1961)
41. Griffiths, R.B.: Consistent histories and the interpretation of quantum mechanics. J. Stat. Phys. **36**(1), 219–272 (1984)
42. Guerlin, C., Bernu, J., Deleglise, S., Sayrin, C., Gleyzes, S., Kuhr, S., Brune, M., Raimond, J.M., Haroche, S.: Progressive field-state collapse and quantum non-demolition photon counting. Nature **448**(7156), 889–893 (2007)
43. Haag, R.: Local Quantum Physics. Springer, Berlin (1996)

44. Haag, R., Kastler, D.: An algebraic approach to quantum field theory. J. Math. Phys. **5**(7), 848–861 (1964)
45. Haagerup, U.: Connes bicentralizer problem and uniqueness of the injective factor of type III1. Acta Math. **158**(1), 95–148 (1987)
46. Heisenberg, W.: Über quantentheoretische Umdeutung kinematischer und mechanischer Beziehungen. In: Original Scientific Papers, pp. 382–396. Springer, Berlin (1985)
47. Hepp, K.: Quantum theory of measurement and macroscopic observables. Helv. Phys. Acta **45**(2), 237–248 (1972)
48. Isham, C.J., Linden, N., Schreckenberg, S.: The classification of decoherence functionals: an analog of Gleason's theorem. J. Math. Phys. **35**, 6360 (1994)
49. Janssens, B., Maassen, H.: Information transfer implies state collapse. J. Phys. A **39**(31), 9845 (2006)
50. Kochen, S.: A reconstruction of Quantum Mechanics (to be published)
51. Kochen, S., Specker, E.P.: The problem of hidden variables in quantum mechanics. J. Math. Mech. **17**(1), 59–87 (1967)
52. Kolmogorov, A.N.: Entropy per unit time as a metric invariant of automorphisms. Dokl. Akad. Nauk SSSR **124**, 754–755 (1959)
53. Lindblad, G.: On the generators of quantum dynamical semigroups. Commun. Math. Phys. **48**(2), 119–130 (1976)
54. Lüders, G.: Über die Zustandsänderung durch den Meßprozeß. Ann. Phys.-Leipzig **443**(5–8), 322–328 (1950)
55. Maassen, H.: Quantum probability and quantum information theory. In: Quantum Information, Computation and Cryptography, pp. 65–108. Springer, New York (2010)
56. Maassen, H., Kümmerer, B.: Purification of Quantum Trajectories. Lecture Notes-Monograph Series, pp. 252–261 (2006)
57. Mott, N.F.: The wave mechanics of alpha-ray tracks. Proc. R. Soc. Lond. Ser. A **126**(800), 79–84 (1929)
58. Neveu, J.: Martingales à Temps Discret. Masson, Paris (1972)
59. Omnès, R.: The Interpretation of Quantum Mechanics. Princeton University Press, Princeton (1994)
60. Penrose, R.: Wavefunction collapse as a real gravitational effect. In: Mathematical Physics, 2000, pp. 266–282. Imperial College Press, London (2000)
61. Peres, A.: Quantum Theory: Concepts and Methods. Springer, New York (1995)
62. Roepstorff, G.: Quantum dynamical entropy. In: Chaos-the Interplay Between Stochastic and Deterministic Behaviour, pp. 305–312. Springer, New York (1995)
63. Rose, B.: Ad Reinhardt: Art as Art, The Selected Writings of Ad Reinhardt. University of California Press, Los Angeles (1991)
64. Schwinger, J.: The algebra of microscopic measurement. Proc. Natl. Acad. Sci. USA **45**(10), 1542–1553 (1959)
65. Sinai, Ya.G.: On the concept of entropy of a dynamical system. Dokl. Akad. Nauk SSSR **124**, 768–771 (1959)
66. Styer, D.F., Balkin, M.S., Becker, K.M., Burns, M.R., Dudley, C.E., Forth, S.T., Gaumer, J.S., Kramer, M.A., Oertel, D.C., Park, L.H., et al.: Nine formulations of quantum mechanics. Am. J. Phys. **70**(3), 288–297 (2002)
67. Takesaki, M.: Tomita's Theory of Modular Hilbert Algebras and Its Applications. Springer, Berlin/Heidelberg/New York (1970)
68. Takesaki, M.: Conditional expectations in von Neumann algebras. J. Funct. Anal. **9**(3), 306–321 (1972)
69. Takesaki, M.: Theory of Operator Algebras, vol. 1. Springer, Berlin (2002)
70. Takesaki, M.: Theory of Operator Algebras, vol. 2. Springer, Berlin (2003)
71. Tsirelson, B.S.: Some results and problems on quantum Bell-type inequalities. Hadronic J. Suppl. **8**(4), 329–345 (1993)
72. von Baeyer, H.C.: Quantum weirdness? It's all in your mind. Sci. Am. **308**(6), 46–51 (2013)

73. Werner, R.: Arrival time observables in quantum mechanics. Ann. I. H. Poincaré-Phy. **47**(4), 429–449 (1987)
74. Werner, R., et al.: http://tjoresearchnotes.wordpress.com/2013/05/13/guest-post-on-bohmian-mechanics-by-reinhard-f-werner/#comment-3374 (2012)
75. Whecler, J.A., Zurek, W.H.: Quantum Theory and Measurement. Princeton University Press, Princeton (1983)
76. Wigner, E.P.: The Collected Works of Eugene Paul Wigner. Springer, New York (1993)

# Chapter 8
# Can Relativity be Considered Complete? From Newtonian Nonlocality to Quantum Nonlocality and Beyond

Nicolas Gisin

## 8.1 Introduction

Hundred years after Einstein miraculous year and 70 years after the EPR paper [1], I like to think that Einstein would have appreciated the somewhat provocative title of this contribution. However, Einstein would probably not have liked its conclusion. But who can doubt that relativity is incomplete? and likewise that quantum mechanics is incomplete! Indeed, these are two scientific theories and Science is nowhere near its end (as a matter of fact, I do believe that there is no end (in contrast to [2])). Well, actually, I am, of course, not writing for Einstein, but for those readers interested in a (necessarily somewhat subjective) account of the *peaceful co-existence* [3] between relativity and quantum physics in the light of the conceptual and experimental progresses that happened during the last 10 years, set in the broad perspective of physics and nonlocality since Newton (for a lively account of the history of quantum nonlocality and of the people who made it happen, see: [4]).

## 8.2 Non-locality According to Newton

Isaac Newton, the great Newton of Universal Gravitation, was not entirely happy with his theory. Indeed, he was well aware of an awkward consequence of his theory: if a stone is moved on the moon,[1] then our weight, of all of us, here on

---

[1]Using a small rocket, so as to displace the center of mass of the moon.

N. Gisin (✉)
Group of Applied Physics, University of Geneva, 1211 Geneva 4, Switzerland
e-mail: Nicolas.Gisin@unige.ch

© Springer-Verlag Berlin Heidelberg 2015
P. Blanchard, J. Fröhlich (eds.), *The Message of Quantum Science*, Lecture Notes in Physics 899, DOI 10.1007/978-3-662-46422-9_8

earth, is *immediately* modified. What troubled so much Newton was this *immediate* effect, i.e. the nonlocal prediction of his theory. Let's read how Newton described it himself [5]:

> That Gravity should be innate, inherent and essential to Matter, so that one Body may act upon another at a Distance thro a Vacuum, without the mediation of any thing else, by and through which their Action and Force may be conveyed from one to another, is to me so great an Absurdity, that I believe no Man who has in philosophical Matters a competent Faculty of thinking, can ever fall into it. Gravity must be caused by an Agent acting constantly according to certain Laws, but whether this Agent be material or immaterial, I have left to the Consideration of my Readers.

It would have been hard for Newton to be more explicit in his rejection of non-locality! Note that this indicates that the no-signalling principle (see Sect. 8.10.1) is part of Newton's world view, not of relativity. However, most physicists didn't pay much attention to this aspect of Newtonian physics. By lack of alternative, physics remained nonlocal until about 1915 when Einstein introduced the world to General Relativity. But let's start 10 years earlier, in 1905.

## 8.3   Einstein, the Greatest Mechanical Engineer

In 1905 Einstein introduced three radically new theories or models in physics. Special relativity of course, but more relevant to this section are his descriptions of Brownian motion and of the photo-electric effect. Indeed, both descriptions show Einstein's deep intuition about mechanics. Brownian motion is explained as a complex series of billiard-ball-like-collisions between a visible molecule—the particle undergoing Brownian motion—and invisible smaller particles. The random collisions of the latter explaining the erratic motion of the former. Likewise, the photo-electric effect is given a mechanistic explanation. Light beams contain little billiard-balls whose energy depends on the color, i.e. wavelength, of the light. These light-billiard-balls (today called photons and recognized as not at all billiard-ball-like) hit the electrons on metallic surfaces and mechanically kick them out of the metal, provided they have enough energy to do so.

General relativity can also be seen as a mechanical description of gravitation. When a stone is moved on the moon, a bunch of gravitons (in modern terminology) fly off in all directions at a finite speed, the speed of light. Hence, about a second later, the earth is *informed* and only then is our weight affected. This is, I believe, the greatest achievement of Einstein, the greatest mechanical engineer[2] of all times: **Einstein turned physics into a local theory!**

---

[2]My friends know well that in my mouth "engineer" has no negative connotation, quite the opposite. For me, a physicist must be a good theorist and a good engineer! Well, I warned you, dear reader, this is a somewhat subjective article.

## 8.4   Quantum Mechanics Is Not Mechanical

Only about 10 years after general relativity came quantum mechanics. This was quite an extraordinary revolution. Until then, greatly thanks to Newton and Einstein's genius, Nature was seen as made out of many little *billiard-balls* that mechanically bang into each other. Yet, quantum mechanics is characterized by the very fact that it no longer gives a mechanical description of Nature. The terminology quantum *mechanics* is just a historical mistake, it should be called *Quantum Physics* as it is a radically new sort of physical description of Nature.

But this new description let nonlocality back into Physics! And this was unacceptable for Einstein.

It is remarkable and little noticed that since Newton, physics gave a local description of Nature only during some 10 years, between about 1915 and 1925. All the rest of the time, it was nonlocal, though, with quantum physics, in quite a different sense as with Newton gravitation. Indeed, the latter implies the possibility of arbitrarily fast signaling, while the former prohibits it.

## 8.5   Non-locality According to Einstein

In 1935 two celebrated papers appeared in respectable journals, both with famous authors, both stressing the—unacceptable in their authors view—nonlocal prediction of quantum physics [1, 6]. A lot has been written on the EPR "paradox" and I won't add to this. I believe that Einstein's reaction is easy to understand. Here is the man who turned physics local, centuries after Newton wrote his alarming text, he is proud of his achievement and certainly deserves to be. Now, only a few years latter, nonlocality reappears! Today one should add that quantum nonlocality is quite a different concept from Newtonian nonlocality, but Einstein did not fully realize this.

What Einstein and his colleagues saw is that quantum physics describes spatially separated particles as one global system in which the two particles are not logically separated. What they did not fully realize is that this does not allow for signaling, in particular no faster than light communication, hence it is not in direct conflict with relativity. In the next section I'll try to present this using modern terminology.

Most physicists didn't pay much attention to this aspect of quantum physics. A kind of consensus established that this was to be left for future examination, once the technology would be more advanced. The general feeling was that quantum nonlocality was nothing but a laboratory curiosity, not serious physics.

Young physicists may have a hard time to believe that such an important concept, like quantum nonlocality, was, during many decades, not considered as serious. But this was indeed the real state of affairs: ask any older professors, a vast majority of them still believes that it is unimportant. Let me add two little stories that illustrate what the situation was like. John Bell, the famous John Bell of the Bell inequalities and of the Bell states, never had any quantum physics student. When a young physicist would approach him and talk about nonlocality, John's first question was: "Do you have a permanent position?". Indeed, without such a permanent position it was unwise to dare talking about nonlocality! Notice that John Bell almost never published any of his remarkable and nowadays famous papers [7] in serious journals: the battle with referees were too ... time wasting (not to use a more direct terminology). Further, if you went to CERN where John Bell held a permanent position in the theory department and asked at random about John's contributions to physics, his work on the foundation of quantum physics would barely be mentioned (true enough, he had so many other great contributions!).[3]

Anyway, so quantum nonlocality remained for decades in the *curiosity lab* and no one paid much attention. But in the 1990s two things changed. First, a conceptual breakthrough happened thanks to Artur Ekert and to his Ph.D. adviser David Deutsch [9]. They showed that quantum nonlocality could be exploited to establish a cryptographic key between two distant partners and that the confidentiality of the key could be tested by means of Bell's inequality. What a revolution! This is the first time that someone suggested that quantum nonlocality is not only real, but that it could even be of some use. A second contribution came from the progress in technology. Optical fibers had been developed and installed all over the world. And Mandel's group at the University of Rochester (where I held a 1-year post-doc position and first met with optics) applied parametric down-conversion to produce entangled photon pairs [10]. This was enough (up to the detectors) to demonstrate quantum nonlocality outside the curiosity laboratory. In 1997 my group at Geneva University demonstrated the violation of Bell inequalities between two villages, Bernex and Bellevue, around Geneva, see Fig. 8.1, separated by a little more than 10 km and linked by a 15 km long standard telecom fiber [11, 12] (since then, we have achieved 50 km [13]). So quantum nonlocality became politically acceptable! But what is it? (for an elementary introduction see [N. Gisin *Quantum chance, Nonlocality, Teleportation and Other Quantum Marvels*, Springer 2014].) Let me introduce the concept using students undergoing "quantum exams".

## 8.6   Quantum Exams: Entanglement

Assume that two students, Alice and Bob, have to pass some exams. As always for exams, the situation is arranged in such a way that the students can't communicate during the exam. Clearly however, they are allowed, and even encouraged, to

---

[3]Another story happened to me while I was a young post-doc eager to publish some work. In a paper [8] I wrote "A quantum particle may disappear from a location A and simultaneously reappear in B, without any flow in-between". The referee accepted the paper under the condition that this outrageous sentence is removed. This referee considered his paternalist attitude so constructive that he declared himself to me: "look how helpful I am to you" (admittedly, he was politically correct).

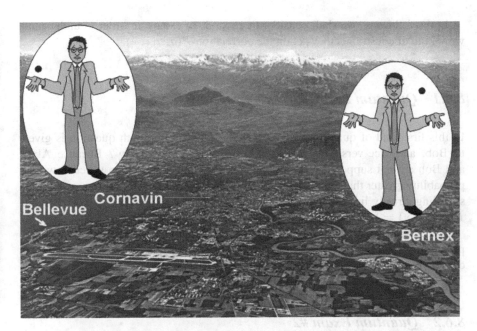

**Fig. 8.1** Bernex and Bellevue are the two villages north and south of Geneva between which our long-distance test of Bell inequality outside the lab was performed in 1997, Sect. 8.5. The *inset* represent two player that toss coins, as explained in Sect. 8.7. In the real experiment the coins were replaced by photons, the players by interferometers, their right and left hands by phase modulators and head/tail by two detectors. The experimental results are similar to that of the game, with weaker but still nonlocal correlations

communicate beforehand. Alice and Bob know in advance the list of possible questions, they also know that this is a kind of exam allowing only for a very limited number of possible answers, often only a binary choice between *yes* and *no*. During the exam Alice receives one question out of the list, let's denote it by $x$; Bob receives question $y$. Finally, denote $a$ and $b$ Alice and Bob's answers, respectively. Hence, an exam is a realization of a random process described by a conditional probability function, often merely called a *correlation*:

$$P(a, b | x, y) \qquad (8.1)$$

Clearly, the choice of questions $x$ and $y$ are under the professor's control. However, as all professors know, the students' answers $a$ and $b$ are not! This is similar to experiments: the choice as to which experiment to perform is under the physicists control, but not the answer given by Nature.

In the following, we shall consider three kinds of exams, in order to understand what kind of constraints they set on the correlation $P(a, b|x, y)$.

## 8.6.1   Quantum Exam #1

In this first kind of quantum exam Alice is asked to tell which question is given to Bob, and vice-versa. This is clearly an unfair exam! Why? Because Alice and Bob are not supposed to communicate. How could they then succeed with a probability greater than mere chance?[4] This simple example shows that prohibiting signaling already limits the set of possible correlation $P(a, b|x, y)$. For example $P(a, b|x, y) = \delta(a = y)\delta(b = x)$ is excluded.

Notice that a correlation $P(a, b|x, y)$ is non-signaling if and only if its marginal probabilities are independent of the other side input: $\sum_b P(a, b|x, y)$ is independent of $y$ and $\sum_a P(a, b|x, y)$ is independent of $x$, see Sect. 8.10.1.

## 8.6.2   Quantum Exam #2

The second kind of quantum exam is closer to standard exams. Alice and Bob are simply requested to provide the same answer whenever they receive the same question. This is clearly feasible: we all expect that good students give the same answer to the same question. It suffices that they prepare for the exam well enough. Note that the quantum exam #2 under consideration here is even easier than standard exams, as there is no notion of correct or incorrect answers. All that is required is that Alice and Bob give consistent answers: it suffices that they jointly decide in advance which answer to give for each of the possible questions.

Now, a central problem: Could Alice and Bob succeed with certainty for such an exam #2 by other means, that is without jointly deciding the answers in advance? Think about it. If you found an alternative trick, then, if you are a student, you should use your trick to pass the next examination: just apply your trick together with the best student, you'll get the same mark as him/her.[5] And if you are a professor and found a trick, then you should stop testing your student with standard exams! Well, of course, there is no other trick, at least none applicable to classical students.

---

[4] Somewhat surprisingly, if there are only two possible questions, then there is a strategy such that the probability that both players succeed is 50 %.

[5] Admittedly, the danger is that both student get the bad mark! But, on average, the poor student improves.

Correlations that satisfy $P(a = b|x = y) = 1$ are necessarily of the form

$$P(a,b|x,y) = \sum_{\lambda} \rho(\lambda)Q(a|x,\lambda)Q(b|y,\lambda) \qquad (8.2)$$

for some probability function $Q$ and some distribution $\rho$ of common strategy $\lambda$. Historically, the $\lambda$ were called "local hidden variables", computer scientist call them "shared randomness"; here the $\lambda$ denote common strategies.

$P(a = b|x = y) = 1$ is but one example of a local correlation, among infinitely many others. Relation (8.2) characterizes all local correlations.

In summary, some exams require common strategies; in other words, some observed correlations can't be explained except by common causes.

### 8.6.3 Quantum Exam #3

The third kind of quantum exam is the most tricky and interesting. For (apparent) simplicity let's restrict the set of questions and answers to binary sets and let us label them by bits, "0" and "1". In this exam Alice and Bob are required to always output the same answer, except when they both receive the question labelled "1" in which case they should output different answers. Note that formally this exam requires that Alice and Bob's data satisfy the following equality, modulo 2: $a + b = x \cdot y$. This time it is not immediately obvious whether they can prepare a strategy that guaranties success.

Assume first that the strategy forces Alice to output an answer that depends only on her input $x$, i.e. Alice's strategy is deterministic. But in such a case, whenever Bob receives the question 1, he can't decide on his output since it should depend on Alice's question. Next, if Alice's output is random, this is clearly of no help to Bob. Consequently there is no way for Alice and Bob to succeed with certainty.

Let us emphasize that successfully completing this exam does not necessarily imply communication between Alice and Bob. Indeed, assume that, somehow, Alice and Bob's data would always satisfy $a + b = x \cdot y$. Would this allow Alice to communicate to Bob, or vice-versa? Well, it depends! If Alice's output $a$ is known to Bob, for instance they decide on $a = 0$ always, then whenever Bob receives $y = 1$, he can deduce Alice's question from $a + b = x \cdot y$ and from his output: $x = b$ in the example. But if Alice's outcome is unknown to Bob, for instance if Alice outcome is merely a random bit, then the relation $a + b = x \cdot y$ is of no help to Bob. We shall come back to this concept of a non-signaling correlation satisfying $a + b = x \cdot y$ in Sect. 8.10.3.

Let us define the mark $M$ of this quantum exam #3 as the sum of the success probabilities [14]:

$$M = P(a + b = xy|x = 0, y = 0)$$
$$+P(a + b = xy|x = 0, y = 1)$$
$$+P(a + b = xy|x = 1, y = 0)$$
$$+P(a + b = xy|x = 1, y = 1) \qquad (8.3)$$

It is not difficult to realize that the optimal strategy for Alice and Bob consists in deciding in advance on a common answer, independent of the questions they receive. With such a strategy they are able to achieve the mark $M = 3$. This is indeed the optimal mark achievable by common strategies:

$$M \leq 3 \qquad (8.4)$$

This is an example of a Bell inequality: a constraint on correlations arising from common strategies. Interesting Bell inequalities are those that can be violated by quantum physics. In the case of (8.4), if Alice and Bob share entangled singlet quantum states, then they can obtain the mark $M_{QP} = 2 + \sqrt{2} \approx 3.41$. Tsirelson proved that this is the highest mark achievable using quantum correlations [15].

Accordingly, quantum theory predicts that some tasks can be achieved that are impossible with any local mechanical model, i.e. some exams are passed with higher marks than classically possible. The fact that such tasks were invented for the purpose of showing the superiority of quantum physics doesn't affect the conclusion. Still, it is only once some useful and natural tasks were found, concretizing the superior power of quantum physics over all possible local strategies, that quantum nonlocality became accepted by the physics community.[6]

## 8.7   Coin Tossing at a Distance

Another way to present nonlocality to non-physicist friends is the following. Imagine two hypothetical players that toss coins. The players are separated in space and toss their coin once per minute. They use their free-will to decide, for each toss, whether to use their right hand or their left hand, independently of each other. And

---

[6]I wish someone establishes the statistics of the occurrences of the words "Bell inequality" and "nonlocality" in Physical Review Letters. I bet that a phase transition happen in the early 1990s, after Ekert's paper on quantum cryptography, see [16]. In 1998 I started a PRL with the sentence [11]: "Quantum theory is nonlocal." and got considerable reactions to what was felt as a provocative statement; today the same statement can be found in many papers, not provoking any reaction.

they mark all results (time, hand and head/tail) in a big black laboratory notebook, see Fig. 8.1.

After thousands of tosses, they get bored. Especially, given that nothing interesting happens: for each of the two players, heads and tails occur with a frequency of 50 %, independently of which hand they use. Hence, the players decide to go for a beer. There, in the bar, they compare their notes and get very excited. Indeed, quickly they notice that whenever at least one of them happened to have chosen his right hand for tossing his coin, both players always obtained the same result: either both head or both tails. But whenever, by mere chance, they both chose the left hand, then they always obtained opposite results: head/tail or tail/head. A very remarkable correlation!

The observation of correlations and the development of theoretical models explaining them is the essence of the scientific method. This is true not only in physics, but also in all other sciences, like geology and medicine for instance. John Bell used to say "correlations cry out for explanations!" [17].

So, why are our two players that excited by the correlation they observe? Note that locally, nothing interesting happens; in particular there is no way for one player to infer from his data which hand the other player chose. Even if one player decides to always use the same hand, this has no effect on the statistics observed by his colleague. Consequently, this game and the observed correlation do not imply any signaling. So, why do we feel that this is impossible? Actually, frankly, I do not know!

Classical correlations are always explained by either of two kinds of causes. The first kind is "signaling", one player somehow informs or influences the other player. This is clearly not the case here, since we assumed the players were widely separated in space (for the physicists we may add "space-like separated"). The second kind of causes for classical correlations is a common cause. For example all hockey players simultaneously stop running, because the umpire whistled. This kind of cause is precisely equivalent to the assumption of a common strategy, as formalized by (8.2) and excluded for the present correlation by Bell's inequality (8.4). Consequently, the correlation observed by our two players is of a different nature. The big surprise is that some sort of cause beyond the two classical causes for correlation exists! This is what Einstein and many others had a hard time to believe. But, today, if one accepts this as a matter of theoretical prediction and experimental confirmation, then the next big question is "why can't the correlation observed by our hypothetical players not be observed in the real world?". Indeed, quantum physics (and tensor products of Hilbert spaces) tell us that Bell's inequality (8.4) can be violated, i.e. not all quantum correlations can be explained by one of the two kinds of classical causes for correlations, but quantum physics does not allow correlations as strong as observed by our hypothetical players. Still, this game is illustrative of quantum nonlocality, as we shall elaborate in Sect. 8.10

## 8.8 Experiments: God Does Play Dice, He Even Plays with Nonlocal Dice

Physics is an experimental science and experiments have again and again supported the nonlocal predictions of quantum theory. All kind of experiments have been performed, in laboratories [18] and outside [11, 12, 19], with photons and with massive particles [20–22], with independent observers to close the locality loophole [11, 12, 19, 23, 24], with quasi-perfect detectors [20, 25, 26] to close the detection loophole, with high precision timing to bound the speed of hypothetical hidden communication [27–31], with moving observers to test alternative models [32] (multi-simultaneity [33] and Bohm's pilot wave [34]).[7] All these results proclaim loudly: **God plays dice**. Note how ironic the situation is: the conclusion "God plays dice" is imposed on us by the experimental evidence supporting quantum nonlocality and by Einstein's postulate that no information can travel faster than light. Indeed, as mentioned in Sect. 8.6.3, a violation of (8.4) with deterministic outputs leads to signaling. Consequently, the experimental violation of (8.4) and the no-signaling principle imply randomness [35–37].

Actually, the situation is even more interesting: Not only does God play dice, but he plays with nonlocal dice! **The same randomness manifests itself at several locations**, approximating $a + b \approx x \cdot y$ better than possible with any local classical physics model.

A very small minority of physicists still refuse to accept quantum nonlocality. They ask (sometimes with anger) *How can these two space-time locations, out there, know about what happens in each other without any sort of communication?* I believe that this is an excellent question! I have slept with it for years [38] (at last, we found the answer, see [31]). I summarize my conclusion in the next section.

## 8.9 Entanglement as a Cause of Correlation

Quantum physics predicts the existence of a totally new kind of correlation that will never have any kind of mechanical explanation. And experiments confirm this: Nature is able to produce the same randomness at several locations, possibly space-like separated. The standard explanation is "entanglement", but this is just a word, with a precise technical definition [39, 40]. Still words are useful to name objects and concepts. However, it remains to understand the concept. Entanglement is a new explanation for correlations. Quantum correlations simply happen, as other things

---

[7]The conclusion that follows from all these experiments is so important for the physicist's world-view, that an experiment closing simultaneously both the locality and the detection loophole is greatly needed.

happen (fire burns, hitting a wall hurts, etc). Entanglement appears at the same conceptual level as local causes and effects. It is a primitive concept, not reducible to local causes and effects. Entanglement describes correlations without correlata [41] in a holistic view [42]. In other worlds, **a quantum correlation is not a correlation between two events, but a single event that manifests itself at two locations**.

Are you satisfied with my explanation of what entanglement is? Well, I am not entirely! But what is clear is that entanglement exists. Moreover, entanglement is incredible robust! The last point might come as a surprise, since it is still often claimed that entanglement is as elusive as a dream: as soon as you try to talk about it, it evaporates! Historically this was part of the suspicion that entanglement was not really real, nothing more than some exotic particles that live for merely a tiny fraction of a second. But today we see a growing number of remarkable experiments mastering entanglement. Entanglement over long distances [11–13, 19, 43], entanglement between many photons [44] and many ions [45], entanglement of an ion and a photon [46, 47], entanglement of mesoscopic systems (more precisely entanglement between a few collective modes carried by many particles) [48–52], entanglement swapping [53–55], the transfer of entanglement between different carriers [56], etc.

In summary: entanglement exists and is going to affect future technology. It is a radically new concept, requiring new words and a new conceptual category. And the time since this was first written amply confirmed the robustness and power of entanglement.

## 8.10   From Quantum Nonlocality to Mere Nonlocality

So far we have seen that quantum physics produces nonlocal correlations. And so what? Ok, this can be used for Quantum Key Distribution and other Quantum Information processes, but that doesn't help much to understand non-locality. Conceptually, one would like to study non-locality without all the quantum physics infrastructure: Hilbert spaces, observables and tensor products. Not too surprisingly, once the existence of non-locality was accepted, the conceptual tools to study it came very naturally. Actually, the tools were already there, in the mathematics [57] and even the physics [35, 36] literature, waiting for a community to wake up! The basic tool is simple, doesn't require any knowledge of quantum physics and allows one, so to say, to study quantum nonlocality "from the outside", i.e. from outside the quantum physics infrastructure.

Let us go back to the quantum exam #3 (Sect. 8.6.3). Assume that Alice and Bob are not restricted by quantum physics, but only restricted by no-signaling. Consequently, they would fail the quantum exam #1. But under this mild no-signaling condition they could perfectly succeed in the quantum exam #3: Alice and Bob would each output a bit which locally looks perfectly random and independent from their inputs—hence there would be no signaling—yet their 2 bits would satisfy $a + b = x \cdot y$, exactly as in the coin tossing game of Sect. 8.7. The hypothetical "machine" that produces precisely this correlation is a basic example of the kind of conceptual tools we need to study nonlocality without quantum physics. Formally, the correlation function is defined by:

$$P(a, b | x, y) = \frac{1}{2} \delta(a + b = x \cdot y) \qquad (8.5)$$

where the $\delta(z_1 = z_2)$ function takes value 1 for $z_1 = z_2$ (modulo 2) and value 0 otherwise.

The correlation (8.5) is often referred to as a PR-box, to recall the seminal work by Popescu and Rohrlich [35, 36], or as a NL-machine (a Non-Local machine[8]). The idea of these terminologies is to emphasize the similarities between quantum measurements on two maximally entangled qubits and the correlation (8.5): in both cases the outcome is available as soon as the corresponding input is given (Alice knows $a$ as soon as she inputs $x$ into her part of the machine and similarly Bob knows $b$ as soon as he inputs $y$, there is no need to wait for the other's input) and in both the quantum and the PR-box cases the "machine" can't be used more than once (once Alice has input $x$, she can't change her mind and give another input). Notice a third nice analogy, neither the quantum nor the NL machines allow for signaling. Indeed, in all cases the marginals are pure noise, independently of any input.

Note that quantum physics is unable to produce the PR correlation (8.5). Indeed, this correlation violates the Bell inequality (8.4) up to its algebraic maximum, $M = 4$, while Tsirelson's theorem [15] states that quantum correlations are restricted to $M \leq 2 + \sqrt{2}$. However, the correlation (8.5) is much simpler than quantum correlations, while sharing many of their essential features. In particular (8.5) is nonlocal but non-signaling.

In order to get some deeper understanding of the power of this hypothetical machine (8.5) as a conceptual tool, let us consider three properties of quantum correlations (many further nice aspects can be found in [59–61]). First we shall consider the so-called quantum no-cloning theorem and see that it is actually not a quantum theorem, but a no-signaling theorem. The next natural step is to analyze quantum cryptography, whose security is often said to be based on the no-cloning theorem, and as we would expect by now, we shall find "non-signaling cryptography". Finally, we consider the question of the communication cost to simulate maximal quantum correlation. But before all this we need to recall some facts about non-signaling correlations.

---

[8]A "machine" is a physicists's terminology for an input-output black-box that is not necessarily mechanical. I believe that this terminology appeared in the quantum physics context with the "optimal cloning machines" introduced by Bužek and Hillery [58].

### 8.10.1   The Set of Non-signaling Correlations

Let us consider the set of all possible bi-partite correlations $P(a, b|x, y)$, where the inputs are taken from finite alphabets $\{x\}$ and $\{y\}$ and similarly for the outputs $\{a\}$ and $\{b\}$, and which are non-signaling[9]:

$$\sum_b P(a, b|x, y) = P(a|x) \text{ is independent of } y \qquad (8.6)$$

$$\sum_a P(a, b|x, y) = P(b|y) \text{ is independent of } x \qquad (8.7)$$

A priori this set looks huge. But it has a nice structure. First, it is a convex set: convex combinations of non-signaling correlations are still non-signaling. Second, there are only a finite number of extremal points (mathematician call such sets *polytopes* and the extremal point *vertices*); accordingly every non-signaling correlation can be decomposed into a (not necessarily unique) convex combination of extremal points. This is analog to quantum mixed states that can be decomposed into convex mixtures of pure states.

Among the non-signaling correlations are the local ones, i.e. those of the form (8.2), analog to separable quantum states. The set of local correlations also forms a polytope, a sub-polytope of the non-signaling one. Moreover all vertices of the local polytope are also vertices of the non-signaling polytope, see Fig. 2 [60]. The facets of the local polytope are in one-to-one correspondence with all tight Bell inequality.

Let us illustrate this for the simple binary case (which is in any case the only one we need in this article), i.e. $a, b, x, y$ are 4 bits. In this case, it is known that there are only eight non-trivial Bell inequalities (i.e. not counting the trivial inequalities of the form $P(a, b|x, y) \geq 0$), i.e. only eight relevant facets of the local polytope. Interestingly, Barret and co-workers [60] demonstrated that the non-signaling polytope has only eight vertices more than the local polytope, exactly one per Bell inequality! Each of these eight vertices is equivalent to the PR correlation (8.5), up to an elementary symmetry (flip an input and/or an output). Although these polytopes live in an eight-dimensional space,[10] their essential properties can be recalled from the simple geometry of Fig. 8.2.

---

[9]Actually, there are at least three different concepts behind this word [62]. (1) There is the mathematical definition given here. (2) No faster than light communication—though light plays no special role in quantum physics. And (3), there is no-signalling as Newton thought of it: no communication without a physical carrier of the information.

[10]More precisely, 8 is the dimension of the space of non-signaling correlations [63].

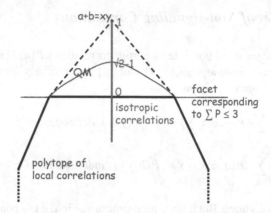

**Fig. 8.2** Geometrical view of the set of correlations. The *bottom part* represents the convex set (polytope) of local correlations, with the upper facet corresponding to the Bell inequality (8.4). The *upper triangle* corresponds to the non-local non-signaling correlations that violate the Bell inequality. The smooth thin curve limits the correlations achievable by quantum physics. The *top of the triangle* corresponds to the unique non-signaling vertex above this Bell inequality, i.e. to the non-local PR machine (8.5). The *thin vertical line* represents the isotropic correlations (8.8) with the indication of some of the values of $p_{NL}$

## 8.10.2   No-Cloning Theorem

Details can be found in [59], as here we would merely like to present the intuition. Let us assume that Alice (input and output bits $x$ and $a$, respectively) shares the correlation (8.5) both with Bob (bits $y$ and $b$) and with Charly (bits $z$ and $c$): $a+b = xy$ and $a + c = xz$. Note that this situation is different from the case where Alice would share one "machine" with Bob and share another independent "machine" with Charly: in the situation under investigation Alice holds a single input bit $x$ and a single output bit $a$. We shall see that the assumption that Alice's input and output bits $x$ and $a$ are correlated both to Bob and to Charly leads to signaling. Hence in a Universe without signaling, Alice can't share the correlation (8.5) with more than one partner: the correlation can't be cloned.

In order to understand this, assume that Bob and Charly come together, input $y = 1$ and $z = 0$, and add their output bits $b + c$. According to the assumed correlations and using the modulo 2 arithmetic $a + a = 0$, one gets: $b + c = a + b + a + c = xy + xz = x$. Hence, they could determine from their data that Alice's input bit is $x$, i.e. Alice could signal to them!

A natural question is how noisy should the correlation (8.5) be to allow cloning? The answer is interesting: as long as the Alice–Bob correlation violates the Bell inequality (8.4), the Alice-Charly correlation can't violate it; if not there is signaling.

We have just seen that the CHSH-Bell inequality (8.4) is monogamous, like well kept secrets. Let's now see that this is not a coincidence!

### 8.10.3   Non-signaling Cryptography

In 1991 Artur Ekert's discovery of quantum cryptography [9] based on the violation of Bell's inequality changed the (physicist's) world: entanglement and quantum nonlocality became respectable. Now, as we shall see in this subsection, the essence of the security of quantum cryptography does not come from the Hilbert space structure of quantum physics (i.e. not from entanglement), but is due to no-signaling nonlocal correlation! The fact that quantum physics offers a way to realize such correlation makes the idea practical. However, if one would find any other way to establish such no-signaling nonlocal correlations (a way totally unknown today), then this would equally well serve as a mean to establish cryptographic keys [64–66].

Let us emphasize that the goal is to assume no restriction on the adversary's power, i.e. on Eve, except no-signaling[11] (for an independent but related work see: [67]). Obviously, if one assumes additional restrictions on Eve, like restricting her to quantum physics, then Alice and Bob can distill more secret bits from their data [68]. But qualitatively, the situation would remain unchanged.

Assume that two partners, Alice and Bob, hold devices that allow them to each input a bit (make a binary choice of what to do, e.g. which experiment to perform) and each receives an output bit (e.g. a measurement result). This can be cast into the form of an arbitrary correlation: $P(a,b|x,y)$, with $a,b,x,y$ four bits. Assume furthermore that the devices held by Alice and Bob do not allow signaling. This simple and very natural assumption suffices to give a nice structure to the set of correlations $P(a,b|x,y)$: as we recall in Sect. 8.10.1 this set is convex and has a finite number of extreme points, called vertices. The nice property is that any correlation $P(a,b|x,y)$ can be decomposed into a convex combination of vertices, hence once one knows the vertices one knows essentially everything. If the correlation is local, i.e. of the form (8.2), then it is not useful for cryptography; indeed the adversary Eve may know the strategy $\lambda$. Hence, let's assume that $P(a,b|x,y)$ violates the Bell inequality (8.4). Consequently $P(a,b|x,y)$ lies in a well defined corner of the general polytope, a sub-polytope. Barrett and co-workers found that this sub-polytope has only nine vertices [60], eight local ones for which $M = 3$ and only one nonlocal vertex, that corresponding to our conceptual tool, i.e. to $a + b = xy$, for which $M = 4$, see Fig. 8.2.

In the case that Alice and Bob are maximally correlated (maximally but non-signaling!), i.e. their correlation correspond to the nonlocal vertex of Fig. 8.2, it is intuitively clear that the adversary Eve can't be correlated neither to Alice, nor to Bob, by the no-cloning argument sketched in the previous subsection. Hence, in such a case Alice and Bob receive from their apparatuses perfectly secret bits.

---

[11]No-signaling should be understood here as in the previous sub-section on the no-cloning theorem. That is, even if two parties joint, for example Eve and Bob come together, then they should not be able to infer any information about the third party's input, e.g. Eve and Bob should not have access no Alice's input.

However, these bits are not always correlated: when $x = y = 1$ they are anti-correlated. But this can be easily fixed by the following protocol. After Alice and Bob received their output bits, Alice announces publicly her input bit $x$ and Bob changes his output bit to $b' = b + xy$. Now Alice and Bob are perfectly correlated and Eve still knows nothing about $a$ and $b'$.

Consider now that Alice and Bob are non maximally correlated:

$$P(a,b|x,y) = \frac{1 + p_{NL}}{2}\frac{1}{2}\delta(a + b = x \cdot y) + \frac{1 - p_{NL}}{2}\frac{1}{4} \qquad (8.8)$$

For $p_{NL} > 0$ this correlation violates the inequality (8.4), for $p_{NL} \leq \sqrt{2} - 1$, it can be realized by quantum physics. Can Alice and Bob exploit such a correlation for cryptographic usage secure against an arbitrary adversary who is not restricted by quantum physics, but only restricted by the no-signaling physics? The full answer to this fascinating question is still unknown. However, there is an optimistic answer if one assumes that Eve attacks each realization independently of the others, the so-called individual attacks. In such a case, one may assume that Eve does actually distribute the apparatuses to Alice and Bob. Some apparatuses are ordinary local ones, for these Eve knows exactly the relation between the input and output bits, both for Alice and for Bob. For example, Eve sends to Alice an apparatus that always outputs a 0, and to Bob an apparatus that outputs the input bit: $b = y$. In this example Eve knows Alice's bit $a$, but doesn't know Bob's bit. For some local pairs of apparatuses Eve knows both $a$ and $b$, or $b$ but not $a$. But, if the Alice–Bob correlation (8.8) violates the Bell inequality (8.4), i.e. if $p_{NL} > 0$, then Eve must sometimes send to Alice and Bob the apparatuses that produce the maximal nonlocal correlation $a + b = xy$,[12] in which case she knows nothing about Alice and Bob's output bits $a$ and $b$. A detailed analysis can be found in [65]. Here we merely recall the result. For $p_{NL} > 0.318$ the Shannon mutual information between Alice and Bob is larger than the Eve-Bob mutual information [65]. Hence for $p_{NL} > 0.318$ Alice and Bob can distil a cryptographic secret key out of their data, secure even against an hypothetical post-quantum adversary, provided this adversary is still subject to no-signaling.

Actually, in [65] we worked out a 2-way protocol for key distillation valid down to $p_{NL} > 0.09$. There, it is also proven that the intrinsic information is positive for all positive $p_{NL}$. It is thus tempting to conjecture that secret key distillation is possible if and only if the Bell inequality is violated.[13]

---

[12]One may think that Eve should sometimes send a weakly non-local machine. But all such correlations are convex combinations of local and fully non-local NL-machines. Hence, it is equivalent for Eve to always send either a local or a NL-machine, with appropriate probabilities.

[13]In [59] we proved that a correlation $P(a, b|x, y)$ is nonlocal iff any possible non-signaling extensions $P(a, b, e|x, y, z)$ has positive Alice–Bob condition mutual information, conditioned on Eve, $I(A, B|E)$, i.e. has positive intrinsic information. This nicely complements the similar result that holds for entangled quantum states and purifications [69]. In [65] we proved that the same relation between nonlocality and positive intrinsic information does also hold when Alice

Another beautiful result is the observation of an *information gain versus disturbance* relationship, very similar to that of quantum physics, based on Heisenberg's uncertainty relations (Scarani, Private communication). Let us analyze separately the cases where Alice announces $x = 0$ and $x = 1$, and denote the respective Alice–Bob error rates $QBER_x$ and the Eve-Bob mutual informations $I_x(B, E)$, i.e. $QBER_x = \sum_y P(a \neq b)|x, y)$ and $I_x(B, E) = H(B|x) - H(B|E, x)$. Remarkably, $I_0(B, E)$ is a function of only $QBER_1$ and $I_1(B, E)$ of $QBER_0$[14]: information gain for one input necessarily produces errors for the other input, in analogy with the quantum case where information gain on basis necessarily perturbs information encoded in a conjugated basis!

To conclude this subsection, let us emphasize that the distribution of the correlation (8.8) by quantum means requires a protocol that differs from the famous BB84 protocol [70]. Indeed, the data obtained by Alice and Bob following the BB84 protocol do not violate any Bell inequality, hence the BB84 protocol is not secure against a non-signaling post-quantum adversary. Indeed, even the noise-free BB84 data can be obtained from quantum measurements on a separable state in higher dimension. The additional dimension could, for the example of polarization coding, be side-channels due to accidental additional wavelength coding. Consequently, standard security proofs [71, 72] must make assumptions about the dimension of the relevant Hilbert spaces (accordingly, no security proof of quantum key distribution is unconditional, contrary to widespread claims). But it is easy to adapt the BB84 protocol, it suffices that Alice measures the physical quantities corresponding to the Pauli matrices $\sigma_z$ or $\sigma_x$, depending on her input bit value 0 or 1, respectively, exactly as in BB84, and Bob measures in the diagonal bases: $\sigma_{+45^o}$ and $\sigma_{-45^v}$ for $y = 0$ and $y = 1$, respectively. In this way Alice and Bob's data are never perfectly correlated, but they can violate the Bell inequality and be thus exploited to distil a secret key valid even against post-quantum adversaries. Note that the violation of a Bell inequality guarantees that no side channels accidentally leak out information. Furthermore, in this protocol, that we like to call the CHSH-protocol, in honor of the four inventors [14] of the most useful Bell inequality (actually equivalent to (8.4)), Alice announces her input bit $x$, i.e. her basis as in BB84, but Bob doesn't speak, he always accepts and merely flips his bit in case $x = y = 1$. In summary, in the CHSH protocol Alice and Bob use all the raw bits, however their data are initially noisier than in the BB84 protocol.

---

announces her input and Bob adapts his output in such a way as to maximize his mutual information with Alice. Proving this in full generality would be a marvelous result.

[14]Precisely one has: $I_0(B, E) = 2 \cdot QBER_1$ and $I_1(B, E) = 2 \cdot QBER_0$.

## 8.10.4   Cost of Simulating Quantum Correlations

Among the many contributions of computer science to quantum information is the beautifully simple question (actually anticipated by Maudlin [73]): what is the cost of simulating quantum correlations? More precisely, Gilles Brassard, Richard Cleve and their student Alain Tapp [74], and independently Michael Steiner [75], asked the question: *How many bits must Alice and Bob exchange in order to simulate (projective) measurement outcomes performed on quantum systems?* The question concerns the communication during the measurement simulation, clearly there must have been prior agreement on a common strategy. If the systems are in a separable state, no communication at all is needed. On the contrary, if the state allows measurements that violate a Bell inequality, i.e. if the state has quantum nonlocality, then it is impossible to simulate it without some communication or some other nonlocal resources.

For the simplest case of two 2-level systems (2 qubits), this game assumes that Alice and Bob receive as input any possible observable, i.e. any vector $\vec{a}$ and $\vec{b}$ on the Poincaré sphere. And they should output one bit, corresponding to the binary measurement outcome "up" or "down" in the physicist's spin $\frac{1}{2}$ language. A simple way to simulate the quantum measurements is that Alice communicates her input $\vec{a}$ to Bob and outputs a predetermined bit (predetermined by Alice and Bob's common strategy). But communicating a vector corresponds to infinitely many bits! My first intuition was that there is no way to do any better, after all the input space is a continuum, quite the contrary to the case of Bell inequalities where the input space is finite, usually even limited to a binary choice. Yet, Brassard and co-workers came out with a model using only 8 bits of communication! What a surprise: is entanglement that cheap? But this was only a start. Steiner published a model valid only for vectors lying on the equator of the sphere, but this model was easy to generalize to the entire sphere [76]: it uses only 2 bits! 2 bits, like in dense coding and teleportation: that should be the end, I thought! But, yet again, I was wrong. Bacon and Toner produced a model using one single bit of communication [77]. Well, by now we should be at the limit, isn't it? But actually, not quite!

Let's come back to the real central question: How does Nature manage to produce random data at space-like separated locations that can't be explained by common causes? The idea that Nature might be exploiting some hidden communication (hidden to us, humans) is interesting. With my group at Geneva University we spent quite some time trying to explore this idea, both experimentally and theoretically. We could set experimental bounds of the speed of this hypothetical hidden communication [27–31]. We also investigated the idea that each observer sends out hidden information about his result at arbitrary large speeds as defined in its own inertial reference frame [32]. The measured bounds on the speed of the hypothetical hidden communication were very high and the latter assumption contradicted by experiments. Also our theoretical investigation cast serious doubts on the existence of hidden communication. Analyzing scenarios involving three parties we could prove that if all quantum correlations would be due to hidden communication, then one should be able to signal (i.e. the hidden communication do not remain hidden) [78, 79]! Hence, the only remaining alternative is that Nature exploits both hidden communication and hidden variables: each one separately contradicts quantum theory, but both together could explain quantum physics. However, recently, we could prove all such models impossible [31]. Hence, let's face the situation: Nature is able to produce nonlocal data without any sort of communication. But is she doing so using all the quantum physics artillery? Aren't there logical building blocks of nonlocality? A partial answer follows.

Let us come back to the problem of simulating quantum measurements, but instead of a few bits of communication let us give Alice and Bob a weaker resource: one instance of the nonlocal machine $a + b = xy$. That this is indeed a weaker resource follows from the observation that the correlation $a + b = xy$ can't be used to communicate any bit, but that by sending a single bit one can easily simulate the nonlocal correlation (since Alice's input is only a bit $x$, it suffices that she communicates it to Bob). The nice surprise is that this elementary resource is sufficient to simulate any pair of projective measurements on any maximally entangled state of two qubits! For the proof the reader is referred to the original article [80] and to the beautiful account in [81] where the relations between all these models are presented.

The above results are very encouraging. One can get the feeling that, at last, one can start understanding nonlocality without the Hilbert space machinery, that, at last, one can study quantum physics from the outside, i.e. from the perspective of future physical theories (assuming these will keep Einstein's no-signaling constraint) and no longer from the perspective of the old classical mechanical physics. But there is still a lot to be done! For instance, it is surprising (and annoying in my opinion) that one is still unable to simulate measurement on partially entangled states using the nonlocal correlation (actually we could prove that this is impossible with a single instance of the NL-machine, but there is hope that one can simulate partially entangled qubit pairs with two instances [82], see also [83]). Let me emphasize that all of today's known simulation models for partially entangled qubits include some sort of communication[15] [77], let's say from Alice to Bob. Consequently, in all these models Bob can't output his results before Alice was given her input. This contrasts with the situation in quantum measurements where Bob doesn't need to wait for Alice (he does not even need to know about the existence of Alice) and with the simulation model for maximally entangled qubits using the PR-box. It would be astonishing if partially entangled state could not be simulated in a time-symmetric way (For another recent results sustaining the conjecture the partially entangled state are more nonlocal than maximally entangled states see: [85]).

---

[15]Using the reduction of an OT-box (Oblivious Transfer to a PR-box) [84] one can simulate any 2-qubit state with one OT-box.

# 8.11   Conclusion

The history of non-locality in physics is fascinating. It goes back to Newton (Sect. 8.2). It first accelerated around 1935 with Einstein's EPR and Schrödinger cat's papers. Next, it slowly evolved, with the works of John Bell, John Clauser and Alain Aspect among many others, from a mere philosophical debate to an experimental physics question, or even to *experimental metaphysics* as Abner Shimony nicely put it [86]. Now, during the last decade, it has run at full speed. Conceptually the two major breakthroughs were, first Artur Ekert's 1991 PRL which strongly suggests a deep link between non-locality and cryptography, Sect. 8.10.3. The second breakthrough, in my opinion, is the PR-box, Sect. 8.10.1, the understanding that non-signaling correlations can be analyzed for themselves, without the need of the usual Hilbert space artillery, thus providing a simple conceptual tool for the unravelling of quantum non-locality [N. Gisin *Quantum chance, Nonlocality, Teleportation and Other Quantum Marvels*, Springer 2014]. We have reviewed that the no-cloning theorem, the uncertainty relation, the monogamy of extreme correlation and the security of key distribution, all properties usually associated to quantum physics are actually properties of any theory without signaling, Sect. 8.10. In particular we emphasized that the second breakthrough, the PR-box, allows one to confirm the first breakthrough: there is an intimate connection between violation of a Bell inequality and security of quantum cryptography.

And relativity, can it be considered complete? Well, if nonlocality is really real, as widely supported by the accounts summaries in this article, then all complete theories should have a place for it. Hence, the question is: "Does relativity hold a place for non-signaling nonlocal correlations?".

**Acknowledgements** This article has been inspired by talks I gave in 2005 at the IOP conference on Einstein in Warwick, the QUPON conference in Vienna, the *Annus Mirabilis* Symposium in Zurich, le séminaire de l'Observatoire de Paris and the Ehrenfest Colloquium in Leiden. This work has been supported by the EC under projects RESQ and QAP (contract n. IST-2001-37559 and IST-015848) and by the Swiss NCCR *Quantum Science and Technology*.

# References

1. Einstein, A., Podolsky, B., Rosen, N.: Can quantum-mechanical description of physical reality be considered complete? Phys. Rev. **47**, 777–780 (1935)
2. Weinberg, S.: Dreams of a Final Theory. Vintage/Random House, New York (1994)
3. Shimony, A.: In: Kamefuchi, S. (ed.) Foundations of Quantum Mechanics in the Light of New Technology. Physical Society of Japan, Tokyo (1983)
4. Gilder, L.L.: The Age of Entanglement. Knopf Publishing, New York (2006)
5. Cohen, B., Schofield, R.E. (eds.): Isaac Newton, Papers & Letters on Natural Philosophy and Related Documents, p. 302. Harvard University Press, Cambridge (1958)
6. Schrödinger, E.: Naturwissenschaften **23**, 807 (1935)
7. Bell, J.S.: Speakable and Unspeakable in Quantum Mechanics: Collected Papers on Quantum Philosophy. Cambridge University Press, Cambridge (1987/Revised edition 2004)

8. Gisin, N.: J. Math. Phys. **24**, 1779–1782 (1983)
9. Ekert, A.K.: Phys. Rev. Lett. **67**, 661 (1991)
10. Mandel, L.: Optical Coherent & Quantum Optics. Cambridge University Press, Cambridge (1995)
11. Tittel, W., Brendel, J., Gisin, N., Zbinden, H.: Phys. Rev. Lett. **81**, 3563–3566 (1998)
12. Tittel, W., Brendel, J., Gisin, N., Zbinden, H.: Long-distance Bell-type tests using energy-time entangled photons. Phys. Rev. A **59**, 4150–4163 (1999)
13. Marcikic, I., de Riedmatten, H., Tittel, W., Zbinden, H., Legré, M., Gisin, N.: Phys. Rev. Lett. **93**, 180502 (2004)
14. Clauser, J.F., Horne, M.A., Shimony, A., Holt, R.A.: $M \leq 3$ is equivalent to the famous CHSH-Bell inequality. Phys. Rev. Lett. **23**, 880 (1969)
15. Cirel'son, B.S.: Lett. Math. Phys. **4**, 93 (1980)
16. Gisin, N.: Bell inequalities: many questions, a few answers. In: Myrvold, W.C., Christian, J. (eds.) Quantum Reality, Relativistic Causality, and Closing the Epistemic Circle. The Western Ontario Series in Philosophy of Science, pp. 125–140. Springer, Berlin (2009). Essays in honour of Abner Shimony
17. Bell, J.S.: Speakable and Unspeakable in Quantum Mechanics, p. 152. Cambridge University Press, Cambridge (1987)
18. Freedman, J., Clauser, J.F.: Phys. Rev. Lett. **28**, 938–941 (1972); Aspect, A., Grangier, P., Roger, G.: Phys. Rev. Lett. **47**, 460–463 (1981); Ou, Z.Y., Mandel, L.: Phys. Rev. Lett. **61**, 50–53 (1988); Shih, Y.H., Alley, C.O.: Phys. Rev. Lett. **61**, 2921 (1988); Kwiat, P.G., Mattle, K., Weinfurter, H., Zeilinger, A., Sergienko, A.V., Shih, Y.H.: Phys. Rev. Lett. **75**, 4337 (1995); Rarity, J.G., Tapster, P.R.: Phys. Rev. Lett. **64**, 2495–2498 (1990); Brendel, J., Mohler, E., Martienssen, W.: Europhys. Lett. **20**, 575–580 (1992); Tapster, P.R., Rarity, J.G., Owens, P.C.M.: Phys. Rev. Lett. **73**, 1923–1926 (1994)
19. Weihs, G., Reck, M., Weinfurter, H., Zeilinger, A.: Phys. Rev. Lett. **81**, 5039 (1998)
20. Rowe, M.A., et al.: Nature **149**, 791–794 (2001)
21. Matsukevich, D.N., et al.: Phys. Rev. Lett. **100**, 150404 (2008)
22. Barreiro, J.T., Bancal, J.-D.: Nat. Phys. (2013). arXiv:1303.2433
23. Aspect, A., Dalibard, J., Roger, G.: Phys. Rev. Lett. **49**, 1804 (1982)
24. Gisin, N., Zbinden, H.: Phys. Lett. A **264**, 103–107 (1999)
25. Giustina, M., et al.: Nature **497**, 227–230 (2013)
26. Christensen, B.G., et al.: Detection-Loophole-Free Test of Quantum Nonlocality, and Applications. Phys. Rev. Lett. **111**, 130406 (2013)
27. Zbinden, H., Brendel, J., Tittel, W., Gisin, N.: Phys. Rev. A **63**, 022111 (2001); Zbinden, H., Brendel, J., Gisin, N., Tittel, W.: J. Phys. A: Math. Gen. **34**, 7103–7109 (2001)
28. Salart, D., Baas, A., Branciard, C., Gisin, N., Zbinden, H.: Testing the speed of spooky action at a distance. Nature **454**, 861–864 (2008)
29. Cocciaro, B., Faetti, S., Fronzoni, L.: A lower bound for the velocity of quantum communications in the preferred frame. Phys. Lett. A **375**, 379–384 (2011)
30. Yin, J., et al.: Phys. Rev. Lett. **110**, 260407 (2013)
31. Bancal, J.D., et al.: Nat. Phys. **8**, 867–870 (2012)
32. Gisin, N., Scarani, V., Tittel, W., Zbinden, H.: 100 years of Q theory. Proc. Ann. Phys. **9**, 831–842 (2000). quant-ph/0009055; Stefanov, A., Zbinden, H., Gisin, N.: Phys. Rev. Lett. **88**, 120404 (2002); Gisin, N.: Sundays in a quantum engineer's life. In: Bell, J.S. (ed.) Proceedings of the Conference in Commemoration, Vienna, 10–14 November 2000; Scarani, V., Tittel, W., Zbinden, H., Gisin, N.: Phys. Lett. A **276**, 1–7 (2000)
33. Suarez, A., Scarani, V.: Phys. Lett. A **232**, 9 (1997)
34. Bohm, D.: Phys. Rev. **85**, 166–193 (1952); Bohm, D., Hilley, B.J.: The Undivided Universe. Routledge, New York (1993)
35. Popescu, S., Rohrlich, D.: Found. Phys. **24**, 379 (1994)
36. Rohrlich, D., Popescu, S.: quant-ph/9508009 and quant-ph/9709026
37. Pironio, S., et al.: Nature **464**, 1021–1024 (2010)
38. Gisin, N.: Science **326**, 1357–1358 (2009). quant-ph/0503007

39. Werner, R.F.: Phys. Rev. A **40**, 4277 (1989)
40. Terhal, B.M., Wolf, M.M., Doherty, A.C.: Physics Today **56**, 46–52 (2003)
41. Mermin, N.D.: quant-ph/9609013 and quant-ph/9801057
42. Esfeld, M.: Stud. Hist. Philos. Mod. Phys. **35B**, 601–617 (2004)
43. Peng, C.-Z., et al.: Phys. Rev. Lett. **94**, 150501 (2005)
44. Zhao, Z., et al.: Nature **430**, 54–58 (2004)
45. Haeffner, H., et al.: Appl. Phys. B **81**, 151 (2005)
46. Blinov, B.B., Moehring, D.L., Duan, L.M., Monroe, C.: Nature **428**, 153–157 (2004)
47. Volz, J., et al.: quant-ph/0511183
48. Julsgaard, B., Sherson, J., Cirac, J.I., Polzik, E.S.: Nature **432**, 482–486 (2004)
49. Altewischer, E., et al.: Nature **418**, 304 (2002)
50. Fasel, S., et al.: Phys. Rev. Lett. **94**, 110501 (2005)
51. Fasel, S., et al.: New J. Phys. **8**, 13 (2006)
52. Chou, C.W., et al.: Nature **438**, 828–832 (2005)
53. Pan, J.W., Bouwmeester, D., Weinfurter, H., Zeilinger, A.: Phys. Rev. Lett. **80**, 3891 (1998)
54. Jennewein, T., Weihs, G., Pan, J.-W., Zeilinger, A.: Phys. Rev. Lett. **88**, 017903 (2002)
55. de Riedmatten, H., et al.: Phys. Rev. A **71**, 050302 (2005)
56. Tanzilli, S., et al.: Nature **437**, 116–120 (2005)
57. Tsirelson, B.S.: Hadronic J. Suppl. **8**, 329 (1993)
58. Bužek, V., Hillery, M.: Phys. Rev. A **54**, 1844 (1996)
59. Masanes, L., Acin, A., Gisin, N.: Phys. Rev. A. **73**, 012112 (2006)
60. Barrett, J., Linden, N., Massar, S., Pironio, S., Popescu, S., Roberts, D.: Phys. Rev. A **71**, 022101 (2005)
61. van Dam, W.: quant-ph/0501159; Wolf, S., Wullschleger, J.: Proceedings of 2006 IEEE Information Theory Workshop (ITW) (2006); Buhrman, H., Christandl, M., Unger, F., Wehner, S., Winter, A.: Proc. R. Soc. A **462**, 1919–1932 (206); Short, T., Gisin, N., Popescu, S.: Quant. Inf. Proc. **5**, 131–138 (2006); Barrett, J., Pironio, S.: Phys. Rev. Lett. **95**, 140401 (2005); Jones, N.S., Masanes, L.: Phys. Rev. A **72**, 052312 (2005); Barrett, J.: Phys. Rev. A **75**, 032304 (2007)
62. Gisin, N.: Quantum correlations in Newtonian space and time: arbitrarily fast communication or nonlocality. In: Struppa, D.C., Tollaksen, J.M. (eds.) Quantum Theory: A Two-Time Success Story, Yakir Aharonov Festschrift, pp. 185–204. Springer (2013)
63. Collins, D., Gisin, N.: J. Phys. A: Math. Gen. **37**, 1775 (2004)
64. Masanes, L., Acin, A., Gisin, N.: Phys. Rev. A **73**, 012112 (2006)
65. Acin, A., Gisin, N., Masanes, L.: Phys. Rev. Lett. **97**, 120405 (2006)
66. Acin, A., Masanes, L., Pironio, S.: New J. Phys. **8**, 126 (2006)
67. Barrett, J., Hardy, L., Kent, A.: Phys. Rev. Lett. **95**, 010503 (2005)
68. Pironio, S., et al.: New J. Phys. **11**, 1–25 (2009)
69. Gisin, N., Wolf, S.: Phys. Rev. Lett. **83**, 4200 (1999); Gisin, N., Wolf, S.: Proceedings of CRYPTO 2000. Lecture Notes in Computer Science, vol. 1880, p. 482. Springer, Berlin (2000). quant-ph/0005042; Acin, A., Gisin, N.: Phys. Rev. Lett. **94**, 020501 (2005). quant-ph/0310054
70. Bennett, C.H., Brassard, G.: Proceedings of IEEE International Conference on Computers, Systems and Signal Processing, p. 175. IEEE Press, New York (1984)
71. Shor, P.W., Preskill, J.: Phys. Rev. Lett. **85**, 441 (2000)
72. Kraus, B., Gisin, N., Renner, R.: Phys. Rev. Lett. **95**, 080501 (2005). quant-ph/0410215
73. Maudlin, T.: Philos. Sci. Assoc. **1**, 404–417 (1992)
74. Brassard, G., Cleve, R., Tapp, A.: Phys. Rev. Lett. **83**, 1874 (1999)
75. Steiner, M.: Phys. Lett. A **270**, 239 (2000)
76. Gisin, B., Gisin, N.: Phys. Lett. A **260**, 323 (1999)
77. Toner, B.F., Bacon, D.: Phys. Rev. Lett. **91**, 187904 (2003)
78. Scarani, V., Gisin, N.: Phys. Lett. A **295**, 167–174 (2002). quant-ph/0110074
79. Scarani, V., Gisin, N.: Braz. J. Phys. **35**, 328–332 (2005). quant-ph/0410025
80. Cerf, N.J., Gisin, N., Massar, S., Popescu, S.: Phys. Rev. Lett. **94**, 220403 (2005)
81. Degorre, J., Laplante, S., Roland, J.: Phys. Rev. A **72**, 062314 (2005)

82. Brunner, N., Gisin, N., Scarani, V.: New J. Phys. **7**, 1–14 (2005). quant-ph/0412109
83. Brunner, N., Gisin, N., Popescu, S.: Phys. Rev. A **78**, 052111 (2008)
84. Wolf, S., Wullschleger, J.: quant-ph/0502030
85. Acin, A., Gill, R., Gisin, N.: Phys. Rev. Lett. **95**, 210402 (2005). quant-ph/0506225; and for a recent review read: Methot, A., Scarani, V.: Quantum Inf. Comput. **7**, 157–170 (2007)
86. Cohen, R.S., Horne, M., Stachel, J. (eds.): Experimental Metaphysics. Kluwer, Boston (1997)

# Chapter 9
# Faces of Quantum Physics

**Rudolf Haag**

## Preface

We review conceptual structures met in quantum physics and note changes of basic
concepts and language partly due to a maturing process in the 80 odd years since
their first evocation by the founding fathers in Copenhagen, partly demanded or
suggested by the passage from quantum mechanics to relativistic quantum field
theory, local quantum physics and high energy experiments. It is in particular the
concept of "observable" which lost its central role as a description of the measurement of some hypothetical property of a "physical system" under investigation and
shifted to an auxiliary position as referring to a detector whose signals serve for the
reconstruction of a history described in equations like (9.6), (9.7). The primary role
is taken over by the notion of a (microscopic) event constituting the bridge to reality
and to finer features of space-time.

## 9.1   Introduction

Do you understand quantum theory? Confronted with this blunt question I can
neither say yes nor no. Since my student days I was fascinated by it and struggled
with it. Sixty years ago, in 1953, I had the good fortune to spend a year in
Copenhagen. Niels Bohr, then 70 years old, had retired from the activities of the
institute but I did get a chance for a lengthy discussion with him due to the fact
that I had been assistant to Fritz Bopp, who had tried for years to improve the

R. Haag (✉)
Waldschmidt str. 4b, D-83727 Schliersee
e-mail: rudolf.haag@t-online.de

© Springer-Verlag Berlin Heidelberg 2015                                          219
P. Blanchard, J. Fröhlich (eds.), *The Message of Quantum Science*, Lecture Notes
in Physics 899, DOI 10.1007/978-3-662-46422-9_9

interpretation of quantum mechanics and had sent several preprints to Niels Bohr who could not understand their motivation at all. Today I regret and am somewhat ashamed of my behaviour in this encounter with the great man, the principal architect of the first coherent theory of quantum mechanics called the Copenhagen interpretation. With the self assurance of a young post doc who had read the book by J. von Neumann on the mathematical foundations of quantum mechanics I came with the prejudice that Bohr's explanations were too vague. So I understood nothing of his speech, nor the meaning of his parting words: "Of course you can change the mathematics but that changes nothing in the essence of what we learned." It took me many years till I understood some of Bohr's insights. The essence for him was "complementarity", felt as a deep and general principle governing all scientific effort. It asserts that in our attempts to describe nature we have to employ symbols and pictures which can never give a full view. To any such choice there is a complementary one illuminating some other aspect. Prime examples are the uncertainty relations between canonically conjugate variables. Another example is the alternative description of matter in terms of particles or waves. An extreme example is the proclaimed complementarity between space-time and causality. We shall return to this later. Niels Bohr distrusted the reliance on a single chain of linear logical reasoning. One of his favorite lines of poetry was:

Nur die Fülle führt zur Klarheit,
und im Abgrund wohnt die Wahrheit. [1]

I do not dare to translate this.

In the physics community the uncertainty relations and complementarity left in their wake an insecurity about the ground on which we stand. This insecurity is still there as witnessed by the heading: "Mysteries, puzzles and paradoxes in quantum mechanics" chosen as title of a series of high class workshops at Gargnano, Italy, 70 years after the birth of quantum mechanics. There, recent experiments in atomic physics and quantum optics, made possible by an incredible advance in experimental techniques, were presented. Some mystery may remain unexplainable but puzzles can be solved and paradoxes lifted. To gain some firm ground on which to stand let me first try to clarify some points of departure and terminology.

## 9.2   Reality, Individuality, Phenomena

### 9.2.1   Reality

In daily life we mostly take it as evident that we are tiny parts of a huge world which is largely independent of our wishes and perceptions and we regard the impressions in our consciousness as images of parts of this outside world. We do not reflect about the relation of this world to the part we call "I". Experimental physicists have to regard their efforts as a dialogue with a sometimes hostile outside world called nature and the individual observed phenomena as "facts", i.e. irreversible elements of reality, where "nature" and "reality" are essentially synonymous. This outside world is felt to be distinct from the human mind, obeying laws independent of our will. This corresponds to a dualistic picture of the universe, with two co-existing parts: human consciousness and will on one side, nature on the other. The question about the relation between these two parts, known as the "mind-body-problem", has been a topic in philosophy for ages. Attempts at unification by eliminating one of the two sides led to the two extreme positions of idealism and materialism.

The advent of quantum mechanics suggested that physics could tip the balance in this dispute in favor of a primacy of the mind. In particular there is the discrepancy between the deterministic development of wave functions by the Schrödinger equation and the statistical nature of all predictions. J. von Neumann observed that the Schrödinger equation alone could not explain the gain of knowledge by experiments. The process would never end unless the observer decides to end it due to his consciousness. Schrödinger illustrated this by his drastic story of the poisoned cat, who is neither dead nor alive. For such reasons the standard interpretation had withdrawn to the safe position that the task of the theory is fulfilled if it is able to predict the results of experiments. I have always felt that this is too modest a view and does not do justice to the motivation of physicists who hoped to learn about the working of nature.

The reality issue received a new impulse by the observation of persistent long range entanglement and the violation of Bell's inequality. It kindled many debates as to whether reality or locality have to be sacrificed. B. D'Espagnat discussed the pro and contra for the assumption of a "mind-independent reality" [2].

In spite of all this, I maintain that physics cannot contribute to the solution of the mind-body-problem. For the purpose of physics it is not relevant to which ideology (if any) one adheres. The experimentalist is safeguarded against becoming the prey of illusions by the extremely stringent requirements for accepting a phenomenon as real. It needs the agreement with many other observers, possibly attainable by documentation; it needs reproducibility by independent work elsewhere. The essential criterion for accepting an element of consciousness as the cognition of a counterpart in reality is the consensus between many observers, which lifts the impression from one individual consciousness to a collective one. If this is satisfied, the agreement of all people concerned is adequate for treating the said observation "as if" it were an element of an outside world.

The theory is in another situation. If it transcends pure phenomenology it is a creation of the human mind. Not all of its concepts and pictures need to have a counterpart in reality. Since a consensus can only be reached about coarse features of phenomena there remains an open end for extrapolations whose merits can only be judged with criteria like fruitfulness, simplicity, naturalness.

## 9.2.2 Individuality, Division

Any gain of knowledge starts by the distinction of different things: different objects, phenomena, even words in a language, different individuals of any kind. Our ability to distinguish individual elements leads us to the concept of numbers, sets and ultimately to the whole structure of mathematics. Physics begins with the observation of individual phenomena and the perception of individual objects. The projection of phenomena into nature we call "events". The singling out of individual elements demands a division of the world. The least objectionable way of doing this starts from the consideration of simple hypothetical universes, allowed by the theory, which could exist without anything else. Among these there are minimal ones consisting of a single, lonely, stable particle. Here, the term "stable particle" denotes anything which permanently stays together, be it an electron, an atom, a molecule or even a piece of solid material.

In quantum physics the relation of a particle to space-time is somewhat subtle. The lonely particle does not produce an event due to the lack of a partner. For its perception we need a detector. The interaction of the particle with the detector produces an event and this is localized in space-time. Prior to this event we cannot assume any localization of the particle because there exist interference effects described by the wave aspect of matter. They tell us that the same particle may be partly here, partly there. But the property of "staying together" means that in a battery of detectors the particle can excite at most one detector at any given time. A particle, though it has no sharp position, is "permanently singly localized", i.e. it cannot produce any twofold equal time coincidences. This constitutes an operational definition of the concept "particle" and is equivalent to the well known requirement that it is a state with sharp mass value.

In pursuit of the old dream of understanding the variety of appearances in terms of a few elements, the division process was carried further and further. The explanation of the structure of matter in terms of electrons and nuclei, of nuclei in terms of protons and neutrons, constitute the most spectacular triumph of quantum mechanics. However, we must note that in these divisions the individuality of the parts becomes blurred. It disappears already in the example of entangled 2-particle states.

## 9.2.3  Phenomena, Events, Observation

In his contribution to the Einstein centennial symposium in Princeton 1979 John Archibald Wheeler formulated two paradoxa which, taken together, might constitute a clue for the next step in the progress of the theory. Unfortunately, I forgot one of them, but the other one sounded: "No phenomenon is a phenomenon unless it is observed." I thought that he wanted to stress the customary doctrine that quantum theory deals with laws governing the observation of nature as distinct from properties of nature itself. So I asked him later what he meant with this statement in view of the common belief that the evolution of stars preceded by a wide margin the appearance of life and consciousness and that cosmologists use quantum physics in the description of such processes. His answer "It has nothing to do with the mind" surprised and satisfied me. However, there remained a need for interpretation. One might generalize the meaning of the term "observation", dissociating it from human consciousness and replacing it by the existence of a document. Still, there remain many unobserved happenings needed as links between or causes of genuinely observed phenomena. We may call them "unobserved events". Among their attributes must be some approximate localization in space and time and irreversibility marking a transition between a factual past and an open future in the relative present; a jump into reality from a possibility. If one wants, one may evoke the picture of an entrance into a universal consciousness in nature. This suggests the replacement of the picture of a universe existing from eternity to eternity by the picture of an evolutionary universe [3–5].

## 9.3   Observables in Quantum Mechanics

### 9.3.1   Brief Sketch of the Formalism and Interpretation

Observables and states are the central concepts of quantum theory. In quantum mechanics they both refer to a "physical system", i.e. some part of the universe singled out for study. It usually consists of a certain number of electrons and nuclei interacting by Coulomb forces and they may be subjected to external fields (electromagnetic or gravitational) described classically. The notion of photons representing the quantum nature of electromagnetic radiation lies, strictly speaking, outside the domain of non-relativistic quantum mechanics, but it is unavoidable and freely used. Parts of nuclear physics in which the structure of nuclei is explained in terms of mechanical forces between protons and neutrons may be included but will not be considered here. In Quantum Field Theory the notion of "physical system" is different. This demands some reconsideration of standard terminology.

The specific way of describing predictions in quantum mechanics arises from two features. The first was formulated by Niels Bohr in the words: "We cannot assign any conventional attributes to an atomic object." The "conventional attributes", like position and momentum are replaced by "unconventional attributes", called "observables", mathematically represented by self-adjoint operators in a Hilbert space associated to the system. They are used to label the measuring procedures to which the system may be subjected. Numerical attributes arise only after the application of such a measuring procedure to an individual system. As measuring results only spectral values of the corresponding self-adjoint operator can appear and they should not be interpreted as properties of the system existing prior to the measurement. They are created in the interaction process between measuring instrument and system.[1]

---

[1] This aspect of the "Copenhagen interpretation" has been violently embattled, and is still not accepted by some. However, it is inescapable if one is not prepared to sacrifice the mathematical structure supported by many experiments, e.g. [6].

The second feature is the lack of deterministic predictability. To ensure reproducibility of findings it is necessary to consider ensembles of individual systems which are (to the best of our knowledge) equally prepared. We must be satisfied with statistical statements. The so called "state of the system" characterizes such an ensemble. Its mathematical description is a positive operator in Hilbert space, normalized to have unit trace. The probability assignments are derived from Born's rule. If the observable $F$ is measured in the state $\rho$ the probability for finding a result within the spectral range $\Delta$ of $F$ is

$$p = \mathrm{Tr}\rho P_{F,\Delta} \tag{9.1}$$

where Tr denotes the trace, and $P_{F,\Delta}$ is the spectral projector of $F$ for the range $\Delta$.

Obviously, different ensembles may be thrown together in arbitrary proportions to form a new resulting ensemble. This "mixing process" implies that the set of all states of a system is a convex body within the set of all operators. For any subset $\{\rho_i\}$ of states the convex combination

$$\rho = \sum \lambda_i \rho_i; \quad \lambda_i > 0; \quad \sum \lambda_i = 1 \tag{9.2}$$

is again a state. The convex body has extremal points, the "pure states" which cannot be written as convex combinations of others. They are represented by one-dimensional projectors. The salient feature of quantum physics is that the convex body of states is not a simplex. This means that the decomposition of an impure state into pure components, while always possible, is highly non-unique. In physical terms the decomposition of a state into a convex combination corresponds to a decomposition of the ensemble into subensembles. Therefore it is often not possible or meaningful to assume that each individual system is in some pure state. The assignment of a particular pure state to an individual system means only that this system is filed as a member of a particular subensemble whose choice remains rather arbitrary.

Several eminent scientists have expressed dissatisfaction with the scheme sketched above. Albert Einstein could not make his peace with the indeterminacy and lack of reference to reality. We shall return to this later. John Bell in his search for an ontological description of the universe wanted to throw out the notion of observables in favor of "beables" [7].

For a critical assessment of the standard terminology in quantum physics it is useful to look at its origins and at the gradual shifting of emphasis.

### 9.3.2 Origins

The enormous amount of knowledge about atomic structure accumulated in the quarter century preceding the advent of quantum mechanics is documented in the book by Sommerfeld [8]. Two features there should be mentioned because of their decisive influence on subsequent development. First, the treatment of the orbital motion of electrons by the canonical formalism governed by a Lagrange function with the formulation of quantum conditions selecting the stable orbits. Secondly, the formulas for the intensity of radiation due to a quantum jump from a higher level $E_i$ to a lower level $E_k$. In this, there occur doubly indexed quantities: an oscillator strength $f_{ik}$ and dipole moment $d_{ik}$ obeying sum rules. This was the decisive cue for Heisenberg leading him to "matrix mechanics" by postulating that all canonical variables should be replaced by matrices whose kinematical relations are expressed by matrix multiplication [9].

Quite different were the observations which led Schrödinger to his formulation of wave mechanics. There was de Broglie's relation between momentum and wavelength supplementing Planck's relation between energy and frequency; there were the interference phenomena found by Davisson and Germer. Centuries earlier such interference phenomena for light had led to the replacement of the corpuscular picture of light by a wave theory.

In this analogy between optics and mechanics the light rays of geometric optics in a medium correspond to the trajectories of a particle in mechanics. This allowed Schrödinger to guess the form of a wave equation describing the propagation of a particle in a potential. Its application to the discussion of atomic structure was heralded by the title of his decisive paper: "Quantisierung als Eigenwertproblem" [10]. For this, one additional idea was needed: the natural boundary conditions at infinity expressed by the square integrability of wave functions. It introduced the notion of Hilbert space into theoretical physics, which has become one of the essential mathematical concepts in quantum physics. The quantities energy and momentum of classical mechanics become operators acting in Hilbert space.

### 9.3.3  Discussion

The creators of matrix mechanics and wave mechanics had to consider only a few basic observables: position, momentum, energy, angular momentum. But in the development of a general, mathematically elegant theory it was assumed that every self-adjoint operator in Hilbert space represents some observable. This generalization is in itself harmless and useful

It leads, however, to the often voiced pseudo-problem: "How can we construct a measuring apparatus for a given self-adjoint operator?" and it may veil the central point that all measurements in atomic physics ultimately end by the detection of an event with its localization in space and time. The observed phenomena are dots in a photographic emulsion, flashes on a scintillation screen, clicks of a Geiger counter or signals from some other localized detector. The great variety of different observables arises only due to the possibility of manipulating the system between state preparation and ultimate detection. In the simplest but important case of the manipulation by an external field this demands the solution of the Schrödinger equation which leads to a unitary transformation of the initial wave function. In the measurement of a spin orientation we determine the position after the deflection by a Stern-Gerlach magnet. A momentum measurement either uses the relation to a mean velocity or one studies the contrast in an interference pattern after the passage of the particle through a lattice. Of particular interest is the measurement of energy levels of atoms. If we separate off the center of mass motion, the isolation of the atom ensures that we shall find it in its ground state. An ensemble of such atoms is automatically described by a pure state, namely the ground state; an isolated atom is (almost) a "beable" as desired by John Bell. The level differences are typically measured by absorption or emission of photons whose energy is determined by the deflection in passing through a spectrometer. The ground state energy itself is determined by studying the ionization process and measuring the momentum of a projectile.

At this point, we should also consider the difference between measurement and detection. The former leads to a measuring result giving a number belonging to the spectrum of the observable. The latter registers a fact or, in the case of a battery of detectors, it offers the choice between various mutually exclusive alternatives. Mathematically, the former corresponds to the self-adjoint operator itself with its spectrum, the latter to a set of orthogonal spectral projectors representable by the Abelian algebra of functions of the observable. There is no natural numbering for detector signals; this is different for preferred observables, like energy and momentum, where such a natural numbering exists. Then, the bridge from the position of the registered event to the spectral value has to be established.

### 9.3.4  Continuous Propagation, Discrete Events

Let us now discuss in some detail the simple but illuminating case: manipulation of a single particle by a classically described external field. Here we see most clearly the division of the process into two stages described and idealized in different ways. First, the (relatively mild) interaction of the particle with the external field. This is described by the Schrödinger equation leading to a unitary transformation of the initial wave function. It is reversible and conserves coherence but by itself it cannot produce any gain in knowledge. All possibilities remain open. Secondly, the catastrophic interaction in the detector leading to an unresolved, irreversible fact. One obvious difference distinguishing the interaction in the first stage from that in the second is that an external field is by definition inert. It acts on the atomic system but does not encounter any appreciable back reaction. Of course we cannot expect that this idealization is perfect but typically it is well satisfied in countless interference experiments with electrons, X-rays, slow neutrons. For interference experiments with much heavier particles see [11]. Reversibility is demonstrated in a so called "quantum eraser experiment" [12]. It shows conservation of coherence in beam splitting, various polarization changes and recombination of beams. It is important to stress here, that the quantum eraser cannot erase any facts. In the intermediate stage when a state is reached for which we can predict with certainty the result of some specific measurement, there remains the essential difference as to whether this measurement has or has not been performed. If the detection process is not performed, the state remains in the coherent reversible stage. If it is performed, an irreversible fact is created. Included in this list of coherent processes may even be experiments in which an atomic beam is crossed by laser light forcing the atoms to oscillate between the ground state and an excited state. After several such encounters an interference between parts of the atomic beam having undergone different histories can still be observed [13]. The reason in this case is that the laser beam may be idealized as an external field. It contains almost infinitely many photons so that absorption or emission of one photon has no effect on it. Still another example is the technique of "parametric down conversion" by the passage of a photon beam through a nonlinear crystal which splits one photon into a coherent superposition of two photons with half energy. This has been extensively used in entanglement experiments, which will be discussed later.

Let us turn now to the detection process. To fix the ideas, let us consider the detection of a charged particle moving against a battery of detectors, say Geiger counters. The transversal coherence length of the wave function of the particle shall be far extended beyond the battery of detectors. The ensuing history can again be divided into two stages. First, a microscopic, triggering event, here the ionization of some molecule in one of the detectors. It is followed by a chain reaction in which an exponentially growing avalanche of secondary ionization processes develops, so that a signal perceptible by our senses results. The customary description of the formation of the avalanche is also given in terms of individual ionization processes, each involving one incoming electron and one molecule. There are good reasons to believe that this picture of breaking up the complex total process into a triggering event and many subsequent secondary ionization processes is a very good approximation. On the side of experiments, the efficiency of a detector can be tested and is typically found to be close to 100 %. This means that if any microscopic event occurs within some detector, it will almost certainly be amplified to a visible signal. On the side of the theory the separation of a triggering event from the rest depends on the mean free path between events in relation to the sharpness of localization in space and time which we may assign to this event. This in turn depends on many factors, which have been discussed in [14]. Still a thorough treatment of the detection process under realistic circumstances would be highly desirable.

The above discussion of the detection process implies some change of emphasis in comparison to the standard treatment of measurement theory. In the idealization of the measuring process by J. v. Neumann one starts from an initial state $\rho$ of the system and the "neutral state" of the measuring device $\rho'$. The total state is the tensor product $\rho' \otimes \rho$. After the measurement (if the system survives) a particular result corresponds to a pair $\rho'_i$, $\rho_i$. The total state is represented by the mixture $\sum p_i \rho'_i \otimes \rho_i$, where $p_i$ denotes the probability of occurrence of the respective result. Assuming that the initial states $\rho'$ and $\rho$ are pure, which, though unrealistic, is not forbidden, the process transforms the pure state into a mixture, which is impossible within the known formalism. Much of the effort in measurement theory was devoted to the resolution of this apparent paradox. One strategy is to refer to the macroscopic nature of the measuring device. The interaction process of the system with this device leads to entanglement with more and more degrees of freedom. One shows that some coarse grained distinctions of properties (projections into high dimensional subspaces) approach stationary limits. These are the states $\rho_i'$ appearing as partners of states $\rho_i$ of the system. If one idealizes the notion "macroscopic" by infinite this argument may be put in precise mathematical language invoking the concept of weak convergence.

The appearance of a highly impure final state is evident for many reasons [15]. There is interaction with uncontrollable partners leading to decoherence [16]. Moreover the initial state of the detector is usually far from purity. Furthermore each reaction involving charged particles is accompanied by a Bremsstrahlung of soft photons which escapes detection. The consequences of this for irreversibility have been discussed [17]. In any case this is not the main part of the paradox.

The essential point is, however, that among the various possibilities precisely one is realized in each individual case. A very weak source will generate a temporal sequence of results with a lawfully determined relative frequency of occurrence for each of them, but no knowledge of which event may occur next. The realization of one specific unpredictable result among many alternatives involves a decision, of which we see at least the distinction between clicks of the different detectors. A decision by whom? Einstein wanted to make God responsible,[2] or instead one might say nature, or one may leave it open as unknown and call it just the "principle of random realization" [14]. We emphasize that the decision arises already on the level of a microscopic triggering event, and that the transition to a macroscopic signal plays only the role of "freezing" this result in a document.

### 9.3.5  Persistent Entanglement, Bell Inequalities and Sequels

In 1964 J. Bell presented an inequality which exhibited a quantitative difference between the quantum mechanical prediction for some probabilities and any explanation of the process in terms of "hidden variables" [19]. He discussed the following process mentioned a few years earlier by D. Bohm: A spin-0 particle decays into two spin-1/2 particles moving in opposite directions for a long time till one of them enters the lab of Alice, the other one the lab of Bob. In both cases the arriving particle is subjected to a measurement of the spin orientation by a Stern-Gerlach arrangement. This can yield two possible outcomes: parallel or antiparallel to the

---

[2]Einstein to James Franck: "Schlimmstenfalls kann ich mir noch vorstellen, daß Gott eine Welt hätte schaffen können, in der es keine natürlichen Gesetze - also kurz gesagt: ein Chaos - gibt. Aber daß es statistische Gesetze mit endgültigen Lösungen geben soll, d. h. Gesetze, die Gott in jedem einzelnen Fall zwingen zu würfeln, das finde ich im höchsten Maße unangenehm." quoted from [18].

orientation of the Stern-Gerlach magnet. We denote this result by $(\mathbf{a}, \alpha)$, where $\mathbf{a}$ is the unit vector describing the direction of the magnet; $\alpha = \pm 1$ distinguishes the two possible results. The spin-part of the two-particle wave function after the decay is a singlet state and this will remain so practically unchanged up to the detection process. This singlet state is a pure two-particle state which cannot be decomposed into any mixture, in particular not into a convex combination of products of single-particle states. This is meant by the term "entanglement" or "non-separability" and this suggests that not only it is impossible to assign any "conventional attributes" ("hidden variables") but even no quantum state to the individual particles. The former impossibility has been demonstrated by Bell, the latter by Clauser, Horne, Shimony and Holt [20]. We shall follow here the arguments by Clauser et al. in deriving the inequality which delimits the second impossibility.

The ensemble of all particles received by Bob may be described by an impure one-particle quantum state $\rho_B$. Since the twin particles are correlated due to their common birth it is not surprising that the probability for a particular measuring result of Bob depends on the result of Alice's measurement on the twin. However, entanglement is more than ordinary correlation.

Suppose now that a particle is endowed with some objective property $\lambda$ and the joint probability in the ensemble of pairs of particles is given by a distribution function $\rho(\lambda_1, \lambda_2)$ which describes ordinary correlation between $\lambda_1$ and $\lambda_2$. We assume further that $\lambda$ determines the probability $w(\lambda; \mathbf{a}, \alpha)$ for the measuring result $(\mathbf{a}, \alpha)$, yielding for the expectation value

$$\langle \mathbf{a}; \lambda \rangle = w(\lambda; \mathbf{a}, +) - w(\lambda; \mathbf{a}, -).$$

We note that $w(\lambda; \mathbf{a}, +) + w(\lambda; \mathbf{a}, -) = 1$ because in the measurement $\mathbf{a}$, one of the alternatives $\pm 1$ must occur. The joint probability for $(\mathbf{a}, \alpha; \mathbf{b}, \beta)$ is then

$$W(\mathbf{a}, \alpha; \mathbf{b}, \beta) = \int d\lambda_1 \, d\lambda_2 \, \rho(\lambda_1, \lambda_2) \, w(\lambda_1; \mathbf{a}, \alpha) \, w(\lambda_2; \mathbf{b}, \beta). \qquad (9.3)$$

For the expectation value in the joint measurement, which is defined by

$$\langle \mathbf{a}; \mathbf{b} \rangle \equiv w(\mathbf{a}, +; \mathbf{b}, +) - w(\mathbf{a}, +; \mathbf{b}, -) - w(\mathbf{a}, -; \mathbf{b}, +) + w(\mathbf{a}, -; \mathbf{b}, -)$$

one obtains the representation

$$\langle \mathbf{a}; \mathbf{b} \rangle = \int d\lambda_1 \, d\lambda_2 \, \rho(\lambda_1, \lambda_2) \, \langle \mathbf{a}, \lambda_1 \rangle \langle \mathbf{b}, \lambda_2 \rangle. \qquad (9.4)$$

From this, together with the positivity and normalization of the distribution function $\rho(\lambda_1, \lambda_2)$ one obtains inequalities between expectation values for combinations of measurements with different orientations of the apparatuses,

$$|\langle \mathbf{a}; \mathbf{b} \rangle + \langle \mathbf{a}; \mathbf{b}' \rangle + \langle \mathbf{a}'; \mathbf{b} \rangle - \langle \mathbf{a}'; \mathbf{b}' \rangle| \leq 2. \qquad (9.5)$$

The experimentally observed violation of this inequality shows that the assumption of an ordinary correlation between assumed properties $\lambda_1$, $\lambda_2$ is not tenable. Instead one has the following situation: If Bob receives the full information from Alice about what she has done and found in her measurements, he can split his ensemble into two subensembles according to Alice's measuring result $\alpha = \pm 1$ on the twin. Then these subensembles define two orthogonal pure states which depend on the orientation of Alice's device. It must be stressed that this has nothing to do with any physical effect of Alice's measurement on the particles received by Bob. Nor is it important how fast the information is transmitted. Bob and Alice can get together leisurely after the experiments are finished to evaluate their records. They only have to establish the correct pairing of the events, which can be found for example from the records of the arrival times. No witchcraft is involved. It shows, however, that the pure state of the particle has no objective significance. It does not describe a property of an individual particle but only the defining information about the subensemble in which the particle is filed. And here this is determined by the result of Alice's measurement on the twin.

This implies an enhancement of Bohr's tenet mentioned in the introduction. Not only can we "not assign any conventional attribute to an atomic object" but we cannot even assign any individual quantum state to the particle. It puts in question our traditional picture of the reality of "atomic objects" (particles). Nicholas Maxwell has coined the term "Propensiton" for such an object [21]. It propagates according to a deterministic law such as a Schrödinger equation which is invariant under time reversal. But it does not represent any phenomenon. It is the carrier of propensity contributing to probability assignments.

## 9.4 Field Theory and High Energy Experiments

### 9.4.1 Quantization and Second Quantization

The way from classical mechanics to quantum mechanics discovered by Heisenberg became the prototype for the development of quantum theories in various regimes. This method called quantization is described in the terminology introduced by Dirac as the replacement of "c-numbers" by "q-numbers". The former denote the conventional variables of the classical theory which have numerical values. The latter denote elements of a noncommutative algebra which can be represented by operators in a Hilbert space. In the case of Hamiltonian mechanics this correspondence has a group theoretic background. The Poisson brackets of the c-number theory are replaced by the commutators of the corresponding q-numbers. The former define the Lie algebra of the group of canonical transformations, the other the Lie algebra of unitary transformations in the Hilbert space of wave functions. Since one can show that these two groups are not isomorphic, this formal correspondence can only hold for some subset of preferred variables. Among these are the generators of the geometric invariance group which, in the non-relativistic case, is the inhomogeneous Galilei group consisting of translations in space and time, rotations and proper Galilei transformations, i.e. transitions to a uniformly moving coordinate system. By Noether's theorem the generators of these relate to momentum, energy, angular momentum and the position of the center of mass at time $t = 0$. The irreducible ray representations of this group can be found in the book by Hammermesh [22]. They describe the quantum theory of a single particle with Heisenbergs commutation relations between position and momentum.

*This work was motivated by the seminal work by Wigner, who classified all irreducible, unitary ray representations of the inhomogeneous Lorentz group and found that those with positive energy correspond to the quantum theory of a single particle in empty space. They are distinguished by the value of mass and spin [23].*

Already in 1930 Heisenberg and Pauli applied the rules of quantization to electrodynamics [24]. It took, however, 20 years with numerous modifications and other ideas till a viable theory, quantum electrodynamics, was established. We must restrict us here to a few comments.

1. The starting point for quantization must be the Dirac Maxwell system, considered as a classical field theory. It involves the complex Dirac field $\Psi(x)$ representing charged matter and a vector potential $A_\mu(x)$ for the Maxwell field. Since the Dirac field originally appeared as the wave function of an electron in quantum mechanics its quantization led to the unfortunate expression "second quantization". This added fuel to mystifications surrounding the concept of quantization lifting it from the position of a heuristic crutch to the level of a fundamental principle. The c-number version of the Dirac field is on the same level as Maxwell's theory. Both describe continuous waves. The quantum aspect does not arise until this is combined with the picture of discrete individual particles, interpreting the wave function as a probability amplitude of an electron resp. a photon.

2. There is a common geometric background of the c-number and q-number versions. They both describe a fiber bundle over space-time. The value of the Dirac field at a base point is recorded on the respective fiber, the fundamental group being $U(1)$ which changes the phase. The vector potential establishes the connection between fibers.

## 9.4.2  Field Theory

Comparing the physical interpretation of quantum mechanics with that of quantum field theory we see one important change: The notion of "physical system" has disappeared or rather it was absorbed in the notion of state. The object of observation is no longer materially defined, but it is the condition of some region of space-time. In quantum electrodynamics one might consider as the basic observable the electric or magnetic field strength at a point or rather their average values in a small region.

Bohr and Rosenfeld undertook it to study the possibility of such measurements and to verify that the ensuing uncertainties are in accordance with the commutation relations [25]. The proposed method appears, however, to be rather adventurous. Concerning realistically available detectors we note that one cannot expect to know the assignment of a specific operator to a given detector. From this we conclude that we should not aim at giving a mathematical counterpart to a single detector but describe the set of all detectors indicating an event in a specific space-time region $\mathcal{O}$. This leads to a correspondence between a space-time region $\mathcal{O}$ and a noncommutative algebra $\mathcal{A}(\mathcal{O})$ specified by a net of algebras $\mathcal{O} \mapsto \mathcal{A}(\mathcal{O})$, satisfying a few natural requirements. This so-called "algebraic approach" or "Local Quantum Physics" has been shown to carry sufficient information to determine the particle content and collision cross sections of the theory [26].

### 9.4.3   High Energy Experiments

In the reaction area of a storage ring high intensity high energy beams of electrons and positrons may cross each other and the results of possible collision processes are registered in arrays of detectors. From these the individual histories of such processes are reconstructed such as

$$e^+ + e^- \to \mu^+ + \mu^- \, ; \qquad \mu^{\pm} \to e^{\pm} + \nu + \bar{\nu} \qquad (9.6)$$

or

$$e^+ + e^- \to q + \bar{q} \, ; \qquad (9.7)$$

The quarks $q$ and $\bar{q}$ transform into hadronic jets in opposite directions. The detector signals do not refer to any of the processes indicated by the arrow in (9.6) and (9.7). They indicate the interaction of one of the charged particles listed on the right hand side of (9.6) and (9.7) with molecules in one of the detectors. Therefore the observables (detectors) play only an auxiliary role and the purpose of the experiment is not the measurement of any observable; it is a reconstruction of a sequence of events. The concept of observable does not fit here and the central role is taken over by the concept of "event".

## 9.5   Concluding Remarks and Outlook

In the foregoing section we have emphasized the need to regard the notion of "event" as a fundamental, primary concept, ultimately replacing the concept of observable. It establishes the bridge to reality and space-time. For many purposes it suffices to understand by the term "event" just a detector signal. But, aiming at a fundamental theory, one can push this to finer distinctions: Invisible processes like microscopic triggering events or elementary reactions as described in (9.6) and (9.7) which are reconstructed from many secondary detector signals. If the occurrence of events is not governed by strict deterministic laws but left to random realizations, this necessarily leads to a bipartition of the theory. On one side there must be the characterization of possible events with their attributes, on the other side the assignments of probability for their occurrence, which is subsumed in the notion of "quantum state". These probability assignments change continuously with the passage of time between events and may be pictured as the propagation of "Propensitons" establishing links between events. The realization of a specific event implies a discontinuous change of the quantum state.

One may note the parallelism between this bipartition and Bohr's principle of complementarity of wave aspects vs. particle aspects. The former refer to the continuous change of the quantum state between events, the latter to the discrete realization of an event. One may even recognize some similarity to Bohr's somewhat mystical complementarity between space-time and causality which was hailed by Heisenberg as the deepest insight. If "causality" is understood as strict determinism (e.g. the Schrödinger equation) and space-time stands for one of the essential attributes of an event, namely its localization, then this complementarity refers to the same bipartition. There is, however, one essential difference. We do not regard the bipartition as a complementarity leaving us the choice to focus on the one or the other aspect. Rather both are needed in succession.

Our discussion here concerned language, concepts and interpretation referring to existing theory. This remains on a qualitative level. What is missing in all this is a precise mathematical representation of possible events which would be needed for a self contained formulation of the theory in terms of these concepts. It must include the essential attributes of an event, namely: the final resulting impure state and the localization in time as well as in space. M. Toller has reviewed existing attempts at defining a time of occurrence and proposed a definition of localization in terms of "positive operator valued measures" [27].

This addresses at least some part of the problem, though it remains far from giving a satisfactory answer. A full solution of this problem could open a wide perspective: the representation of possible histories as a category whose objects are events and whose directed arrows are propensitons describing causal links between events. It would include the description of possible space-times generated by the processes and dependent on the energy-momentum flow

**Acknowledgements** I want to thank Berge Englert, Dieter Haidt, Klaus Hepp and Heide Narnhofer for helpful discussions. I am greatly indebted to Erhard Seiler and Vanessa Paulisch for essential help without which this manuscript could not have been completed.

# References

1. Schiller, F.: Sprüche des Konfuzius, Musen-Almanad 1796, Neustrelitz
2. d'Espagnat, B.: Quantum Physics and Reality (2011). arXiv: 1101.4545v1
3. Whitehead, A.N.: Process and Reality. Cambridge University Press, Cambridge (1929)
4. Weizsäcker, C. F. v.: Probability in quantum mechanics. Br. J. Philos. Sci. **24**, 321 (1973)
5. Haag, R.: An evolutionary picture for quantum mechanics. Commun. Math. Phys. **180**, 73 (1996)
6. Rauch, H.: Quantum physics with neutrons: from spinor symmetry to Kochen-Specker phenomena. Found. Phys. **42**, 153 (2012)
7. Bell, J.: Speakable and Unspeakable in Quantum Mechanics. Cambridge University Press, Cambridge (1987)
8. Sommerfeld, A.: Atombau und Spektrallinien. F. Vieweg und Sohn, Braunschweig (1921)
9. Born, M., Heisenberg, W., Jordan, P.: Zur Quantenmechanik II. Z. Physik **35**, 577 (1926)
10. Schrödinger, E.: Quantisierung als Eigenwertproblem. Ann. der Physik **385**, 437 (1926)
11. Arndt, M., et al.: Wave particle duality of C60 molecules. Nature **401**, 680 (1999)
12. Walborn, S.P., Terra Cunha, M.O., Padua, S., Monken, C.H.: Double-slit quantum eraser. Phys. Rev. A **65**, 033818 (2002)
13. Riehle, F., Witte, A., Kisters, Th., Helmcke, J.: Interferometry with Ca atoms. Appl. Phys. B **54**, 333 (1992); Sterr, U., Sengstock, K., Müller, J.H., Bettermann, D., Ertmer, W.: The magnesium ramsey interferometer: applications and prospects. Appl. Phys. B **54**, 341 (1992)
14. Haag, R.: On the sharpness of localization of individual events in space and time. Found. Phys. **43**(11), 1295–1313 (2013)
15. J. Fröhlich, see contribution in this book
16. Zeh, H.D.: On the interpretation of measurement in quantum theory. Found. Phys. **1**, 69 (1970); Zurek, W.H.: Decoherence, chaos, quantum-classical correspondence, and the arrow of time. Acta Phys. Pol. **29**, 3689 (1998)
17. Buchholz, D., Roberts, J.E.: New light on infrared problems: sectors, statistics, symmetries and spectrum (2013). arXiv:1304.2794
18. A. Einstein to J. Franck, quoted from Einstein sagt Piper Verlag (1999)
19. Bell, J.: On the Einstein Podolsky Rosen paradox. Physics **1**, 195–200 (1964)
20. Clauser, F., Horne, M.A., Shimony, A., Holt, R.A.: Proposed experiment to test local hidden variable theories. Phys. Rev. Lett. **49**, 1804 (1969)
21. Maxwell, N.: Is the quantum world composed of propensitons. In: Suarez, M. (ed.) Probabilities, Causes and Propensities in Physics, pp. 221–243. Springer, Dodrecht (2011)
22. Hammermesh, M.: Group Theory and Its Application to Physical Problems. Reading, Addison-Wesley (1964)
23. Wigner, E.P.: On unitary representations of the inhomogeneous Lorentz group. Ann. Math. **40**, 149 (1939)
24. Heisenberg, W., Pauli, W.: Zur Quantentheorie der Wellenfelder. Z. Phys. **59**, 1 (1929)
25. Bohr, N., Rosenfeld, L.: Zur Frage der Messbarkeit der elektromagnetischen Feldgrößen. Kgl. Danske Vidensk. Selskab. Math.-Fys. Medd. **12**(8), 3–65 (1933)
26. Haag, R.: Local Quantum Physics. Springer, Heidelberg (1996)
27. Toller, M.: Localization of events in space-time (1998). arXiv:quant-ph/9805030v

# Chapter 10
# Computation Through Neuronal Oscillations

K. Hepp

"Allerdings" ("Certainly". Poem by Goethe (G) to a Physicist (P)):

| | | | | |
|---|---|---|---|---|
| P: | Ins Innre der Natur - | | P: | Into the interior of Nature - |
| G: | O du Philister! - | | G: | Oh you Philistine! - |
| P: | dringt kein erschaffner Geist. | | P: | no created mind can enter. |
| | .. | | | |
| P: | Glückselig, wenn sie nur | | P: | Happy, if she only |
| | die äussre Schale weist! | | | shows the external shell! |
| | .. | | | |
| G: | Alles gibt sie reichlich und gern. | G: | All she gives richly and eagerly. |
| | Natur hat weder Kern | | | Natur has neither core |
| | noch Schale. | | | nor shell. |
| | Alles gibt sie mit einem Male. | | | Everything she gives at once. |
| | .. | | | |

## 10.1 Introduction

Some of us believe that natural sciences are governed by simple and predictive
general principles. This hope has not yet been fulfilled in physics for unifying
gravitation and quantum mechanics. Epigenetics has shaken the monopoly of the
genetic code to determine inheritance [4]. It is therefore not surprising that quantum

K. Hepp (✉)
Institute for Theoretical Physics, ETHZ, Zürich, Switzerland
e-mail: khepp@phys.ethz.ch

© Springer-Verlag Berlin Heidelberg 2015
P. Blanchard, J. Fröhlich (eds.), *The Message of Quantum Science*, Lecture Notes
in Physics 899, DOI 10.1007/978-3-662-46422-9_10

mechanics does not explain consciousness or more generally the coherence of the brain in perception, action and cognition. In an other context, others [105] and we [66, 67] have strongly argued against the absurdity of such a claim, because consciousness is a higher brain function and not a molecular binding mechanism. Decoherence in the warm and wet brain is by many orders of magnitude too strong. Moreover, there are no efficient algorithms for neural quantum computations. However, the controversy over classical and quantum consciousness will probably never be resolved (see e.g. [50, 53]).

Are there new and powerful coherence generating mechanisms in the brain, based on classical physics? This is the central question of this essay. Is Goethe's poem relevant for modern neuroscience?

The human brain is the most complex 'stand-alone computer' on our planet. Each $mm^3$ of the gray matter in neocortex contains $\sim 10^5$ neurons and $\sim 4\,km$ wire. Humans have $\sim 10^{11}$ neurons and each neuron is synaptically connected to $\sim 10^4$ other neurons. We shall mainly deal with the vertebrate neocortex with $\sim 80\,\%$ pyramidal neurons [102] and $\sim 20\,\%$ interneurons [39]. The colocalization of processing and memory provides an architecture for efficient parallel processing. Parallel processing requires synchronization, a conventional wisdom from computer science applied to the brain by Singer [100]:

'The brain is a highly distributed system in which numerous operations are executed in parallel and that lacks a single coordinating center. This raises the questions of (i) how the computations occurring simultaneously in spatially segregated processing areas are coordinated and bound together to give rise to coherent percepts and actions, (ii) how signals are selected and routed from sensory to executive structures without being confounded, and finally (iii) how information about the relatedness of contents is encoded. One of the coordinating mechanisms appears to be the synchronization of neuronal activity by phase locking of self-generated network oscillations.'

Looking back on the history of seven decades of investigating neural oscillations in the brain, starting with [2] in the olfactory bulb and then brought to prominence by Gray and Singer [46] in the visual cortex, we will critically contrast neuronal oscillations in the brain with analogous operations in the Josephson quantum computer (JQC).

A quantum computer using Josephson junction qubits is, from the point of the microscopic electron-phonon interaction, a macroscopic device which obtains its current-flux nonlinearity from a phase transition in the thermodynamic limit of two weakly coupled BCS system ([8, 60]; caricaturized by Hepp [52]). By combining such nonlinear inductors with other circuit elements one obtains a classical Hamiltonian system which, when quantized, leads after truncation to qubits (see [95]). These 2-level systems can be controlled by external electromagnetic fields via the Jaynes-Cummings model of electron-photon interactions. Clearly, the many approximations in this construction necessitate complicated error corrections, with the hope that there is no macroscopic conspiracy of errors (see [29]).

On the 'top-down' level the JQC belongs to a well defined computational framework [1]. The neglected degrees of freedom of the underlying nonrelativistic quantum electrodynamics (QED) only enter as 'noise', producing decoherence. The most active research is 'middle-out', on the device level. The 'bottom-up' approach to the JQC is absurd: nobody can describe the implementation of the Shor algorithm to factor 15 on a JQC [73] in the formalism of microscopic QED. The three levels are tightly coupled by common physical laws.

The human brain, on the contrary, is a construct shaped by the devious course of evolution. The brain has many different scales. The microscale of $O(1/1,000\,mm)$ involves individual neurons, glia cells and synapses. On the mesoscopic scale of $O(mm)$ one deals with groups of hundred thousand neurons and their circuits. On the macroscopic level one would like to understand the whole brain with $O(10^{11})$ neurons. Before dealing with neuronal oscillations on the mesoscopic scale in Chap. 4 we have to deal with the macroscopic scale in Chap. 2 and the microscale in Chap. 3. The impatient readers can avoid the technical and highly incomplete elaborations in Chaps. 2, 3 and 4 and jump directly to the conclusions in Chap. 5 or to their favorite articles in the References.

It is impossible to explain here systematically the basic neurophysiological notions. Fundamental neuroscience is treated in excellent textbooks [62, 103]. In my essay I will refer to them and to the many easily accessible reviews in the Scholarpedia on the Internet. I will reformulate in less technical language the careful wordings in the abstracts of several peer-reviewed papers, while being less formal when providing the general background. This 'picture gallery' will convey to a physicist a flavor of modern neuroscience.

## 10.2   Connectome

"I am my connectome!" [94]

Neuroanatomy is the basis for formulating and testing ideas about how computations are performed by neural circuits. Connectomes, complete wiring diagrams of brains' are scarce, the most famous and only example being that of the 302 neurons of the roundworm *Caenorhabditis elegans (C. elegans)*.

Helmstaedter [51] present for $\sim0.001\,mm^2$ of the mouse inner plexiform layer (the main computational region in the mammalian retina) the dense reconstruction of 950 neurons and their mutual contacts. This was achieved by applying a combination of manual annotation using human experts and machine-learning-based volume segmentation to serial block-face electron microscopy data. They found a new type of retinal bipolar cell and many violations of Peter's Rule (saying that the synaptic connections in the brain are determined essentially by geometrical proximity). Measured by this heroic technical effort the new results seem to be meager. If achievable, however, connectomics of mammalian brains will probably become as important as the human genome has become for the molecular biology of the cell [4].

However, even if one had complete access to the connectome of the human retina with its $\sim10^6$ ganglion cells one would not understand how this important input station of the visual system works. Synaptic connections in the brain are constantly changing, showing context sensitivity and time- and activity dependent effects. These are manifest over a vast range of time scales, from synaptic depression lasting a few milliseconds to long-term potentiation lasting weeks. In short, connectivity is as transient, adaptive, and context sensitive as brain activity *per se*. Therefore, it is unlikely that the characterization of the static connectome alone will furnish deep insights about the dynamic processing of the brain. Helmstaedter's next target (private communication) is the dense reconstruction of layer 4 of a rodent's cortical barrel column. It will take many years to see what is microscopically true in the claims by Hill et al. [54] and taken up in Chaps. 3 and 5 that Peter's Rule and the morphologies of cortical pyramidal neurons and interneurons ensure a robust and invariant set of distributed inputs and outputs between specific pre- and postsynaptic populations of neurons in a cortical column.

Neuroscience, however, can thrive with a macroscopic connectome provided by light-microscope neuroanatomy [28]. Recent work in Henry Kennedy's laboratory started from a parcellation of the macaque neocortex into 93 areas (see Fig. 10.1, [74])

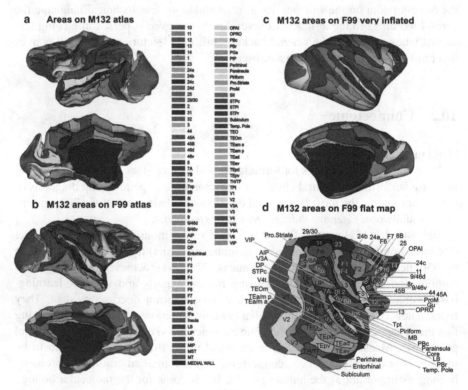

**Fig. 10.1** Reconstruction of the cortex of monkey M132. (**a, b**) are side views from the outside and inside of the two hemispheres. (**c**) is an inflated view and (**d**) a surface-area preserving map of the parcellation into 93 areas. With permission from [74]

Retrograde tracers were injected into 29 areas distributed over six regions in the occipital, temporal, parietal, frontal, prefrontal, and limbic cortex. These regions contained the areas, in which neuronal oscillations will be analyzed in Chap. 4: The areas V1 [23], V2, V4 and TEO of the ventral visual stream, the areas MT, MST, DP and LIP in the dorsal visual stream, and the frontal eye fields (FEF), areas 8l and 8m [104].

The neocortex can be divided 'horizontally' into six layers, the so-called supragranular layers L1, L2 and L3, the granular layer L4, which receives the main projections from the subcortical thalamus, and the infragranular layers L5 and L6. Strongly simplified, feedforward (FF) anatomical pathways that connect different areas of cortex originate mainly from supragranular layers and terminate in L4 in hierarchically higher cortical areas, while feedback (FB) pathways originate mainly in infragranular layers and L1 in higher areas and avoid L4 in hierarchically lower areas. In the visual system a hierarchy can be established using the 'SLN value'. The SLN value s(i,j) of a projection from source i to target j is the fraction of labeled neurons located in the superficial layers of area j after tracer injections in a single area i. By fitting SLN to the parcellation of Fig. 10.1 [9] and [75] determined hierarchical levels of the areas in the ventral and dorsal visual system (see Fig. 10.2).

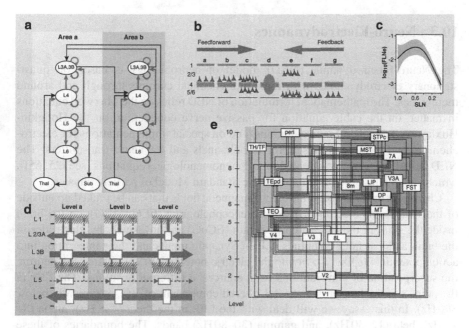

**Fig. 10.2** (a) Canonical microcircuit [31] connecting 'vertically' two neocortical areas. (b) Cartoon of the laminar distributions in a cortical mid-level area. (d) Cartoon of cortical counterstreams. (e) Hierarchical organization of visual cortical areas using SLN as a distance measure. *Left* ventral, *right* dorsal visual stream. With permission from [75]

It was a new discovery that the foveal area 8l of the FEF (dedicated to the highest density of receptors on the retina) has the same hierarchical level as V4.

Systematic injections in 29 typical areas of the parcellation determined a $29 \times 29$ matrix M of elements $s(i,j)$ for each area-to-area pathway. The full macroscopic connectome of the parcellation would be the $91 \times 91$ matrix, which is presently unknown. The graph-theoretical analysis of M has brought many important insights [76]. The data reveal high-density cortical graphs in which economy of connections is achieved by weight heterogeneity and distance-weight correlations. These properties define a model that predicts many binary and weighted features of the cortical network. FB and FF pathways between areas implement a dual counterstream organization, and their integration into local circuits constrains cortical computation.

The functional interpretation of the anatomical hierarchy in Fig. 10.2 is still unclear. In the visual system of the macaque the shortest latency for the onset of a salient target determines another ordering in the dorsal stream: the latency in response to a sudden visual stimulus of the fastest neurons in V1 is ~35 ms, of MT ~40 ms, of LIP ~45 ms and of FEF ~50 ms. In Chap. 4 we will analyze footprints of neural oscillations in the macaque during active vision, where information will propagate along the macroscopic connectome.

## 10.3  Neuro-Electrodynamics

The Neuro-Electrodynamic (NED) model of neurobiology is based on quasi-stationary electrodynamics with ionic and chemical currents through and around membranes. The mathematical formulation of NED relies on the Maxwell equations in matter, on the cable equation for passive nerve conduction, on the Hodgkin-Huxley excitable membrane equations, and on special implementations for electro-chemical and electric synapses, for ion channels and for neuro-transmitters. The NED model is a well-documented set of phenomenological equations (see [25, 65]), which is much less fundamental than the Standard Model of particle physics.

Charges in the brain generate electromagnetic fields that can be picked up outside of the skull in the electro- or magneto-encephalogram (EEG, MEG), on the brain inside the skull in the electro-corticogram (ECoG) and invasively by electrodes in the brain as local field potentials (LFP). Synaptic inputs and their integration into action potentials (spikes) generate the electric potential $V(x,t)$ outside the neuron, the superposition of contributions from sources at different positions. The LFP at the point x is the low-pass temporal filtered component of $V(x,t)$ (typically below 300 Hz). In this essay we will deal with the LFP in the theta (4–8 Hz), alpha (8–12 Hz), beta (12–30 Hz), and gamma (30–80 Hz) bands. The boundaries of these frequency bands vary between different studies.

The LFP is the mesoscopic manifestation of the activity of possibly millions of neurons, depending on the task. It is relatively easy to measure using a multicontact electrode, even in alert human subjects subject to ethical constraints. Hence the LFP is important, but it is difficult to interpret. By an array of intracellular and extracellular electrodes [7] found that extracellular fields induced nonsynaptically mediated changes in the somatic membrane potential that were less than 0.5 mV under subthreshold conditions for spike generation. Despite their small size, these fields could strongly entrain action potentials, particularly for slow (<8 Hz) fluctuations of the extracellular field. One can boost slow oscillations in the brain by applying from the outside of the scull time-dependent (0.75 Hz) external fields to the brain and potentiate memory during sleep [78]. It is unknown in what form the LFPs have to be included in the equations of NED.

Estimates about the reach the LFP vary strongly depending on measurement: Katzner et al. [63] found that >95 % of the LFP signal originated within 0.25 mm of the recording electrode. On the other hand, [61] estimated by direct measurements that LFPs spread passively to sites more than a centimeter from their origins. These findings appeared to be independent of the frequency content of the LFP.

What information is carried by the LFP signal? An important visual stimulus is a plaid pattern. Consider two superimposed gratings moving in different directions. This plaid may be perceived either as two surfaces, one being transparent and sliding on top of the other (component motion) or as a single pattern whose direction of motion is intermediate to the component vectors (pattern motion). The degree of transparency, and hence the perception, can be manipulated by changing the luminance of the grating intersection. Khawaja et al. [64] studied the transformation from V1 to MT to MST for gratings and plaids and found that the LFPs measured simultaneously with the spikes often exhibited properties similar to that of the presumptive feedforward input to each area: in the high-gamma frequency band, the LFPs in MST were as component selective as the spiking outputs of MT, and in MT the LFPs have plaid responses that were similar to the spiking outputs of V1. Such and similar findings suggest (wrongly, see below) that one can obtain from the LPF the synaptic inputs to a local area and from the action potentials the output.

Compared to the rate coding of spikes the LFP signal can be rather coarse: Lashgari et al. [69] found in V1 that LFP and single unit activity (SUA) had similar stimulus preferences for orientation, direction of motion, contrast, size, temporal frequency, and spatial phase. However, the average SUA had 50 times better signal-to-noise relation, 20 % higher contrast sensitivity, 45 % higher direction selectivity, and 15 % more tuning depth than the average LFP.

LFPs are everywhere in the brain: Gamma oscillations can be *voluntarily* generated. Engelhard et al. [34] trained monkeys to specifically increase low-gamma power in selected sites of motor cortex to move by 'pure thought' (driving a robot via a brain-machine-interface) a cursor and obtain a reward. The monkeys learned to robustly generate oscillatory gamma waves, which were accompanied by a dramatic increase of spiking synchrony of highly precise spatiotemporal patterns, possibly by attention to movement control [37]. On the other side LFPs were recorded in the brain of a *dying* rat. Borjigin et al. [15] identified a transient surge of synchronous gamma oscillations that occurred within the first 30 s after cardiac arrest and preceded the isoelectric electro-encephalogram. Gamma oscillations during cardiac arrest were global and highly coherent. This frequency band exhibited a striking increase in anterior–posterior connectivity and tight phase-coupling to both theta and alpha waves. High-frequency neurophysiological activity in the near-death state exceeded levels found during the conscious [45] waking state. The relation to human consciousness reported in near-death experiences is, however, far fetched.

Giving its usefulness it is important to understand the LFPs from first principles. Reimann et al. [88] simulated the LFP in a model of the rodent neocortical column composed of >12,000 reconstructed, multi-compartmental, and spiking cortical L4 and L5 excitatory pyramidal neurons and inhibitory basket cells, including five million dendritic and somatic compartments with voltage- and ion-dependent currents, realistic connectivity, and probabilistic AMPA, NMDA, and GABA synapses. They found that the LFP reflects local and cross-layer processing. Membrane currents dominate the generation of LFPs, not synaptic ones. Spike-related currents can be visible in the LFP not only at higher frequencies but also below 50 Hz. In the [88] model of a neocortical column layers 1, 2, 3 and 6, glia cells (see [84]) and the back-reaction of the LFP on the membrane processes [7] have been omitted, and the connectome has been dealt as schematically as in [54]. Still the conclusion of the paper is valid, the genesis of LFPs has to be re-evaluated.

In this example we see the problems with bottom-up large scale simulations of the brain: size, intransparency of the computational steps and choice of neglected degrees of freedom. A computer simulation of a brain area is a very unconstrained activity, because one can never have all the data and because the theory of NED is only a phenomenological framework. Therefore, whatever results emerge, many are simply an artefact of the willful choice of approach and data. Thus the simulation results will raise many illusory questions that will lead away from deeper understanding and will never exhibit general laws. Simulations of the brain have no universality like gravitation in Newton's equations.

An alternative approach is to study the behavior of neural circuits in 'wet-ware' using controlled perturbations. Optogenetics [36] allows genetically marked neural populations in an alert behaving animal to be stimulated by light. Cardin et al. [24] showed in the cortex of an alert mouse that light-driven activation of fast-spiking parvalbumin (PV) interneurons selectively amplified gamma oscillations with a resonance at gamma frequency. In contrast, pyramidal neuron activation amplified only lower frequency oscillations. They found that the timing of a sensory input relative to a gamma cycle determined the amplitude and precision of evoked responses of a pyramidal neuron.

This approach has recently been applied to the olfactory bulb (OB) in alert behaving mice. A mouse has ~1,000 different genetically encoded odor receptor proteins, which are distributed in the olfactory sensory neurons (OSN) across the nasal receptor surface. Every OSN of one odorant projects to a unique receptor-specific glomerulus in the OB, whose excitatory output neurons, the mitral (M) cells, are coupled dendro-dendrically to inhibitory granule (G) cells (see [97]). The mitral-granule circuit generates gamma oscillations in the awake mouse. Lepousex and Lledo [70] showed that gamma oscillations were amplified, or abolished, after optogenetic activation or selective lesions to the M-neurons. In response to pharmacological perturbations, long-range gamma synchronization was selectively enhanced while mean firing activities remained unchanged. This change of the oscillatory state of the network impaired odor discrimination in an olfactory learning task: gamma oversynchronization coupled in the network M-neurons related to different odor components with a loss of smell acuity. Gamma oscillations with (negative!) behavioral consequences for an alert mammal could be shown to be dynamic properties of a neural microcircuit. Necessary and sufficient for gamma was the reciprocal pyramidal-interneuronal (PING) coupling between M- and G-neurons. However, the entire circuit of the mouse OB is much more complicated and a simulation based on connectomics and the equations of NED is still impossible.

PING is not the only microscopic mechanism of neuronal oscillations. Courtin et al. [27] showed by single neuron recordings and optogenetic manipulations in alert mice that fear expression was causally related to the phasic inhibition (not excitation!) of PV interneurons in the prefrontal (PF) cortex. The disinhibition of PF pyramidal neurons projecting to the amygdala, a structure that encodes associative fear memories, synchronized these neurons by the resetting of local theta oscillations. These two complementary mechanisms, disinhibition in the PING circuit and synchronization by resetting theta oscillations, led to fear expression. Many more microscopic mechanisms for neuronal oscillations are proposed by theoreticians (see e.g. [107]) or seen in reduced preparations.

## 10.4 Manifestations of Neural Oscillations in Active Vision

Correlations, correlations, correlations...

James [56] had already characterized attention in its most modern form: 'Every one knows what attention is. It is the taking possession by the mind, in clear and vivid form, of one out what seem several simultaneously possible objects or trains of thought. Focalization, concentration of consciousness are of its essence. It implies withdrawal from some things in order to deal effectively with others,...'

In this section we will discuss the contribution of oscillatory processes in the alert behaving macaque monkey during active vision, namely vision, attention and saccadic eye movements [38]. We will concentrate on the ventral and dorsal visual stream, which have been most intensely studied on the cellular and LFP level. Nonhuman primates are necessary to interpolate between mice and humans to understand cognition and brain dysfunctions.

After the discovery of beta oscillations in the motor cortex by Murthy and Fetz [81] neuronal oscillations were found in many areas of the brain of insects and vertebrates, and very prominently in humans. What are their functions? Today thousands of abstracts emerge when one searches for alpha, beta, gamma or theta oscillations in 'Medline'.

In the active vision system several testable and refutable hypotheses have emerged. The correlation theory of brain functions by von der Malsburg [111] (often criticized, e.g. by Shadlen and Movshon [96]) and the finding of gamma oscillations in the cat visual cortex by Gray and Singer [46] and Eckhorn et al. [32] has led to the *binding by synchrony* (BBS) hypothesis [99]:

Spatially segregated neurons should exhibit synchronous response episodes if activated by
a single stimulus or by stimuli that can be grouped together into a single perceptual object.

The first data in the alert monkey from visual area MT supported BBS: Kreiter
and Singer [68] found that cells with overlapping receptive fields (RF: Alonso
and Chen [6]), but different preferences for directions of motion, could engage in
synchronous activity, if they were stimulated with a single moving bar. In contrast,
if the same cells were activated with two different bars, each moving in the direction
preferred by the cells at the two respective sites, responses showed no or much fewer
synchronous epochs.

The BBS hypothesis was questioned by many experiments: Thiele and Stoner
[106] manipulated the apparent transparency of plaid patterns. Although the direc-
tional tuning of plaids correlated highly with perceptual coherence, coherent plaids
elicited significantly less synchrony than did non-coherent gratings. Roelfsema et al.
[90] tested the BBS hypothesis in V1 of monkeys engaged in a contour-grouping
task and found that synchrony was unrelated to contour-grouping. Rate covariation depended on perceptual grouping, as it was strongest between neurons that
responded to features of the same object. Palanka and DeAngelis [83] generalized
the approach of [68]. By sampling more broadly and employing stimuli that contain
partially occluded objects, they conducted a more incisive test of the BBS in area
MT. They found that synchrony in spiking activity showed little dependence on
this task, whereas gamma band synchrony in field potentials could be significantly
stronger when features were grouped. However, these changes in gamma were small
relative to the variability of synchrony across recording sites and did not provide a
robust population signal for feature grouping.

The ultimate 'disproof' of the implementation of BBS in V1 or MT came from
Wolf Singer's laboratory, again by using plaid stimuli. Lima et al. [72] found in
alert behaving monkeys that responses to the single components from gratings
exhibited strong and sustained gamma oscillations (30–65 Hz). The superposition of
the second component, however, led to profound changes in the temporal structure
of the responses, characterized by a drastic reduction of gamma oscillations in the
spiking activity and systematic shifts to higher frequencies in the LFP.

Despite these negative findings in V1 the interest in neural oscillations increased.
Research on V1 brought several intriguing results:

Ray and Maunsell [86] discovered that gamma increased monotonically with
contrast. Changes in stimulus contrast over time lead to a reliable gamma frequency
modulation on a fast timescale. Further, large stimuli whose contrast varied across
space generated gamma at significantly different frequencies in simultaneously
recorded neuronal assemblies separated by as little as 0.4 mm. Chalk et al. [26]
found that directing attention toward a visual stimulus at the RF of the recorded
neurons decreased LFP gamma power and gamma-spike coherence. Jia et al. [57]
discovered that the spatial extent of gamma and its relationship to local spiking
was stimulus dependent. Burns et al. [19] and Xing et al. [115] interpreted gamma
to be 'filtered broadband noise' (but: Nicolic et al. [82]). They saw identical

temporal characteristics of gamma activity in both awake and anesthetized monkeys, including large variability of peak frequency, brief oscillatory epochs (<100 ms on average), and stochastic statistics of the incidence and duration of oscillatory events. Jia et al. [58] found that there was no fixed relationship between LFP gamma power and peak frequency, and neither was related to the strength of spiking activity.

Not all investigations aimed at 'disproving' the functionality of gamma in V1: Womelsdorf et al. [114] found that the mutual influence among neuronal groups depended on the phase relation between rhythmic activities within the groups. Phase relations supporting interactions between the groups preceded those interactions by a few milliseconds, consistent with a mechanistic role. These effects were specific in time, frequency, and space. They therefore proposed that the pattern of synchronization flexibly determined the pattern of neuronal interactions.

A counterstream (see Fig. 10.2d) of neuronal oscillations in the visual cortex was discovered: Buffalo et al. [17] compared the magnitude and latency of attentional enhancement of firing rates in V1, V2, and V4 in the same animals performing the same task. They found a reverse order of attentional effects, such that attentional enhancement was larger and earlier in V4 and smaller and later in V1, with intermediate results in V2. Buffalo et al. [18] found that spike-field coherence in the gamma frequency range (40–60 Hz) was largely confined to the superficial layers, whereas the deep layers showed maximal coherence at low frequencies (6–16 Hz), which included the alpha range. In the superficial layers of V2 and V4, gamma synchrony was enhanced by attention, whereas in the deep layers, alpha synchrony was reduced by attention. Spaak et al. [101] found a robust coupling between alpha phase in the deeper layers and gamma amplitude in granular and superficial layers. Moreover, the power in the two frequency bands was anticorrelated.

Roelfsema et al. [91] and van Kerkoerle [108] found that gamma activity started in input layer 4 and propagated to the other layers, in accordance with a feedforward information flow. In contrast, alpha oscillations were initiated in feedback recipient layers 1 and 5 and propagated towards layer 4 in accordance with a feedback effect. Alpha flowed in the feedback direction, from V4 to V1, whereas gamma propagated in the feedforward direction. Microstimulation of V1 increased the gamma oscillations in V4. Conversely, microstimulation in V4 caused an increase in the V1 alpha oscillations while suppressing the gamma rhythm. These results, taken together, indicate reciprocal mechanisms of alpha and gamma oscillations in monkey visual cortex. They suggest that the early visual system gamma rhythm is involved in the feedforward processing of information, while the alpha rhythm reflects recurrent interactions. Vezoli et al. [109] and Bastos et al. [10] implanted ECoG grids on macaque hemispheres covering the 8 visual areas V1, V2, 8l, V4, TEO, DP, 8m and 7A shown in Figs. 10.1 and 10.2. The monkeys performed a visual discrimination task during which the authors recorded LFP correlations between all 28 pairs of sites with clear beta, gamma and theta bands. Analysing their directionality by Granger causality (how much signals at one site are related to the recent past in another area) they found that the gamma and theta bands carried a feedforward and beta a feedback signature. These directed manifestations of a temporal neural coding could be grouped into a functional hierarchy which corresponded astonishingly well with the anatomical hierarchy of Fig. 10.2. The question arises how these signals are integrated through the synchronized firing patterns in different cortical layers, as it has been shown by Roelfsema et al. [91] for the connection between V1 and V4. Are the alpha and beta rhythms between V1 and V4 in Roelfsema et al. [91] and Vezoli et al. [109] just band overlaps or are they different mechanisms acting on the neuronal firing rates?

These directed LFP correlations led to a new functional interpretation of neural oscillations, to the *Communication by Coherence* (CTC) hypothesis [11, 42]:

> Activated neuronal groups oscillate and thereby undergo rhythmic excitability fluctuations that produce temporal windows for communication. Only coherently oscillating neuronal groups can interact effectively, because their communication windows for input and for output are open at the same [delayed-corrected] times.

In view of the difficulties with oscillatory coding in V1 it was not obvious that CTC could operate between V1 and V2 on the neuronal level. However, Jia et al. [59] found that visual stimuli that induced a strong, coherent gamma rhythm resulted in enhanced V1 synchrony of spikes in pairs of neurons. This was associated with stronger coupling of V1 and V2 by spiking activity, in a retinotopically specific manner. Coupling was more strongly related to the gamma modulation of V1 firing than to the downstream V2 rhythm. Roberts et al. [89] discovered a possible pathway for CTC: Although gamma frequency increased with gratings of varying contrast in V1 and V2 (by ~25 Hz), V1–V2 gamma coherence was maintained for all contrasts. Moreover, while gamma frequency fluctuated spontaneously by ~15 Hz during constant contrast stimulation, this fluctuation was highly correlated between V1 and V2. The strongest coherence connections showed a layer-specific pattern, matching feedforward anatomical connectivity, based on Granger causality.

Note that gamma followed the feedforward projection from the supragranular layers of V1 to the granular layer of V2, and this was modeled by Roberts et al. [89] using neurons obeying the Hodgkin-Huxley equations. One should, however, keep in mind that the projections from V1 to V2 are highly diverse [33]. V2 is thought to use its input from V1 as the basis for computations that are important for visual form processing, such as signaling angles, object borders, illusory contours, and relative binocular disparity. Neuronal oscillations alone can only change the mesoscopic gain of this subtle information transfer effected by the spiking activity of small neuronal populations.

V4 is the gateway of the ventral stream and has been intensively scrutinized for neural oscillations. Fries et al. [41] discovered that neurons activated by the attended stimulus showed increased gamma frequency (35–90 Hz) synchronization but reduced low frequency (<17 Hz) synchronization compared with neurons at nearby V4 sites activated by distractors. Naively, CTC should need a significant spike-spike correlation for an efficient output of V4 to the next downstream neurons in TEO, since without spikes there is no fast long distance communication. Such correlations could not be found in a direct analysis of spike trains in the time domain. However, when first multiplying the spike trains with discrete prolate spheroidal sequence tapers and then transforming this signal into the frequency domain Fries et al. [43] found in the same data zero-phase gamma coherence among spike trains of V4 neurons. This synchronization was particularly evident during visual stimulation and enhanced by selective attention, thus confirming the pattern inferred from LFP power and spike-LFP coherence, and it was rapid enough for the interactions of top-down spatial and feature attention with bottom-up saliency. Attention inside the receptive field of the recorded neuronal population enhanced gamma synchronization and strongly reduced alpha (9–11 Hz) synchronization in the prestimulus period. Figure 13 in [43] shows how a small effect (0.01 vs. 0.03 % coherence) can be exhibited by sophisticated statistical methods. It is still unclear by what circuits in the receiving areas the tapered V4 output could be decoded.

The statistical approach to spike-LFP correlations has recently been much more perfected. Vinck et al. [110] studied how visual attention is orchestrated by the activity of excitatory and inhibitory cells. They tentatively identified these neurons as broad (BS) and narrow spiking (NS) cells and analyzed their synchronization to the LFP in two macaque monkeys performing a selective visual attention task. Across cells, gamma phases scattered widely and were unaffected by stimulation or attention. During stimulation, the phases of NS cells lagged BS cells on average by ~60° and gamma synchronized twice as strongly. Attention enhanced and reduced the gamma locking of strongly and weakly activated cells, respectively. During a prestimulus attentional cue period, BS cells showed weak gamma synchronization, while NS cells gamma-synchronized as strongly as with visual stimulation. These analyses reveal cell-type-specific gamma patterns in macaque visual cortex and suggest that attention affects neurons differentially depending on cell type and activation level. This study is an important first step to study neuronal oscillations in active vision on the circuit level. A pandemonium of correlations emerged which often had no cogent interpretation in terms of CTC. For instance, taking the LFP as a clock, gamma phases of individual neurons scattered widely and were unaffected by stimulation or attention. In the future optogenetic manipulations of subpopulations in V4 might clarify this picture after having solved many more hard questions, as they were already discussed in Chap. 3.

Some oscillatory patterns in V4 correlated well with theoretical expectations. Bichot et al. [13] recorded from neurons in area V4 of monkeys freely scanning a complex array to find a target defined by color, shape, or both. Throughout the period of searching, neurons gave enhanced responses and synchronized their activity in the gamma range whenever a preferred stimulus in their receptive field matched a feature of the target, as predicted by parallel models for visual search. Neurons also gave enhanced responses to candidate targets that were selected for saccades, reflecting a serial component of visual search. Womelsdorf et al. [113] showed that behavioral response times to a stimulus change could be predicted by the degree of gamma-band synchronization among those neurons in monkey V4 that were activated by the relevant stimulus. When there were two visual stimuli and monkeys had to detect a change in one stimulus while ignoring the other, their reactions were fastest when the relevant stimulus induced strong gamma synchronization before and after the change in stimulus. This enhanced gamma synchronization was also followed by shorter neuronal response latencies on the fast trials. Conversely, the monkeys' reactions were slowest when gamma-band synchronization was high in response to the irrelevant distractor.

These papers show convincing correlations between neuronal oscillations and behavior on a mesoscopic scale, but not on the level of neural circuits.

There are manifestations of CTC between V1 and V4: Bosman et al. [16] used two stimuli, activating disjoint V1 sites in the ECoG, and both activating a V4 site equally strongly. When one of those stimuli activated one V1 site, its gamma synchronized to V4. When the two stimuli activated two V1 sites, primarily the cued site gamma synchronized to V4. Frequency bands of gamma activities showed substantial overlap containing the band of interareal coherence. Gamma-mediated interareal influences were predominantly directed from V1 to V4, based on Granger causality. Only LFP correlations were tested in this permanently implanted ECoG array. Similarly Grothe et al. [49] found that gamma-band phase correlation of the local V4 population with spatially disjoint sub-populations of its V1 input was differentially modulated by attention. It was high with the input sub-population representing the attended stimulus, while simultaneously very low between the same V4 population and the other input-providing sub-population representing the irrelevant stimulus. Also here no spike-spike correlations could be tested.

Is CTC necessary for visual object recognition along the ventral stream? The standard approach to model the rapid recognition of objects despite substantial variation in appearance is based on a $\sim$50 ms interval rate code without relying on neuronal oscillations [30, 85]. Optogenetic experiments in V1 of alert mice by Histed and Maunsell [55] showed that behavioral effects of weak inputs to a visual RF depended only on the number of spikes over a period of 100 ms in the corresponding neuronal population, regardless of the temporal pattern of optogenetic stimulation by which the visual input was 'simulated'. No behavioral effects were seen for beta- or gamma-patterned light pulses. This linear summation in V1 of weak inputs showed the effectiveness of rate coding. However, other tasks like attention in primates [79] or conscious perception [87] could be enhanced by coherent neuronal oscillations.

An important test for CTC is attention in the counterstream between V4 and TEO in the ventral visual stream: Saalmann et al. [92] mapped the visual RF in the thalamus, in the Pulvinar (PU), relative to those in V4 and TEO and recorded spikes and field potentials from these interconnected sites in monkeys performing a visuo-spatial attention task. The PU synchronized activity between V4 and TEO according to attentional allocation. Their interpretation of the data, namely that the pathway V4 to PU to TEO relied on alpha oscillations, is surprising in view of the strong evidence that alpha and gamma are anti-correlated in V4. However, Saalmann et al. [92] and Schafer et al. [93] also found significant gamma synchronization between V4, PU and TEO. Both synchronizations could be used, since the brain usually does not rely on only one mechanism and since alpha in PU could be different from alpha between V1 and V4. The correlational evidence is ambiguous and [92, 93] disagree in their interpretation of the data.

Interesting findings have been made the dorsal visual stream on the interaction between FEF and LIP with V4 (see Fig. 10.2). In an early fundamental paper [20] found that the frontal cortex was the first to reflect the target location during top-down attention, whereas the parietal area LIP signaled the target earlier during bottom-up attention to 'pop-up' stimuli. Synchrony between frontal and parietal areas was stronger in lower frequencies during top-down attention and in higher frequencies during bottom-up attention. This result suggested that top-down and bottom-up signals arose from the frontal and sensory cortex, respectively, and different modes of attention could synchronize at different frequencies.

In the FEF the neuronal responses in the memory-guided saccade task can be classified according to their visual and/or saccade-related activity: visual neurons have only visual responses at onset of the visual stimulus, movement neurons have no visual responses, but have saccade-related activity, and visuo-movement neuron show both patterns. V4 and FEF are on the same hierarchical level and even form a closed driving loop in the scheme of [75]. Here the functionality of CTC can be seen on the cellular level: Gregoriou et al. [47] found that attention to a stimulus in their joint receptive field leads to enhanced oscillatory coupling between V4 and FEF, particularly at gamma frequencies. This coupling appeared to be initiated by FEF and was time-shifted by about 8–13 ms across a range of frequencies, the estimated propagation delay from FEF to V4. Gregoriou et al. [48] found that in the attention task, only visual and visuo-movement FEF neurons showed enhanced responses, whereas movement cells were unchanged. Importantly, visual, but not movement or visuo-movement cells, showed enhanced gamma frequency synchronization with activity in V4 during attention. Within FEF, beta synchronization was increased for movement cells during attention but was suppressed in the saccade task, consistent with the findings by Murthy and Fetz [81].

Neuronal oscillations in many more anatomically connected areas have been explored in the alert behaving monkey and have provided a fairly consistent interpretation of the macroscopic connectome. In humans Moratti et al. [80] studied using MEG the effective coupling of visual (VIS) and parietal (PA) areas during the recognition of a 'Gestalt' in two-tone images. They applied a statistical test (DCM, Friston et al. [44]) to the spectral densities of these sources in the gamma range. Thereby they tested different model circuits between VIS and PA and inferred a reverse hierarchical processing in this task (i.e. from top to bottom in Fig. 10.2e), correlated with gamma oscillations. This is an important discovery, but much more is needed to anchor in neuronal circuit dynamics the correlational evidence of this chapter.

## 10.5   Conclusion

Synergy between the billions of neurons in the mammalian brain, where every nerve cell alone is a sophisticated 'microprocessor', is obviously necessary for sensorimotor coordination and cognition. In absence of a central clock synergy is implemented in different circuits and tasks differently by neuronal firing rates and rhythms (see [3]). Coherent neuronal oscillations are one of the manifestations of the microscopic dynamics of the brain. There is a superficial analogy with the coherent quantum superpositions in the JQC: The computational power of a quantum computer is lost by decoherence beyond error correction, and similarly the highly optimized performance of neural circuits deteriorates when one perturbs coherence by pharmacological or optogenetic means (see e.g. [70] in Chap. 3). However, as Hirsted and Maunsell [55] found (see Chap. 4) fine temporal, and more specifically oscillatory codes are not always necessary to explain behaviour on the circuit level. Similarly, in the JQC computations by classical algorithms are still possible, when decoherence is too strong.

There is accumulating evidence that dysfunctions of the human nervous system can be due to the degradation of neural processes whose footprint are neural oscillations. For instance, gamma oscillations are linked to resonances implemented by PV interneurons (see [22, 24, 112]), and one of the causes of schizophrenia has been traced to alteration in PV interneurons which contribute to gamma disturbances and cognitive deficits [71]. Another crippling dysfunction of the brain is Parkinson's disease. Here the tremor frequency is about half that of beta oscillations, which are overexpressed in the brainstem subthalamic nucleus (RN). Again, as in the OB, neuronal oversynchronization is dysfunctionally related to behavior, and neuronal oscillations are the footprint of large-scale interactions in the motor system [98]. What is their microscopic basis? This awaits further study.

It is much easier to construct a JQC based on known physical laws than to find out experimentally the combination of possible mechanisms which synchronize multiareal neural activity in a mammalian brain. Synthesis is simpler than analysis, even if we are still uneasy about the foundations of quantum mechanics.

Several large-scale international collaborations promise new tools. One tool is connectomics as discussed in Chap. 2. Another is to capture nerve activity in an entire mammalian brain 'spike by spike' [5]. One can estimate the amount of data and their processing in this approach from a recent paper of [12]. Here recordings from only 512 channels were implemented. The analysis of spikes and LFPs from 256 simultaneously recorded microelectrodes was possible only because decades of research on the hippocampus [21] have established a meaningful mesoscopic picture, in terms of which these neuronal firing patterns could be analysed. For a billion neurons this is as absurd as trying to understand the implementation of the Shor algorithm on a JQC from microscopic QED.

One could also envisage to simulate a mammalian brain bottom up on a network of supercomputers. In 2011 the ambitious "Blue Brain Project" (BBP, Markram [77]) promised in its submission to the ETH Board "to simulate the human brain neuron-for-neuron within 10 years in order to understand mental diseases and to develop new computers". The inside report about the proposal and its reviews was published only 2 years later [35]. The 'proof of feasibility' in this proposal was based on the same schematic 'neocortical column' as the [88] simulation described in Chap. 3. The claims of the BBP created strong criticism among neuroscientists in a meeting in Bern 2012. The majority of the participants believed that the promises of the BBP will never be fulfilled (Google "Perspectives of High Power Computing in Neuroscience"). In 2013 the "Human Brain Project" has been accepted by the European Union as 'Flagship' for the next 10 years, with the BBP as 'Brain Simulation Platform' and as a leading house. Clearly the outlandish claims (discussed in this essay) in the 2011 proposal to the ETH were instrumental in the Swiss government's decision to spend 160 Million Euros on the BBP. For me, 'Flagships' have to sail on real water.

Goethe's advice in "Allerdings" was motivated by his disputes with the physicists on color vision. The great technical advances in the neurosciences show that today there is no separation between the "Schale", the sensorimotor periphery, and the "Kern", the immense interneuronal connectome. Neuronal oscillations, albeit mostly based on correlational evidence, have provided a new view on systems neuroscience and so have connectomes, multi-array spike sampling and computer simulations. The scope of neuroscience has become wider: The links to psychiatry and neurology are immediate. More distant is neurolinguistics (see [40]) studying human languages and music as they are implemented in the brain in a highly distributed manner. For their understanding the contribution of the humanities is essential. I think that Goethe would be happy today.

This essay was almost completed in 2013, when the abstracts of the annual meeting of the 'Society of Neuroscience' (SfN) appeared on the Internet. It will probably be obsolete before the appearance our book [14]. However, the many references in my review could be useful for young physicists to overcome their 'Higgs hangover'. There are wide open spaces in neurobiology where some of the basic question can be investigated with new techniques and theories, often inspired by physics. "Why are you still lacing, when others are racing?" (Advertisement for a Swiss ski boot, when laces became obsolete).

**Acknowledgements** I am grateful to my colleagues in the Physics Department of the ETHZ, Jürg Fröhlich, Hans-Ruedi Ott and Thomas Schulthess, for listening to my concerns about the BBP. I have learnt most about neuroscience from the late neurologist Volker Henn and from collaborations in the Institute of Neuroinformatics in Zürich. Constructive remarks by Pascal Fries and Kevan Martin have been very helpful, but all misrepresentations are mine.

# References

1. Aaronson, S.: Quantum Computing Since Democritus. Cambridge University Press, Cambridge (2013)
2. Adrian, E.D.: Olfactory reactions in the brain of the hedgehog. J. Physiol. **100**, 459–473 (1942)
3. Ainsworth, et al.: Rates and rhythms: a synergistic view of frequency and temporal coding in neuronal networks. Neuron **75**, 572–583 (2012)
4. Alberts, B., et al.: Molecular Biology of the Cell, 5th edn. Garland, New York (2008)
5. Alivisatos, A.P., et al.: The brain activity map. Science **339**, 2084–2085 (2013)
6. Alonso, J.-M., Chen, Y.: Receptive field. Scholarpedia **4**(1), 5393 (2009)
7. Anastassiou, C.A., et al.: Ephaptic coupling of cortical neurons. Nat. Neurosci. **14**, 217–223 (2011)
8. Bardeen, J., et al.: Theory of superconductivity. Phys. Rev. **108**, 1175–1204 (1957)

9. Barone, P., et al.: Laminar distribution of neurons in extrastriate areas projecting to visual areas V1 and V4 correlates with the hierarchical rank and indicates the operation of a distance rule. J. Neurosci. **20**, 3263–3281 (2000)
10. Bastos, A.M., et al.: Visual areas exert feedforeward and feedback influences through distinct frequency channels. Neuron **85**, 1–12 (2015)
11. Bastos, A.M., et al.: Communication through coherence with interareal delays. Curr. Opin. Neurobiol. **31**, 173–180 (2015)
12. Berényi, A., et al.: Large scale, high density (up to 512 channels) recording of local circuits in behaving animals. J. Neurophysiol. **111**, 1132–1143 (2013)
13. Bichot, N.P., et al.: Parallel and serial mechanisms for visual search in macaque area V4. Science **308**, 529–534 (2005)
14. Blanchard, P., Fröhlich, J.: Message from Quantum Science. Springer, Heidelberg (2015)
15. Borjigin, J., et al.: Surge of neurophysiological coherence and connectivity in the dying brain. Proc. Natl. Acad. Sci. **110**, 14432–14437 (2013)
16. Bosman, C.A., et al.: Attentional stimulus selection through selective synchronization between monkey visual areas. Neuron **75**, 875–888 (2012)
17. Buffalo, E.A., et al.: A backward progression of attentional effects in the ventral stream. Proc. Natl. Acad. Sci. **107**, 361–367 (2010)
18. Buffalo, E.A., et al.: Laminar differences in gamma and alpha coherence in the ventral stream. Proc. Natl. Acad. Sci. **108**, 11262–11267 (2011)
19. Burns, S.P., et al.: Is gamma-band activity in the local field potential of V1 cortex a "clock" or "filtered noise"? J. Neurosci. **31**, 9658–9664 (2011)
20. Buschman, T.J., Miller, E.K.: Top-down versus bottom-up control of attention in the prefrontal and posterior parietal cortices. Science **315**, 1860–1862 (2007)
21. Buzsaki, G.: Hippocampus. Scholarpedia **6**(1), 1468 (2011)
22. Buzsaki, G., Wang, X.-J.: Mechanisms of gamma oscillation. Ann. Rev. Neurosci. **35**, 203–225 (2012)
23. Carandini, M.: Area V1. Scholarpedia **7**(7), 12105 (2012)
24. Cardin, J.A., et al.: Driving fast-spiking cells induces gamma rhythm and controls sensory responses. Nature **459**, 663–666 (2009)
25. Carnevale, N.T., Hines, M.L.: The NEURON Book. Cambridge University Press, Cambridge (2006)
26. Chalk, M., et al.: Attention reduces stimulus-driven gamma frequency oscillations and spike-field coherence in V1. Neuron **66**, 114–125 (2010)
27. Courtin, J., et al.: Prefrontal parvalbumin interneurons shape neuronal activity to drive fear expression. Nature **505**, 92–96 (2014)
28. Da Costa, N.M., Martin, K.A.C.: Sparse reconstructions of brain circuits: or, how to survive without a microscopic connectome. NeuroImage **80**, 27–36 (2013)
29. Devoret, M.H., Schoelkopf, R.J.: Superconducting circuits for quantum information: an outlook. Science **339**, 1169–1174 (2013)
30. DiCarlo, J.J., et al.: How does the brain solve visual object recognition? Neuron **73**, 415–434 (2012)
31. Douglas, R.J., Martin, K.A.: Neural circuits of the neocortex. Ann. Rev. Neurosci. **27**, 419–451 (2004)
32. Eckhorn, R., et al.: Coherent oscillations: a mechanism of feature linking in the visual cortex? Biol. Cybern. **60**, 121–130 (1988)
33. El-Shamayleh, Y., et al.: Visual response properties of V1 neurons projecting to V2 in macaque. J. Neurosci. **33**, 16594–16605 (2013)
34. Engelhard, B., et al.: Inducing gamma oscillations and precise spike synchrony by operant conditioning via brain-machine interface. Neuron **77**, 361–375 (2013)
35. ETH Board Blue Brain Project: Internationale Begutachtung www.ethrat.ch/en/node/1361 (2013)
36. Fenno, L., et al.: The development and application of optogenetics. Ann. Rev. Neurosci. **35**, 389–412 (2011)

37. Fetz, E.E.: Volitional control of cortical oscillations and synchrony. Neuron **77**, 216–218 (2013)
38. Findlay, J., Walker, R.: Human saccadic eye movements. Scholarpedia **7**(7), 5095 (2012)
39. Freund, T., Kali, S.: Interneurons. Scholarpedia **3**(9), 4720 (2008)
40. Friederici, A.D.: The brain basis of language: from structure to function. Physiol. Rev. **91**, 1357–1392 (2011)
41. Fries, P.: Modulation of oscillatory neuronal synchronization by selective visual attention. Science **291**, 1560–1563 (2001)
42. Fries, P.: A mechanism for cognitive dynamics: neuronal communication through neuronal coherence. Trends Cogn. Sci. **9**, 474–480 (2005)
43. Fries, P., et al.: The effects of visual stimulation and selective visual attention on rhythmic neuronal synchronization in macaque area V4. J. Neurosci. **28**, 4823–4835 (2008)
44. Friston, K.J., et al.: DCM for complex-valued data: cross spectra, coherence and phase delays. NeuroImage **59**, 439–455 (2012)
45. Gaillard, R., et al.: Converging intracranial markers of conscious access. PLoS Biol. **7**, e1000061 (2009)
46. Gray, C.M., Singer, W.: Stimulus-specific neuronal oscillations in the cat visual cortex: a cortical functional unit. SfN Abstr. **13**, 404.3 (1987)
47. Gregoriou, G.G., et al.: High-frequency, long-range coupling between prefrontal and visual cortex during attention. Science **324**, 1207–1210 (2009)
48. Gregoriou, G.G., et al.: Cell-type-specific synchronization of neural activity in FEF with V4 during attention. Neuron **73**, 581–594 (2012)
49. Grothe, I., et al.: Switching neuronal inputs by differential modulations of gamma-band phase-coherence. J. Neurosci. **32**, 16172–16180 (2012)
50. Hameroff, S., Penrose, R.: Conciousness in the universe: A review of the 'Orch OR' theory. Phys. Life Rev. **11**, 39–78 (2013)
51. Helmstaedter, M., et al.: Connectomic reconstruction of the inner plexiform layer in the mouse retina. Nature **500**, 168–174 (2013)
52. Hepp, K.: Two models for Josephson oscillators. Ann. Phys. **90**, 285–294 (1975)
53. Hepp, K.: Coherence and decoherence in the brain. J. Math. Phys. **53**, 095222 (2012)
54. Hill, S.L., et al.: Statistical connectivity provides a sufficient foundation for specific functional connectivity in neocortical neural microcircuits. Proc. Natl. Acad. Sci. **109**, 16772–16773 (2012)
55. Histed, M.H., Maunsell, J.H.R.: Cortical neural populations can guide behavior by integrating inputs linearly, independent of synchrony. Proc. Natl. Acad. Sci. **111**, E178–E187 (2013)
56. James, W.: The Principles of Psychology, p 403. H. Holt, New York (1890)
57. Jia, X., et al.: Stimulus selectivity and spatial coherence of gamma components of the local field potential. J. Neurosci. **31**, 9390–9403 (2011)
58. Jia, X., et al.: No consistent relationship between gamma power and peak frequency in macaque primary visual cortex. J. Neurosci. **33**, 17–25 (2013)
59. Jia, X., et al.: Gamma and the coordination of spiking in early visual cortex. Neuron **77**, 762–774 (2013)
60. Josephson, B.: Possible new effects in superconductive tunneling. Phys. Lett. **1**, 251–253 (1962)
61. Kajikawa, Y., Schroeder, C.E.: How local is the local field potential? Neuron **72**, 847–858 (2011)
62. Kandel, E.R., et al.: Principles of Neural Science, 5th edn. McGraw-Hill, New York (2013)
63. Katzner, S., et al.: Local origin of field potentials in visual cortex. Neuron **61**, 35–41 (2009)
64. Khawaja, F.A., et al.: Pattern motion selectivity of spiking outputs and local field potentials in macaque visual cortex. J. Neurosci. **29**, 13702–13709 (2009)
65. Koch, C.: Biophysics of Computation: Information Processing in Single Neurons. Oxford University Press, New York (1999)
66. Koch, C., Hepp, K.: Quantum mechanics in the brain. Nature **440**, 611–612 (2006)

67. Koch, C., Hepp, K.: In: Chiao, R.J., Cohen, M.L., Legget, A.J., Phillips, W.D., Harper, jr C.L. (eds.) Visions of Discovery: New Light on Physics, Cosmology, and Consciousness. Cambridge University Press, Cambridge (2011)
68. Kreiter, A.K., Singer, W.: Stimulus-dependent synchronization of neuronal responses in the visual cortex of the awake macaque monkey. J Neurosci. **16**, 2381–2396 (1996)
69. Lashgari, R., et al.: Response properties of local field potentials and neighboring single neurons in awake primary visual cortex. J. Neurosci. **32**, 11396–11413 (2012)
70. Lepousez, G., Lledo, P.-M.: Odor discrimination requires proper olfactory fast oscillations in awake mice. Neuron **80**, 1–15 (2013)
71. Lewis, D.A.: Inhibitory neurons in human cortical circuits: Substrate for cognitive dysfunction in schizophrenia. Curr. Opin. Neurobiol. **26**, 22–26 (2014)
72. Lima, B., et al.: Synchronization dynamics in response to plaid stimuli in monkey V1. Cereb. Cortex **20**, 1556–1573 (2010)
73. Lucero, E., et al.: Computing prime factors with a Josephson phase qubit quantum processor. Nat. Phys. **8**, 719–723 (2012)
74. Markov, T., et al.: A weighted and directed interareal connectivity matrix for macaque cerebral cortex. Cereb. Cortex **24**, 17–36 (2013)
75. Markov, T., et al.: The anatomy of hierarchy: feedforward and feedback pathways in macaque visual cortex. J. Comp. Neurol. **522**, 225–259 (2013)
76. Markov, T., et al.: Cortical high-density counterstream architectures. Science **342**, 1238406 (2013)
77. Markram, H.: The blue brain project. Nat. Rev. Neurosci. **7**, 153–160 (2006)
78. Marshall, L., et al.: Boosting slow oscillations during sleep potentiates memory. Nature **444**, 610–613 (2006)
79. Miller, E.K., Buschman, T.J.: Cortical circuits for the control of attention. Curr. Opin. Neurobiol. **23**, 216–222 (2013)
80. Moratti, S., et al.: Dynamic gamma frequency feedback coupling between higher and lower order visual cortex underlies perceptual completion in humans. Neuroimage **86**, 470–479 (2013)
81. Murthy, V.N., Fetz, E.E.: Coherent 25- to 35-Hz oscillations in the sensorimotor cortex of awake behaving monkeys. Proc. Natl. Acad. Sci. **89**, 5670–5674 (1992)
82. Nicolic, D., et al.: Gamma oscillations: precise temporal coordination without a metronome. Trends Cogn. Sci. **17**, 54–55 (2013)
83. Palanca, B.J.A., DeAngelis, G.C.: Does neural synchrony underlie visual feature grouping? Neuron **46**, 333–346 (2005)
84. Pannasch, U., Rouach, N.: Emerging role for astroglial networks in information processing: from synapse to behavior. Trends Neurosci. **36**, 405–417 (2013)
85. Poggio, T., Serre, T.: Models of visual cortex. Scholarpedia **8**(4), 3516 (2013)
86. Ray, S., Maunsell, J.H.R.: Differences in gamma frequencies across visual cortex restrict their possible use in computation. Neuron **67**, 885–898 (2010)
87. Rey, H.G., et al.: Timing of single-neuron and local field potential responses in the human medial temporal lobe. Curr. Biol. **24**, 299–304 (2014)
88. Reimann, M.W., et al.: A biophysically detailed model of neocortical LFPs predicts the critical role of active membrane currents. Neuron **79**, 375–390 (2013)
89. Roberts, M.J., et al.: Robust gamma coherence between macaque V1 and V2 by dynamic frequency matching. Neuron **78**, 523–536 (2013)
90. Roelfsema, P.R., et al.: Synchrony and covariation of firing rates in the primary visual cortex during contour grouping. Nat. Neurosci. **7**, 982–991 (2004)
91. Roelfsema, P.R., et al.: Alpha and gamma oscillations as markers of feedforward and feedback processing in areas V1 and V4 of monkey visual cortex. SfN Abstr. **1**, 623.03 (2012)
92. Saalmann, Y.B., et al.: The Pulvinar regulates information transmission between cortical areas on attention demands. Science **337**, 753–756 (2012)
93. Schafer, R.J., et al.: Visual and attentional functions of the lateral pulvinar. SfN Abstr. 673.10 (2012)

94. Seung, S.: I am my connectome (2010). Video on TED.com
95. Shafranjuk, S.E., Ketterson, J.B.: Principles of josepson-junction-based quantum computation. In: Bennemann, K.H., Ketterson, J.B., (eds.) Superconductivity, vol 1. Spinger, Berlin (2008)
96. Shadlen, M.N., Movshon, J.A.: Synchrony unbound: a critical evaluation of the temporal binding hypothesis. Neuron **24**, 67–77 (1999)
97. Shepherd, G., (ed.): The Synaptic Organization of the Brain. Oxford University Press, Oxford (2004)
98. Shimamoto, S.A., et al.: Subthalamic nucleus neurons are sychronized to primary motor cortex local field potentials in Parkinson's disease. J. Neurosci. **33**, 7220–7233 (2013)
99. Singer, W.: Time as coding space in neocortical processing: a hypothesis. In: Gazzaniga, M.S. (ed.) The Cognitive Neurosciences, pp. 91–104. MIT, Cambridge, MA (1997)
100. Singer, W.: Binding by synchrony. Scholarpedia **2**(12), 1657 (2007)
101. Spaak, E., et al.: Layer-specific entrainment of gamma-band neural activity by the alpha rhythm in monkey visual cortex. Curr. Biol. **22**, 2313–2318 (2012)
102. Spruston, N.: Pyramidal neuron. Scholarpedia **4**(5), 6130 (2009)
103. Squire, L.R., et al.: Fundamental Neuroscience. Elsevier, Amsterdam (2013)
104. Squire, R.F., et al.: Frontal eye field. Scholarpedia **7**(10), 5341 (2012)
105. Tegmark, M.: Importance of quantum decoherence in brain processes. Phys. Rev. E **61**, 4194–4206 (2000)
106. Thiele, A., Stoner, G.: Neuronal synchrony does not correlate with motion coherence in cortical area MT. Nature **421**, 366–370 (2003)
107. Traub, R.D., et al.: Single-column thalamocortical network model exhibiting gamma oscillations, sleep spindles, and epileptogenic bursts. J. Neurophysiol. **93**, 2194–2232 (2005)
108. Van Kerkoerle, et al.: Alpha and gamma oscillations characterize feedback and feedforward processing in monkey visual cortex. Proc. Natl. Acad. Sci. **111**, 14332–14341 (2014)
109. Vezoli, J., et al.: Extracting structure from function: Inter-areal causal interactions at gamma and beta rhythms reveal cortical hierarchical relationships. SfN Abstr. 723.10 (2012)
110. Vinck, M., et al.: Attentional modulation of cell-class-specific gamma-band synchronization in awake monkey area V4. Neuron **80**, 1077–1089 (2013)
111. Von der Malsburg, C.: The correlation theory of brain function. In: Domany, E., Van Hemmen, J.L., Schulten, K. (eds.) MPI Biophysical Chemistry, Internal Report 81-2. Reprinted in Models of Neural Networks II (1994). Spinger, Berlin (1981)
112. Wang, X.-J.: Neurophysiological and computational principles of cortical rhythms in cognition. Physiol. Rev. **90**, 1195–1268 (2010)
113. Womelsdorf, T., et al.: Gamma-band synchronization in visual cortex predicts speed of change detection. Nature **439**, 733–736 (2006)
114. Womelsdorf, T., et al.: Modulation of neuronal interactions through neuronal synchronization. Science **316**, 1609–1612 (2007)
115. Xing, D., et al.: Stochastic generation of gamma band activity in the primary visual cortex of awake and anesthetized monkeys. J. Neurosci. **32**, 13873–13880 (2012)

# Chapter 11
# Local Properties, Growth and Transport of Entanglement

**Roland Omnès**

## 11.1 Introduction

To describe the nature of entanglement, Schrödinger considered as an example the case of two quantum systems $A$ and $B$, initially independent, which begin to interact at some time and separate again after some more time [1]. Both $A$ and $B$ are initially in a pure state but, whereas this is still true of the compound system $AB$ after their interaction, it is no more true for each one of them separately. Schrödinger viewed this property of lasting entanglement between wave functions as the most characteristic aspect of quantum mechanics, which estranges it farther from classical physics than anything else and leads finally to a famous contradiction between the quantum principles and the uniqueness of measurement data [2].

A peculiarity of entanglement is to be an intrinsic property of the state of a compound system (an eigenfunction of the density matrix $\rho_{AB}$ is not a product of an eigenfunction of $\rho_A$ and an eigenfunction of $\rho_B$). This feature distinguishes entanglement from the "physical properties", which were associated by von Neumann with projection operators in Hilbert space and, on the contrary, tended to diminish the gap between classical and quantum physics [3]. This is probably the main reason why some apparently natural questions become so obscure when entanglement is concerned, such as asking if entanglement can grow as time goes on and reach wider regions, or whether one can give a measure for the amount of entanglement between two distant subsystems sharing interactions with a third one (see for instance [4]).

This type of question will be the topic of the present discussion. One will deal with the locality, growth and transport of entanglement, as well as with

R. Omnès (✉)
Laboratoire de physique théorique, Université de Paris-Sud, Orsay, France
e-mail: roland.omnès@th.u-psud.fr

© Springer-Verlag Berlin Heidelberg 2015                                                      257
P. Blanchard, J. Fröhlich (eds.), *The Message of Quantum Science*, Lecture Notes in Physics 899, DOI 10.1007/978-3-662-46422-9_11

local measures of its amount, but not through usual approaches using Schmidt factorization [5] or algorithmic entropy, which has too strong limitations [4, 6].

The first step, introduced in Sect. 11.2, consists in pointing out a topological and dynamical character of entanglement, rather than dealing only with sums of tensor products of some state vectors, which is its the usual expression of entanglement and remains purely algebraic. This new approach is explained for convenience in a case where System $A$ is an energetic alpha particle and System $B$ a Geiger counter containing an argon gas. Using Feynman histories or perturbation theory, one notices that the property of some atom in $B$ to be already entangled (or not yet entangled) with $A$ at some time $t$ has a topological meaning in terms of graphs. As a matter of fact, it refers to some clustering between this atom and the alpha particle $A$, which occurred (or did not occur) before time $t$ in some histories or some perturbation graphs [7–11].

This recognition of a topological character of entanglement is used in Sect. 11.3 to rid it from an apparent restriction to perturbation theory and express it through a convenient refinement of the Schrödinger equation, in which the partial and local characters of entanglement are taken into account. This is indeed a refinement and not a modification because it does not imply any change in standard wave functions and only recognition of the presence of more information in their mathematical expressions as functions of atoms positions. The corresponding mathematical aspects are considered in Sect. 11.4. The approach is then extended to quantum fields in Sect. 11.5, to make clear both its generality when dealing with quantum measurements and its restriction to non-relativistic and macroscopic measuring devices.

This approach through field theory provides in a rather simple way a local "measure of entanglement" $f_1(x, t)$, expressing explicitly the relative amount of entanglement at time $t$ in a small macroscopic space region around a point $x$. Using this macroscopic meaning together with standard arguments from kinetic physics, one shows then in Sect. 11.6 that $f_1(x, t)$ satisfies a rather simple *nonlinear* partial differential equation, which is solved in special cases. The most remarkable consequence of non-linearity is that entanglement grows behind the front of a wave moving at a computable velocity and extending from the region in which the $A - B$ interactions occurred. Finally, some possible consequences, including some already proposed earlier in terms of algorithmic entropy [12], are briefly indicated in Sect. 11.8.

## 11.2 A Topological Aspect of Entanglement

One will deal mostly with the case of an alpha particle $A$ having a straight-line trajectory and entering into the Geiger counter $B$ at some time 0 Fig. 11.1 shows a perturbation graph for the events occurring inside the detector before some time $t$. The same figure could represent as well a Feynman history during the same interval of time, except that the straight lines representing propagation of the alpha particle

**Fig. 11.1** The clustering topology of entanglement

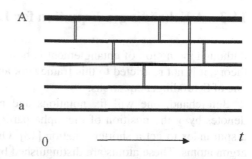

and of the argon atoms would have to be highly wiggling. The heavy horizontal line represents propagation of the alpha particle and the light horizontal lines represent propagation of atoms. We assume that all the interactions are governed by potentials with a very short range for a pair of argon atoms and a less short range for the alpha-atom interactions, these interactions being represented by vertical lines.

No explicit calculation is needed to see that, in this graph, some atoms have become entangled with $A$ at time $t$ because they interacted directly with $A$ or they interacted with an atom that had already interacted with $A$ previously, or they interacted with an atom that had interacted with an atom that had interacted, and so on. In Fig. 11.1, the atom denoted by $a$ is not entangled with the alpha particle $A$ at time $t$, according to this graph.

This rather trivial expression of entanglement has several significant consequences. First of all, it is clearly a topological property, either of perturbation graphs or of Feynman histories. Secondly, it shows that the mechanism of entanglement is very similar to a contagion: an atom can catch entanglement directly from the alpha particle, but it can also catch it from an already entangled atom.

This topology of entanglement can also be viewed as a form of clustering if one considers that the set of entangled particles constitutes a cluster in the sense of graph theory. In that sense, the idea is not new and it already occurred in other domains of physics where it played a significant part. It first occurred in quantum statistical physics where it was used to establish the proportionality to volume of extensive thermodynamic quantities [7, 8]. It occurred also in $S$-matrix theory [9] where it was often expressed intuitively as the property according to which an experiment in Geneva is insensitive to another experiment in Brookhaven. Here it would mean that an atom that is far enough from the alpha track will feel very little influence from the alpha particle, at least for some time. The formulation these cluster properties has also been thoroughly expressed in Weinberg's book on quantum field theory [10], where it is used as a basic property in the construction of effective quantum fields. One encounters also some other forms of clustering in the scattering theory of several particles [11].

One could be tempted to give a special name to these local and topological properties of entanglement, which are familiar to everybody. The name "intricacy", for instance, would seem proper, but it could also become more an impediment than a help in a first introduction and we shall therefore keep the name of "entanglement" with a slightly refined meaning.

## 11.3 A Schrödinger Equation for Local Entanglement

Although the notion of entanglement is best understood by means of perturbation theory, it is not restricted to this framework and can be cast directly into the wider one of Schrödinger equations.

Beforehand, one will fix notations and make the model more precise. One denotes by $y$ the position of the alpha particle, including tacitly in this notation a spin index to get a shorter notation [13]. One denotes by $N$ he total number of argon atoms. These atoms are distinguished by an index $n$ going from 1 to $N$, the position of an atom is denoted by $x_n$ and the set of all these positions (i.e., a point in the configuration space) is denoted simply by $x$. The initial state of the alpha particle is supposed a pure state with wave function $\chi(y)$. The initial state of the Geiger counter $B$, which is necessarily mixed, is expressed by a density matrix $\rho_B$ and one will concentrate on one of its eigenfunctions, denoted by $\psi(x)$. Initially, the $A - B$ state under consideration is therefore

$$\Psi(x, y; 0) = \psi(x; 0)\chi(y; 0). \tag{11.1}$$

Its evolution is governed by the Schrödinger equation

$$i\hbar \mathrm{d}\Psi/\mathrm{d}t = (K_A + K_B + U + V)\Psi, \tag{11.2}$$

where $K_A$ and $K_B$ denote the kinetic energies of the alpha particle and of the argon atoms. For the time being, we assume that all the interactions can be represented by a potential so that the potential $U$ is a sum of interactions of the alpha particle with the various atoms, and $V$ is the sum of potential interactions between pairs of atoms. Distinguishing these atoms by an index $n$, one has:

$$U = \sum_n U(y, x_n), V = \sum_{(n,n')} V(x_n, x_{n'}). \tag{11.3}$$

We must now describe individual entanglement of atoms and, labeling an atom by a label $n$, one introduces an entanglement index taking the value 1 (in the case of entanglement) or 0 (in the case of no entanglement). Perturbation theory showed that the direct generation of entanglement by the alpha particle leads to a change of index $0 \rightarrow 1$. An interaction between two atoms leads to the possible changes $00 \rightarrow 00$, $01 \rightarrow 11$, $10 \rightarrow 11$ or $11 \rightarrow 11$, and all these transitions can be expressed by using $2 \times 2$ matrices

$$P_{n0} = \begin{pmatrix} 0 & 0 \\ 0 & 1 \end{pmatrix}, \quad P_{n1} = \begin{pmatrix} 1 & 0 \\ 0 & 0 \end{pmatrix}, \quad S_n = \begin{pmatrix} 0 & 1 \\ 0 & 0 \end{pmatrix} \tag{11.4}$$

$P_{n0}$ and $P_{n1}$ are projection operators conserving respectively a state of non-entanglement or of entanglement whereas $S_n$ brings a state with no entanglement into an entangled state. These are the only such operators we need because the entanglement of one atom is obtained irreversibly under an interaction, so that although $S_n$ is not hermitian, its hermitian conjugate can never occur in an history of entanglement. These matrices could also be written in terms of Pauli matrices (dropping the index $n$) by

$$P_0 = (I - \sigma_z)/2, \quad P_1 = (I + \sigma_z)/2, \quad S = \sigma_+^{\mp}(\sigma_x + i\sigma_y)/2 \qquad (11.5)$$

Using again perturbation theory, one can express algebraically the generation and contagion of entanglement by means of a simple replacement in the potentials insuring automatically these properties, namely for generation:

$$U(y, x_n) \to U_n \equiv \quad U(y, x_n) \quad A_n, \text{ with } A_n = (S_n P_{n0} + P_{n1}). \qquad (11.6)$$

The first term in the $2 \times 2$ matrix $A_n$ describes the generation of entanglement from a non-entangled atom $n$, which is recognized as non-entangled by $P_{n0}$, and brought to entanglement by $S_n$. The second term expresses that, when the alpha particle interacts with an already entangled atom, it conserves this entanglement, recognized and maintained by $P_{n1}$.

The conservation or contagion of entanglements in the interaction of two atoms $n$ and $n'$ is similarly expressed algebraically through the replacement

$$V(x_n, x_{n'}) \to V_{nn'} \equiv V(x_n, x_{n'}) \quad O_{nn'}; \qquad (11.7)$$

$$\text{with } O_{nn'} = P_{n0} \otimes P_{n'0} + P_{n1} \otimes P_{n'1} + S_n P_{n0} \otimes P_{n'1} + P_{n1} \otimes S_{n'} P_{n'0} \qquad (11.8)$$

It becomes clear then that the evolution of entanglement is not intrinsically linked with perturbation theory but has a wider meaning. The first step along that direction consists in characterizing the state of entanglement for the $N$ atoms by strings $q$ consisting of $N$ bits of entanglement indices having the value 0 or 1. There are $2^N$ such strings. To each string, one associates a wave function $\Phi_q(x, y; t)$. The set of these wave functions can be considered as a $2^N$-dimensional vector $\Phi$ depending on $(x, y, t)$ for which one writes down an evolution equation

$$i\hbar d\Phi/dt = H'\Phi. \qquad (11.9)$$

This evolution expresses the evolution of the wave functions $\Phi_q$ as well as generation and contagion of entanglement or, more explicitly

$$i\hbar d\Phi_q/dt = (K_A + K_B)\Phi_q + \sum_{q'} (U_{qq'} + V_{qq'})\Phi_{q'}, \qquad (11.10)$$

with the notations

$$K_A = -(\hbar^2\nabla_y^2)/2m_A, \quad K_B = -\sum_n \hbar^2\nabla_{x_n}^2/2m_a,$$

$$U_{qq'} = \sum_n U_n I_{qq'}^{(n)}, V_{qq'} = \sum_{nn'} V_{nn'} I_{qq'}^{(nn')}. \qquad (11.11)$$

In these formulas, the matrix $I_{qq'}^{(n)}$ acts through the $2 \times 2$ matrix $A_n$ on the entanglement index $\mu$ of atom $n$ and does not affect the entanglement indices of other atoms, which are therefore the same in $\Phi_q$ and $\Phi_{q'}$ Similarly, the matrix $I_{qq'}^{(nn')}$ acts through the $4 \times 4$ matrix $O_{nn'}$ on the entanglement indices of the pair of atoms $(nn')$ and does not affect the entanglement of other atoms.

The set of equations (11.10) is of course still more involved than the standard Schrödinger equation (11.2), since it contains in principle more information. The differential matrix operator $H'$ in (11.9) is not self-adjoint, because the generation and the contagion of entanglement are not time-reversible. Nevertheless, if one assumes convenient bounds on the potentials and their derivatives, Eq. (11.9) has essentially the same existence properties as the Schrödinger equation (11.2), as can be shown for instance by microlocal analysis [14].

To understand better the relation between this detailed description of entanglement and the global one in the Schrödinger equation (11.2), one notices that the action of a matrix $A_n$ in (11.12) brings out always an entangled state of atom $n$, whatever could have been the previous state of entanglement of this atom. More generally, if one introduces a $2^N \times 2^N$ matrix $\Pi$ as the tensor product of the $N$ matrices $A_n$:

$$\Pi = \prod_n^{\otimes} A_n, \tag{11.12}$$

the action of this matrix on any string $q$ of entanglement indices brings it onto the string of complete entanglement $q_0 = (1111111\ldots)$. If one defines therefore the Schrödinger wave function by

$$\Psi q_0 = \Pi \Phi, \tag{11.13}$$

one finds that this function satisfies the standard evolution (11.2) as a consequence of (11.9). In terms of wave functions, the relation (11.13) is of course equivalent to $\Psi = \sum_q \Phi_q$.

To conclude on this point, one may say that a rather detailed description of the generation and contagion of entanglement is perfectly compatible with the quantum principles. One can also add a last remark concerning indistinguishable atoms. The argon atoms in our example are indistinguishable and the Schrödinger wave function $\Psi$ is therefore invariant under their permutations. In place of the $2^N$ functions $\Phi_q$, one can deal therefore with a set of $N + 1$ functions $\Xi_r (r = 0, 1, \ldots, N)$ in which $r$ atoms are entangled with the alpha particle and $N - r$ are not. Such a function is symmetric under permutations of the entangled atoms as well as under permutations of the non-entangled atoms. One has furthermore in that case

$$\Psi = \sum_r \Xi_r. \tag{11.14}$$

## 11.4 Some Mathematical Aspects of the Construction

It could seem at first sight that this refinement in the idea of entanglement goes against the familiar Hilbert space formulation of quantum mechanics. This is not so however and the present idea is presumably a special case of a wider one, which would be that there is more in wave functions and the details of their evolution than in the algebra of Hilbert space. The local properties of entanglement and their evolution would probably appear more obvious if one could actually perform explicit computations for a large system and look at a wave function in detail at every step of its evolution. One will try briefly to say a little more on that point now.

It will be convenient to consider this question in the case of the symmetric functions $\Xi_j$. The evolution equation (11.9) is linear and one can therefore formulate it by means of some vector spaces. One may consider for instance that a wave function $\Xi_j$. is associated abstractly with a vector $|\Xi_r\rangle$ in a Hilbert space $E_r$ Two different such spaces, corresponding to different steps of entanglement $r \neq r'$, are not orthogonal usually ($\langle \Xi_r | \Xi_{r'} \rangle$ does not generally vanish). The set $E'$ of these $N$ vector spaces could be considered as a sheaf in the sense of sheaves theory [15], but it would not be of great help and one could consider it rather as a vector space $E$ by allowing addition between vectors $|\Xi_r\rangle$ belonging to different spaces $E_r$, as in (11.14).

But the algebraic meaning of this summation needs more attention. Every Hilbert space $E_r$ is isomorphic to the standard Hilbert space $E$ to which the standard wave function $\Psi$ in (11.2) belongs. Said otherwise, $E_r$ is a copy of $E$. Every $E_r$ is therefore a Hilbert space by itself, but two of them are not intrinsically related. All these vector spaces have only the vector zero in common and this meeting point is significant: The evolution equation (11.9) is continuous in time. At time 0 before entanglement, there is only a state vector $|\Xi_0\rangle$ with $r = 0$. The evolution of entanglement under (11.9) goes always later on through the birth of more and more new levels of $r$, starting at an infinitesimal level during an infinitesimal lapse of time from a lower level of $r$. One can expect that after enough time (to be considered in Sect. 11.6), a completely entangled state $|\Xi_N\rangle$ will survive alone and evolve according to the standard Schrödinger equation.

The Hilbert space $E_N$ appears then as a reference in the process of evolution in entanglement and the geometric meaning of (11.14) consist then in projecting all the vector spaces $E_r$ on $E_N$ by means of the matrix $A$, and thus regaining explicitly the standard Hilbert space formulation of quantum mechanics.

The process is not reversible on the other hand: One cannot construct the functions in the sheaf $E'$ from the function $\Psi$ by means of projection operators in the standard Hilbert space $E$, which means that the development of entanglement cannot be considered as a Von Neumann's "physical property" [3]. This development *belongs to the history* of $\Psi$ and not on its value $\Psi(t)$ at a specific time $t$. Whether or not this feature might call attention to a comparison between the respective meanings of Feynman histories and of the Heisenberg–Schrödinger formulation of quantum mechanics could be a subject of more reflection, but I shall not venture along that direction.

## 11.5    An Approach Using Quantum Field Theory

Two points need further consideration in the previous discussion: The first one is a more explicit account for the evolution of entanglement among undistinguishable atoms. The second one is concerned with a measure of entanglement: The idea of a probability of entanglement for a specific atom is neither of much practical nor of conceptual interest. One would be much more interested in getting local properties of entanglement. For instance, one would like to dispose of a measure of entanglement for all the atoms in some region of space inside a detector. One expects that, when an alpha particle has just crossed rapidly the gas, there is much more entanglement near its track than farther away. On the other hand, the growth of clustering with time must imply that, gradually, the gas becomes more entangled until entanglement with $A$ is complete. One would like to make these features more precise.

These questions are directly concerned with locality and they become clearer when the framework of quantum field theory is used [13]. To begin with, we shall still ignore entanglement and recall a few points concerning this approach: The atoms are described by a field $\varphi(x)$ where the notation $x$ involves again the position of an atom and eventual spin indices. The field satisfies either commutation or anti-commutation properties according to the spin value, but the two cases are very similar and we shall retain only for illustration the case of Bose–Einstein statistics, with commutation relations

$$[\varphi(x), \varphi(x')] = 0 \quad [\varphi(x), \varphi^\dagger(x')] = \delta(x - x'). \tag{11.15}$$

Denoting the vacuum state by $|0\rangle$, a state of the gas with $N$ atoms with wave function $\psi(\{x\})$ is given by the expression

$$|\psi\rangle = \int \{dx\}\psi(\{x\}) \prod_{N=1}^{N} \phi^\dagger(x_r) |0\rangle. \tag{11.16}$$

(Notice the difference between the notation $x$ for localization of the field $\varphi(x)$ and the notation $\{x\}$ for all the variables in the wave function.)

The Hamiltonian de is given by

$$H = \int dx\varphi^\dagger(x)(-\nabla^2/2m)\varphi(x) \tag{11.17}$$

$$+ (1/2) \int dxdx'\varphi^\dagger(x)\varphi^\dagger(x')V(x, x')\varphi(x)\varphi(x'),$$

where the factor $1/2$ in the last term is due to the fact that a pair of atoms with positions $x$ and $x'$ occurs twice in this expression with the orderings $(x, x')$ and $(x', x)$.

To describe entanglement, one introduces two fields $\varphi_0(x)$ and $\varphi_1(x)$, standing respectively for no entanglement and entanglement. Both of them, together with the adjoint fields, satisfy the commutation relations (11.15) and moreover commute together and with the adjoint fields (for instance: $[\varphi_0(x), \varphi_1^\dagger(x')] = 0$). The evolution operator is then $H_0 + H_1 + H_{01} + D_0 + D_1$ where in this sum, $H_0$ and $H_1$ represent the independent evolutions of non-entangled and entangled atoms; they have the same form as (11.17) after a convenient replacement of $\varphi(x)$ by $\varphi_0(x)$ or $\varphi_1(x)$. The coupling $H_{01}$ represents the contagion of entanglement is given by

$$H_{01} = \int dx dx' \varphi_1^\dagger(x) \varphi_1^\dagger(x') V(x, x') \varphi_0(x) \varphi_1(x'). \tag{11.18}$$

The generation of entanglement by the alpha particle and its absence of renewal when the alpha particle interacts again with an already entangled atom are described respectively by the two terms $D_0$ and $D_1$. They involve a field $\alpha(y)$ describing the alpha particle and are given by

$$D_0 = \int dx dy \alpha^\dagger(y) \varphi_1^\dagger(x) U(x, y) \alpha(y) \varphi_0(x), \tag{11.19}$$

$$D_1 = \int dx dy \alpha^\dagger(y) \varphi_1^\dagger(x) U(x, y) \alpha(y) \varphi_1(x). \tag{11.20}$$

We shall not write down explicitly the corresponding evolution equations for the set of wave functions $\{\Xi\}$ and only add a comment: Whereas the relation between the set of entangled wave functions $\Xi_r(t)$ and the standard wave function $\Psi(t)$ is trivial, since it amounts to a summation, there is no simple relation between $\varphi(x)$ and the pair of fields $\varphi_0(x)$ and $\varphi_1(x)$. This may be interpreted as a warning: Apparently, and as could have been expected, the existence of local properties of entanglement is closely linked with the consideration of macroscopic *objects*, whose elements behave in a non-relativistic way in a reference frame. This is by of course in agreement with the relation of the growth and transport of entanglement with kinetic physics, which will be considered in the next section. Whether or not there could be a wider theoretical framework allowing an extension of these methods is another question and we shall not try to look for it presently.

A great power of the field approach is nevertheless its flexibility: It applies to every kind of particles, either fermions or bosons, to systems containing a unique species of atoms or different species, to relativistic as well as non-relativistic behavior for the measured quantum systems. Whereas the approach in Sect. 11.3 holds when all the atoms in the detector are in their ground state, the present approach works also when there are excited states, ions, free electrons or photons. In a solid system or subsystem, one can use again fields to describe phonons or conducting electrons. In that sense, one may presume that an approach, in which entanglement is considered as a topological clustering property has a much wider validity domain than can be explored in the present first encounter.

## 11.6   Growth and Transport of Entanglement

As a next step, we shall go from the quantum description of entanglement to its macroscopic behavior implying local growth and transport inside the detector $B$. It will be convenient for this purpose to cover $B$ by a set of macroscopic (though rather small) Gibbs cells $C_\beta$. Denoting by $x$ the center of one of these cells, by $V_\beta$ its volume, by $\rho_\beta$ the corresponding density matrix and by $n$ the average number of atoms per unit volume, one can define two local measures for entanglement $f_1(x)$ and for no entanglement $f_0(x)$ through

$$f_j(x) = (nV_\beta)^{-1} \int_{C_\beta} Tr\{\rho_\beta \varphi_j^\dagger(y)\varphi_j(y)\}d^3y, \tag{11.21}$$

with $j$ equal to 1 or 0.

These quantities are positive. We shall assume furthermore that they satisfy the relation

$$f_1(x) + f_0(x) = 1. \tag{11.22}$$

Notice however that a justification of this equation is typical of the gap existing between kinetic theory and fundamental quantum theory. The evolution operator $H_0 + H_1 + H_{01} + D_0 + D_1$ in the previous section commutes with the sum

$$\int_B dx\{\varphi_1^\dagger(x)\varphi_1(x) + \varphi_2^\dagger(x)\varphi_2(x)\}, \tag{11.23}$$

so that the sum of average numbers of entangled and of non-entangled atoms in the detector is a constant of motion, equal to the total number $N$ of atoms. But the postulated equation (11.22) assumes a similar conservation in every Gibbs cell, i.e., at a local level. It is suggested by the short range of interactions and can be expected to be valid, up to small fluctuations, but a rigorous proof is lacking. It can yield on the other hand quite a few interesting and suggestive consequences, which we now examine.

When making a transition from quantum elementary effects to a kinetic behavior at larger scales, one must always rely ultimately on a classical description of atoms or other carriers undergoing a random motion, whereas elementary exchanges in transported quantities are derived from quantum mechanics [16]. Although entanglement is in some sense a paradigm of quantum physics (since it is not even associated with an observable) its exchange properties are remarkably simple because of the contagion in these exchanges.

Assuming the gas in the detector at thermal equilibrium with temperature $T$, the average velocity of an atom is $v = (3k_BT/2m)^{1/2}$. One introduces also the mean free time $\tau$ and mean free path $\lambda = v\tau$ of the atoms. If one considers the motion of atoms as a random walk, the consideration of entanglement takes two significant aspects: On one hand, an entangled atom keeps its entanglement under collisions with other atoms, whether these atoms are themselves entangled or not. The corresponding evolution of the measure of entanglement $f_1(x,t)$ is governed in that case by a diffusion equation

$$(\partial f_1/\partial t)_{diffusion} = D\nabla^2 f_1, \tag{11.24}$$

where one the diffusion coefficient as given by $D = (1/6)\lambda^2/\tau$.

There is in addition a local growth in entanglement. Its probability of occurrence for an individual non-entangled atom in the cell $C_\beta$ during a short time interval $\delta t$ is $f_1(x)\delta t/\tau$ whereas the measure for these non-entangled atoms is $f_0(x)$. Since contagion makes such an atom entangled, the corresponding increase in local entanglement is given by

$$(\partial f_1/\partial t)_{contagion} = f_1 f_0/\tau. \tag{11.25}$$

When the two effects of diffusion and contagion are put together with account of the relation (11.22), one obtains a nonlinear partial differential equation for the evolution of entanglement, which is

$$\partial f_1/\partial t = f_1(1 - f_1)/\tau + D\nabla^2 f_1 \tag{11.26}$$

In principle, one should add a source term in this equation to account for entanglement from direct collisions with the alpha particle, but one can leave this effect out when considering the transport of entanglement at some distance from the alpha track, or when one looks only for the mathematical consequences of (11.26).

Simple numerical estimates in one dimension show that it would be impossible to obtain a solution satisfying everywhere the conditions $1 \geq f_1(x,t) \geq 0$. One needs therefore boundary conditions for Eq. (11.26) and, fortunately, these conditions come out easily from the basic property of contagion: In a one-dimensional model (in which case the thermal velocity is $v' = 3^{-1/2}v$), one notices that after a collision between an entangled atom and a non-entangled one, both atoms are entangled after collision and therefore, entanglement is carried away on average in both directions of positive and negative directions of $x$ at the velocity $v'$. This is also true in three-dimensional space where one expects that entanglement is carried away from its origin (on the alpha track) at the velocity $v'$ (which coincides with the velocity of sound in the case of a random walk of atoms for which also $D = (1/6)\lambda^2/\tau$).

The corresponding boundary condition is then that $f_1(x,t)$ should vanish outside a space region including the source of entanglement on the track and having a boundary moving away from the track at the velocity v', which is the velocity of sound. Said otherwise, $f_1(x,t)$ is positive and non-vanishing behind a wave front moving at that velocity. Farther away, beyond the front, $f_1(x,t)$ must vanish. The geometric behavior of the contagion of entanglement is therefore very similar to the behavior of self-maintaining waves, which are known typical of some nonlinear equations of evolution, as the case is for (11.26) [17].

## 11.7  An Example

As an example, one considered a model where the source of entanglement stands on a plane, located at a large distance on the left ($x < 0$) and the measure of entanglement $f_1(x, t)$ behaves accordingly as a function $g(x - v't)$, far enough from this source. Using again the previous conventions, this function should obey the nonlinear differential equation

$$3^{-1/2}dg/dz + g(1 - g) + (1/6)d^2g/dz^2 = 0, \tag{11.27}$$

where the units of length and time have been taken as the mean free path and the mean free time of atoms. When the front is at $z = 0$ (i.e., $g(0) = 0$) and the boundary condition $g(-\infty) = 1$ is used, the solution of this equation is shown in Fig. 11.2 (Note: One could get the impression when looking at this figure that $g'(0)$ vanishes, but the computation gives only a small number $g'(0) \approx -0.06$).

These properties are presumably general, although they would require some adaptation in various circumstances. When entanglement originates for instance from an external collision on a solid box enclosing the gas, it must be carried by phonons and grows initially through phonon-phonon collisions (in which case, it move again at the velocity of sound). In the case of entanglement for an electric signal in a conductor, the wave front will move at the Fermi velocity. When entanglement is carried by photons, the velocity of entanglement waves becomes $c$. More cases have been considered and, generally, they suggested combinations of different waves of entanglement with different heights moving at different velocities.

In any case, the final outcome will be a complete entanglement after some lapse of time. Remarkably, this time is not very short and, in the simplest model where

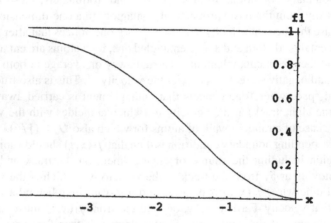

**Fig. 11.2**  An entanglement wave

only the motion and collisions of atoms carry the entanglement, it is of order $L/v'$, where $L$ denotes the largest dimension of the box enclosing the gas.

## 11.8   Conclusions and Perspectives

The main conclusion I would be tempted to draw is the probable existence of a richness of information in the quantum state of a macroscopic system. The present results, even if partial, show that some of this information is not necessarily expressible as a mean value of an operator in Hilbert space and, moreover, wave functions can hold much memory of the past that cannot be extracted by a measurement, even an ideal one [3].

A different question would be to ask whether this kind of information, which is accessible in principle to a Turing machine, could be of some consequence for realistic physical phenomena.

This question leads inevitably to considering quantum measurements, for which entanglement is so essential that any addition to its understanding could turn out valuable. As a matter of fact, the past sections dealt with a measurement though a very special one: a measurement in which the initial state is taken for sure and the alpha particle has probability 1 for entering the detector and the final outcome is anyway predictable. A next step must consist obviously in asking what happens if the initial state of the alpha particle is a superposition

$$\sum_j c_j \, |\alpha, j\rangle \qquad\qquad (11.28)$$

The various channels $j$ in this expression can differ in many ways: They can represent alpha particle states in which the particle hits the detector at different places (in which case a bubble chamber or a wire chamber would be more adequate than a Geiger counter). The particle can also arrive at different times in the different channels, or there is a mute channel in which it does not enter the detector. Moreover, the detection device needs not be unique and can involve several detectors at different places. Although these variants are more or less trivial when seen from the standpoint of abstract measurement theory, they raise interesting variations when one tries understanding better the different histories of entanglement in them.

One will not however discuss these alternatives because it would be too long and a simple reflection using the content of the previous sections is enough for guessing the results in each case. The main conclusion is quite clear anyway and asserts that, *as long as the measuring device is perfectly isolated*, the final state of the system after a complete evolution of entanglement will coincide with its standard description [2, 3, 18] and will keep no trace of locality, even in the topology of wave functions.

The situation is different however when the existence of an environment for the system $AB$ is taken into account and there is decoherence. One knows that, basically, decoherence is a manifestation of entanglement between the $AB$ system and its environment [19] and the question has been raised whether the conjugation of this entanglement with the one occurring through measurement in the $AB$ system could have deep consequences and perhaps as far as leading to collapse [12, 20].

In Zurek's proposal of Quantum Darwinism, a key factor would be the existence of collective effects in the environment involving transport of the entanglement with $AB$. This transport was described however only by using relative entropy between different parts of the environment [12]. Perhaps the present methods, which seem to carry more information, could be of use in this very interesting theory. Another proposal, which was at the origin of the present investigations, concentrates on the contrary over possible effects of the entanglement of the environment with $AB$ inside the detector. The main point would not be the effect of decoherence but the fact that this entanglement is carried by many different traveling waves of which the succession and accumulation would be ultimately responsible for collapse [20]. It could also be that the two ideas are not mutually exclusive, but it would be premature and out of place to add anything more presently.

# References

1. Schrödinger, E.: Discussion of probability relations in separated systems. Proc. Camb. Philos. Soc. **31**, 555 (1935); **32**, 446 (1936)
2. Schrödinger, E.: Die gegenwärtige Situation in der Quantenmechanik. Naturwissensschaften **23**, 807, 823, 844 (1935); reprinted in Wheeler, J.A, Zurek, W.H.: Quantum Mechanics and Measurement. Princeton University Press, Princeton (1983)
3. von Neumann, J.: Mathematische Grundlagen der Quantenmechanik. Springer, Berlin (1932)
4. Laloë, F.: Comprenons-nous vraiment la mécanique quantique? CNRS Editions. EDP Sciences, Paris (2011)
5. Schmidt, E.: Zur theorie der linearen und nichtlinearen Integralgleichungen. Math. Ann. **63**, 161 (1907)
6. Plenio, M.B., Virmani, S.: An introduction to entanglement measures. Quant. Inf. Comput. **7**, 1–51 (2005)
7. Hugenholtz, N.M.: Physica **23**, 481 (1957)
8. Kubo, R.: J. Math. Phys. **4**, 174 (1963)
9. Wichmann, E.H., Crichton, J.H.: Phys. Rev. **132**, 2788 (1983)
10. Weinberg, S.: The Quantum Theory of Fields I, Chapter 4. Cambridge University Press, Cambridge (2011)
11. Faddeev, L.D., Merkuriev, S.P.: Quantum Scattering Theory for Several Particle Systems. Springer, Berlin (1993)
12. Zurek, W.H.: Quantum Darwinism. Nat. Phys. **5**, 181 (2009)
13. Brown, L.S.: Quantum Field Theory. Cambridge University Press, Cambridge (1992)
14. Hörmander, L.: The Analysis of Linear Partial Differential Operators. Springer, Berlin (1985)
15. Godement, R.: Topologie algébrique et théorie des faisceaux. Hermann, Paris (1973)
16. Lifshitz, E.M., Pitaevskii, L.P.: Physical Kinetics. Pergamon, London (1981)
17. Dautray, R., Lions, J.-L.: Mathematical Analysis and Numerical Methods for Science and Technology. Evolution Problems, I, II. Springer, Berlin (2000)

18. Wigner, E.P.: Interpretation of quantum mechanics. Am. J. Phys. **31**, 6 (1963)
19. Joos, E., Zeh, H.D., Kiefer, C., Giulini, D., Kupsch, K., Stamatescu, I.O.: Decoherence and the Appearance of a Classical World in Quantum Theory. Springer, Berlin (2003)
20. Omnès, R.: A quantum approach to the uniqueness of Reality. In: Agazzi, E. (cd.) Turing's Legacy, Proceedings of the meeting in September 2012 of the International Academy of Philosophy of Science, FranceAngeli, Milano (2013). arxiv.org/quant.phys/1302.1750

# Chapter 12
# Unavoidable Decoherence in Matter Wave Interferometry

Helmut Rauch

## 12.1 Introduction

Neutrons are massive particles which exhibit wave and particle features and, therefore, they are proper tools for testing quantum phenomena (e.g. [21]). With perfect crystal interferometers widely separated coherent beams can be produced and modified by nuclear, electroweak and gravitational interactions. Highly efficient detectors, polarisers and spin flippers are available. Most experiments use thermal neutrons from a research reactor with velocities in the order of 2,000 m/s and wavelengths in the order of about 2 Å. The spin state of neutrons can be described by a typical two-level Zeeman system where transitions can be induced by electromagnetic interaction. The two coherent beam paths passing through the interferometer can also be seen as a two-level system and related entanglements can be created and used for novel entanglement and contextuality quantum measurements.

A high degree of coherence, expressed by the visibility of the interference pattern (up to 95 %), and high order interferences (up to 200th) have been observed. The beam in forward direction behind the interferometer consists of wave functions from beam path I and beam path II, which are transmitted, reflected and reflected (trr) and reflected, reflected and transmitted (rrt), respectively (Fig. 12.1). From symmetry considerations follows that both beams are equal in intensity and in phase ($\Psi_{trr} = \Psi_{rrt}$)

$$I_0 \propto \left| \Psi_{trr} + e^{i\chi} \Psi_{rrt} \right|^2 \propto 1 + |\Gamma(\chi)| \cos \chi, \tag{12.1}$$

H. Rauch (✉)
Atominstitut, TU-Wien, 1020 Wien, Austria
e-mail: rauch@ati.ac.at

© Springer-Verlag Berlin Heidelberg 2015
P. Blanchard, J. Fröhlich (eds.), *The Message of Quantum Science*, Lecture Notes in Physics 899, DOI 10.1007/978-3-662-46422-9_12

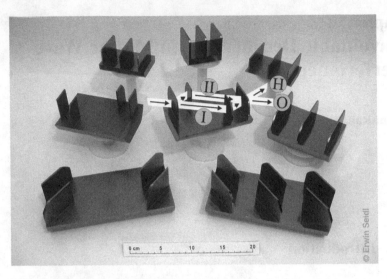

**Fig. 12.1** Photos of several perfect crystal interferometers

where a phase shift $\chi$ between the two beams is present. This phase shift is connected to a spatial phase shift $\Delta$ and is given by the index of refraction $n$ of any material, which depends on the particle density $N$, the coherent scattering length $b_c$, the neutron wave length $\lambda$ and the thickness of the sample $D$

$$\chi = (1 - n)kD = -Nb_c\lambda D = nk.\Delta. \tag{12.2}$$

When wave packets with momentum width $\delta k$ are used to describe the neutron within an interferometer one can define a distinct degree of coherence. The coherence features are usually described in terms of the well-known Glauber [11] formalism. The coherence function is defined as the auto-correlation function of the wave function, which reads for a stationary situation as

$$\Gamma^{(12.1)}(\vec{\Delta}) = \left\langle \psi(\vec{r}) * \psi(\vec{r} + \vec{\Delta}) \right\rangle = \int g(\vec{k}) \, e^{i(\vec{k}\vec{\Delta})} d^3k \,, \tag{12.3}$$

where $\vec{\Delta}$ denotes the spatial shift of the two wave fields and $g(\vec{k})$ the momentum distribution in the related direction. The characteristic widths of these functions define the coherence lengths $\Delta_i^c$ which are related to the widths $\delta k_i$ of the related momentum distributions as: $\Delta_i^c \delta k_i \leq 1/2$; $i = x, y, z$. From measurements of the momentum distributions and the loss of contrast at high order interference, which is described by $|\Gamma^{(1)}(\vec{\Delta})|$, one obtains coherence lengths in the order of about 200–500 Å, which represents also the size of the related wave packets [22]. Figure 12.2 shows typical results of such a measurement.

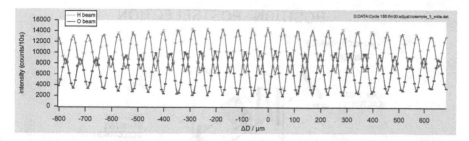

**Fig. 12.2** Neutron interferences measured up to high orders

The complementarity and duality feature of quantum physics is an essential part for understanding it. According to the Englert-Greenberger-Yasin relation the wave-particle feature can be considered as a further "two level" system where pure particle features exist at the north and south poles of the Bloch sphere and pure wave properties along the equator [8, 9, 13]. This relation combines path predictabilities P with wave properties V, which equals the visibility of the interference pattern.

$$P^2 + V^2 \leq 1 \tag{12.4}$$

This relation provides the basis for further entanglement experiments with other two level systems.

Pre- and post-selection experiments apply higher monochromaticity, better collimation, spatial resolution or additional time information and, therefore, they provide more information about a quantum system. Such measurements also demonstrate that coherence persists even when the coherence pattern is smeared out due to finite monochromaticity, i.e. of $|\Gamma(\Delta)|$, and due to inhomogeneities along the beam paths or statistical motions of the interferometer crystal during the time-of-flight the neutron spends within the interferometer ($\delta\Delta$). Pre- and post-selection experiments have verified that coherence features are determined by the parameters of the apparatus, i.e. the monochromaticity and the collimation of the beam which determine $\delta k$. But the interference pattern also includes effects from the imperfection of any experimental set-up and, therefore the interference pattern has to be described by a generalized form of Eq. (12.1)

$$I_{meas} = A + B\,|\Gamma(\Delta)|\cos(\chi + \phi), \tag{12.5}$$

where $A$, $B$ and $\phi$ are characteristic parameters for each set-up influenced by small and local variations of the crystal structure, of the geometry, of small vibrations and small temperature variations.

Various post-selection methods are shown in Fig. 12.3. Pre- and post-selection provides a step towards more quantum complete measurements as it will be described in the following sections.

## POSITION POSTSELECTION

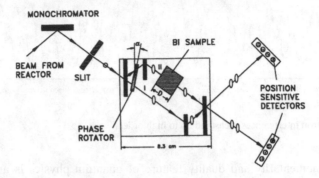

## MOMENTUM POSTSELECTION

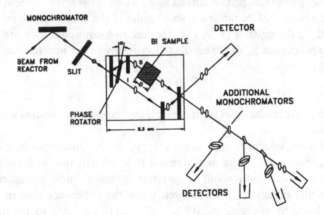

## TIME POSTSELECTION

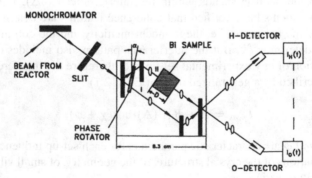

**Fig. 12.3** Various post-selection methods; position post-selection (*above*), momentum post-selection (*middle*) and time post-selection (*below*)

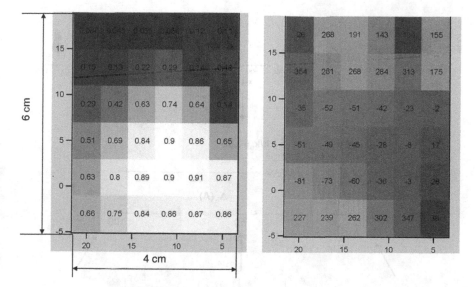

**Fig. 12.4** Position sensitive post-selection of the contrast $B$ (*left*) and of the internal phase $\phi$ (*right*) by means of a position sensitive detector

## 12.2  Position Post-selection

Since the beam cross section is usually much larger than the size of the wave packets one has to average Eqs. (12.1) and (12.5) over the beam cross section. Figure 12.4 shows the results of position sensitive measurements of the contrast and of the beam intensity. One notices that there is a rather marked variation of these quantities. This indicates that there is much more information in the beam than is usually extracted, or much higher degrees of coherence can be achieved when proper measurement methods are applied.

## 12.3  Momentum Post-selection

### 12.3.1  Visibility Loss due to Wavelength Spread

In the course of several neutron interferometer experiments it has been established that smeared out interference properties at high interference order can be restored even behind the interferometer when a proper spectral filtering is applied. Figure 12.5 shows the idealized situation for Gaussian wave packets in ordinary space and the related momentum modulation [19] which reads as:

$$I_0(k) \propto \exp\left[-(k-k_0)^2/2\delta k^2\right] \{1 + \cos(\chi k_0/k)\}. \tag{12.6}$$

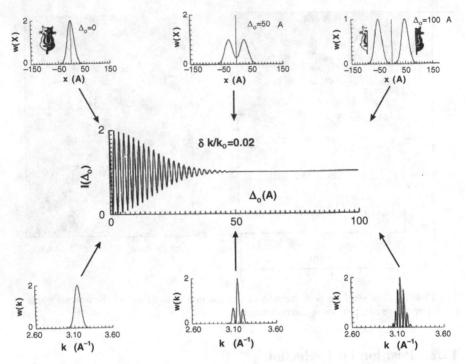

**Fig. 12.5** Wave packets and momentum distributions for various phase shifts at low and high interference order [19]

At high interference orders so-called Schrödinger cat-like states are produced where the neutron occupies spatially separated regions in ordinary space. Related measurements have been performed by Jacobson et al. [15], which show that interference fringes can be restored when a proper momentum filter is applied. Figure 12.6 shows more recent measurements were intensity wiggles have been observed for various phase shifts [3]. In this case the imperfection of the neutron beam (wavelength spread) cause contrast reduction at high order due to $|\Gamma(\Delta)|$.

The increasing modulation at high order is clearly visible. This indicates a forceless beam modulation due to interference. Figure 12.7 shows such a double humped distribution measured with a double loop interferometer [3]. The spatial phase shift $\Delta_1$ in the first loop produces the Schrödinger cat-like state and $\Delta_2$ scans the spatial distribution within the second loop.

As a conclusion of this section, we can say that a loss of visibility of the interference pattern at high order does not mean a loss of coherence but rather a shift of coherence features from ordinary space into momentum space. By a proper post-selection procedure the interference pattern can be retrieved. In all cases a complete retrieval is impossible since unavoidable losses occur at any interaction the system experiences [20].

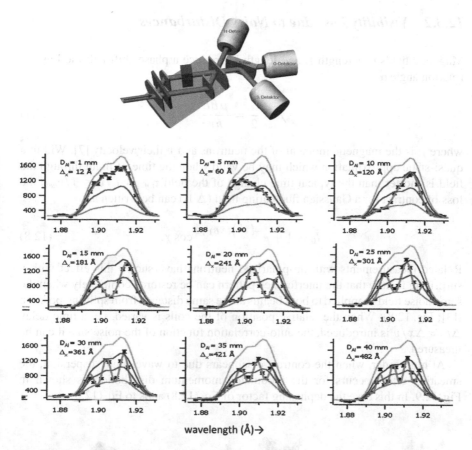

**Fig. 12.6** Experimental arrangement (*above*) with a third crystal as a momentum (wavelength) filter and typical momentum scans at different interference order [3]. The blue curves indicate the momentum distribution without phase shifter

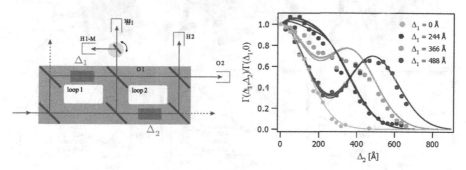

**Fig. 12.7** Double loop interferometer (*left*) and measured spatial distribution for different phase shifts in the first loop [3]

### 12.3.2  Visibility Loss due to Noisy Disturbances

Magnetic fields of strength B and length L produce a phase shift half the Larmor rotation angle $\alpha$

$$\chi = \frac{\alpha}{2} = \frac{\mu BL}{\hbar v}, \tag{12.7}$$

where $\mu$ is the magnetic moment of the neutrons and $v$ their velocity [7]. Within a quasi-static approximation which may be valid when the time-of flight through the field is shorter than the typical time change of the field $\tau_{tof} = L/v \leq 1/v_{field}$ the loss of contrast for a Gaussian fluctuating field ($\Delta$ B) can be written as

$$\bar{I}_0 \propto 1 + e^{-\left(\frac{\mu L}{\hbar v}\right)(\Delta B)^2/2} \cos \chi. \tag{12.8}$$

Related measurements with un-polarized neutrons have shown this effect with a surprising feature that the interference pattern can be restored completely when the same noise field is applied to both beams at the same distance from splitting position ([3]; Fig. 12.8). When the relative position of the coils is varied, i.e. a time delay $\Delta t = \Delta x/v$ is introduced, the auto-correlation function of the noise signal can be measured.

At high order, when the contrast disappears due to wavelength dispersion, the smearing effect occurs for the modulated momentum distribution as shown in Fig. 12.9. In this case the dephasing factor of Eq. (12.8) adds to Eq. (12.6).

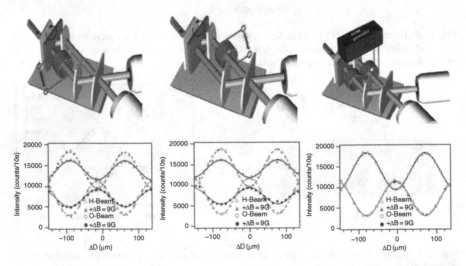

**Fig. 12.8** Interference pattern with (*full lines*) and without magnetic noise field (*dashed lines*) with frequencies between 0 and 20 kHz and a mean amplitude of 9 G [3]. The noise field has been applied to the beams separately (*above and middle*) and to both beams synchronously (*below*)

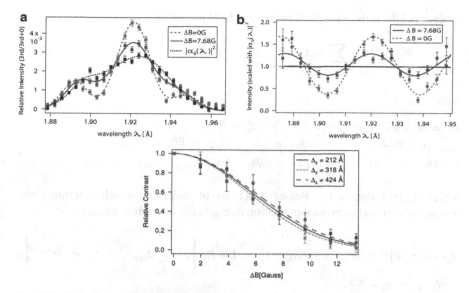

**Fig. 12.9** Reduction of the momentum beam modulation when noisy fields are applied shown on an absolute scale (**a**) and in a modulation corrected way (**b**). The general loss of contrast as a function of the strength of disturbance is shown *below* [26]

Within the experimental errors the loss of contrast is independent of the spatial separation of the Schrödinger cat-like states. The reversibility feature when the same noisy field is applied to both beams indicate that we deal with a dephasing rather than a decoherence effect [23]. The question how "real" decoherence (irreversibility) can be achieved remains open. As long as the same disturbance is applied to both beams the coherence features are preserved and that holds even when absorption processes occur with the same probabilities in both beams [24].

Time-dependent fields, like those discussed in above, cause also photon exchange between field and neutron. In the case of a Rabi resonance flipper an one photon exchange occurs and the neutron polarization changes from spin-up to spin-down and vice versa [2]. For non-resonance magnetic fields multi-photon exchange has been observed [25]. Here we want to see whether these inelastic processes can yield decoherence.

One assumes oscillating fields within a region $0 < x < L$ which are described as

$$\vec{B}(\vec{r},t) = [B_0 + B(t).(\Theta(x) - \Theta(x - L))]\,\hat{z} \qquad (12.9)$$

$$B(t) = \sum_{i=1}^{N} B_i \cos(\omega_i t + \varphi_i).$$

Taking into account that the kinetic energy of the neutrons ($\sim 20\,\text{meV}$) is much higher than the maximal potential barrier ($\mu B_i \sim 0.5\,\text{neV}$) one obtains after some analytical efforts the wave field behind the field region [27]

$$\Psi_{III}(x,t) = \sum_{\vec{n}} J_{n_1}(\beta_1)\ldots\ldots J_{n_N}(\beta_N)e^{-\vec{n}\vec{\eta}}e^{ik_n x}e^{-i\omega_n t} \tag{12.10}$$

with

$$\omega_{\vec{n}} = \omega_0 + \vec{n}\vec{\omega} \qquad \vec{k}_{\vec{n}}^2 = k_0^2 - \tfrac{2m}{\hbar^2}\mu B_0 + \tfrac{2m}{\hbar}\vec{n}\vec{\omega}$$
$$\eta_i = \varphi_i + \tfrac{\omega_i T + \pi}{2} \qquad \beta_i = 2\alpha_i \sin\tfrac{\omega_i T}{2} \qquad \alpha_i = \tfrac{\mu B_i}{\hbar\omega_i} \qquad T = \tfrac{L}{v_0}$$
$$\vec{n} = (n_1\ldots\ldots n_N), \quad \vec{\varphi} = (\varphi_1\ldots\ldots\varphi_N), \quad \vec{\omega} = (\omega_1\ldots\ldots\omega_N), \quad \vec{\eta} = (\eta_1\ldots\ldots\eta_N),$$

where $J_{n_i}(\beta_i)$ denote the Bessel functions of order $n_i$, which determine the transition amplitudes. From this relation one gets the interference pattern as:

$$I_0(x.t) = \tfrac{1}{2}\left|\Psi_I(x,t) + e^{i\chi}\Psi_{III}(x.t)\right|^2 = 1 + Re\left\{e^{i\chi}\sum_{\vec{n}} J_{n_1}\ldots\ldots J_{n_N}.e^{i\vec{n}(\vec{\xi}+\vec{\omega}t)}\right\}$$

$$\text{with}\quad \xi_i = \eta_i - \tfrac{\omega_i x}{v_0} \tag{12.11}$$

When the fundamental frequency of all frequencies is $\omega_f$ the interference pattern can be expressed in a Fourier series

$$I_0(x,t) = \sum_{m=-\infty}^{m=\infty} c_m(x)e^{im\omega_f t}, \tag{12.12}$$

where a comparison with Eq. (12.8) gives, for un-polarized neutrons where only even m-terms remain.

$$c_m = \delta_{m0} + \sum_{\substack{\vec{n};\,\vec{n}\vec{\omega} = m\omega_f \\ \sum_i n_i even}} J_{n_1}(\beta_1)\ldots\ldots J_{n_N}(\beta_N)e^{i\vec{n}\vec{\xi}}\cos\chi, \tag{12.13}$$

which shows that the Fourier coefficient belonging to the frequency $m\omega_f$ contains the same product of Bessel functions as the transition amplitudes for an energy exchange $m\hbar\omega_f$. The argument of the Bessel functions also contains a $\sin(\omega t/2)$-term defining a "resonance"-condition. If the time-of-flight $T = L/v$ through the field region fulfils $\omega_i T = 2\pi l$ ($l = 1, 2, 3\ldots$) no resulting energy exchange occurs.

Related experiments have been performed with a time resolved analysis of the interference pattern, where the periodicity of the applied magnetic field ($\omega_f = 2\pi f_f$ ; $f_f = 1\,kHz$) also manifests itself in the interference pattern [27]. The related energy transfers lie in the order of $1\,peV$ and energy gain and energy loss processes are equal, which is caused by the symmetric sinusoidal fields. The calculated results show good agreement with the measured values (Fig. 12.10). One

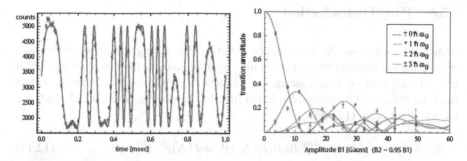

**Fig. 12.10** Example of a time-resolved interference pattern for a two mode field $f_1 = 3\,kHz$ and $f_2 = 3\,kHz$ and $B_1 = 30\,G$ (*left*) and extracted multi-photon transition probabilities (*right*) [27]

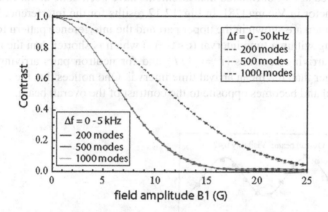

**Fig. 12.11** Calculated loss of contrast for a low frequency and a high frequency field as a function of the mean field amplitude [27]

notices that absorption and emission probabilities are equal which results from the symmetric interaction potential [Eq. (12.9)].

This procedure has been extended experimentally up to a 10 mode field and further by numerical methods up to a 1,000 mode field. When random phases, Gaussian variations of the amplitude and different frequency spectra are employed one can simulate a noisy field. Figure 12.11 shows the results for a rather low frequency and a higher frequency spectrum.

As conclusion of this section it should be mentioned that a loss of visibility in an interference experiment does a priori not indicate a loss of coherence, but that a non adequate measuring method has been applied and it indicates that decoherence is also rather difficult to achieve and probably as difficult as to preserve coherence. More and more complete quantum measurements may reduce decoherence by means of more sophisticated investigations, but some decoherence will always remain as a residual of all interactions experienced by the system and due to unavoidable losses (see Sect. 12.5, [20]).

## 12.4   Time Post-selection

Instead of measuring the interference pattern by scanning the phase shifter and registering the intensity over a certain time interval, one can also measure the arrival time of each neutron, which means one extracts more information. From these data the time dependent intensity correlation function can be extracted, which reads for a Poissonian beam [12, 16],

$$G^{(2.2)}(\Delta, t) = \langle I_1(0,0)I_2(\Delta, t)\rangle = \bar{I}(\Delta)e^{-\bar{I}(\Delta)t}, \qquad (12.14)$$

and gives the probability of registering a neutron at time $t$, if there was another one registered at $t = 0$. Related measurements have been done at our low intensity research reactor in Vienna [28]. In Fig. 12.12 results for the interference pattern of the whole beam are shown in the upper part and the interference pattern for neutron pairs arriving within a time interval ($0 < t < 3\,\text{s}$) which is shorter than the mean time interval of arriving neutrons ($\bar{\tau} = 1/\bar{I}$) and for neutron pairs arriving within a interval larger than the mean arrival time interval. One notices that the contrast can be increased and becomes opposite to the contrast of the overall beam.

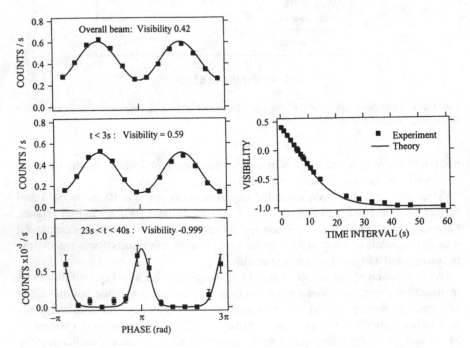

**Fig. 12.12** Interference contrast of the overall beam and of neutron pairs within different time intervals (*left*) and the visibility as a function of the time intervals (*right*) [28]

These results demonstrate that considerably more information can be deduced, even from a Poissonian beam, when the individual arrival times of the neutrons are registered to define pair correlation functions inherent to quantum systems.

## 12.5   Unavoidable Losses

We discuss this aspect with the means of a phase echo experiment done with two opposite phase shifting samples as shown in Fig. 12.13 [5]. The interference contrast disappears at high order for a thick Bi and Ti sample but is restored when both phase shifters are inserted since the phase shift of Bi is negative and that of Ti positive with the sum zero. Nevertheless, quantum theory teaches us that at any barrier of height $\bar{V} = 2\pi\hbar^2 Nb_c/m$ and thickness $L$ there are more wave components than the incident and the transmitted wave. The reflectivity reads as (e.g. [6])

$$R = 1 - T = 1 - \frac{4E(E - \bar{V})}{4E(E - \bar{V}) + \bar{V}^2 \sin^2 kL} \cong \frac{1}{2}\left(\frac{\bar{V}}{2E}\right)^2 \left(1 - \cos^2 kL\right)$$

$$(12.15)$$

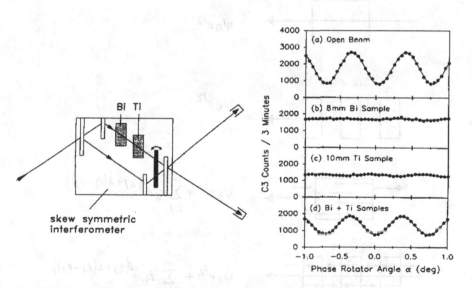

**Fig. 12.13** Phase-echo experiment using phase shifters with positive (Bi) and negative (Ti) coherent scattering lengths causing positive and negative phase shifts. The retrieval of the contrast becomes visible when both phase shifters are inserted and the total phase shift becomes zero [5]

The oscillating term vanishes when $E >> \bar{V}$ and when the barrier produces phase shifts larger than the coherence length of the beam and one gets

$$\bar{R} \cong \frac{1}{2}\left(\frac{\bar{V}}{2E}\right)^2,\qquad(12.16)$$

which is of the order $10^{-10}$ for thermal neutrons and reasonable materials or magnetic potentials but these losses are unavoidable. The transmission and the reflectivity of arbitrarily shaped potentials can be calculated in a similar way leaving the general conclusions unchanged, e.g. Cohen-Tannoudji et al. [6]. For multiple potential barriers the formulae become rather complicated because multiple interference effects have to be taken into account. Figure 12.14 shows that different

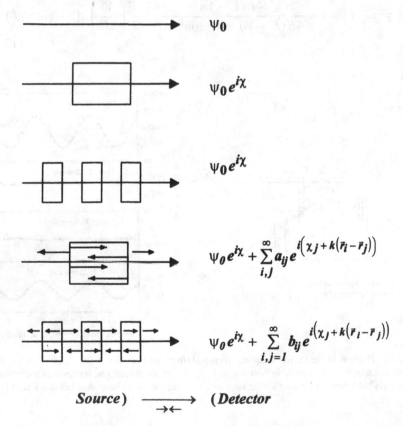

**REVERSIBILITY - IRREVERSIBILITY**

**Fig. 12.14** Approximate and complete wave-functions when differently shaped potential barriers (phase shifters) are used. One notices that the correct wave-functions have much more components than the simplified ones (*above*) which are often used for straight-forwardness

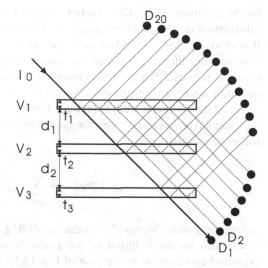

**Fig. 12.15** Parasitic waves from a three-fold barrier and how they can be measured, in principle

wave functions arise for different arrangements of the same potential. In the upper part the simplified wave functions are shown whereas in the lower part the complete wave functions are indicated. There are infinite back and forth reflections even in the case that there is only one source and one detector since there has to be a potential on both sides. The wave function behind the interaction region (and in front of it) contains the full information about the structure and strength of the interaction region and a quantum more complete experiment would have to measure all parasitic beams as shown in Fig. 12.15. The measurement of all parasitic beams gives one also full information about the main beam and can be considered as a weak measurement of that beam [1]. A complete retrieval to the original state appears impossible although only unitary actions have been applied. Thus, decoherence appears as an intrinsic feature of all physical processes starting with the first interaction the quantum system experiences.

## 12.6 Discussion

Neutron interference experiments have shown that a washed out interference pattern can be restored to a high degree when proper post-selection procedures are applied. The newly generated visibility values may be considered to violate the particle-wave duality relation Eq. (12.4), but that is not true because the particle features become changed as well. Even in the case of a multi-mode noise field with multi-photon exchange the contrast can be retrieved when the same noise field is applied to the other beam or a time-resolved measurement is carried out. A retrieval procedure up to a certain degree is possible whenever the reason for dephasing is known, but unavoidable quantum losses of coherence remain from any interaction experienced by the quantum system [20]. More complete measurements show how much more information is inherent in any experiment and usually only a small part is extracted. It also means that a complete isolation from the environment is impossible and a residual entanglement between the quantum object and the environment remains resulting in causal connections between the micro- and macro world [29, 30]. In the quantum measurement process a superposition state between the quantum object $|\psi_0\rangle$ and the apparatus $|\Re_0\rangle$ is produced, which results in an entanglement between the quantum object and the environment. The evolution of the joint system $|\psi_0\rangle |\Re_0\rangle$ follows a unitary operator $\cup$

$$|\psi_0\rangle |\Re_0\rangle \xrightarrow{\cup} \sum_{n=0}^{N} c_n |\psi_n\rangle |\Re_n\rangle , \qquad (12.17)$$

where the Pointer states $\Re_n$ appear according to the probabilities $c_n$ which are determined by the solution of the linear Schrödinger equation. There may be preferred $c_n$-values as in the case of Fig. 12.15 where $c_0$ may be dominant but all other $c_n$-values exist as well. The arrow in Eq. (12.17) can be interpreted as an arrow of time since a complete retrieval of the states is impossible.

This is not in contradiction with the concept that irreversibility is a fundamental feature of nature and that reversibility is only an approximation, a conclusion stated by several authors (e.g. [4, 14, 18]). At the same time it is closely connected to the quantum measurement problem, because a complete retrieval becomes impossible since in any interaction small loss mechanisms are unavoidable and they are caused by boundary and initial conditions [10, 17]. The appearance of entropy associated with decoherencing effects reflects the presence of an arrow in time in quantum theory, i.e. a fundamental irreversibility in the formalism of the theory itself [9, 20]. This irreversibility comes into play only through initial and boundary conditions in our universe.

# References

1. Aharonov, Y., Vaidman, L.: J. Phys. A: Math.Gen. **24**, 2315 (1991)
2. Badurek, G., Rauch, H., Summhammer, J.: Phys. Rev. Lett. **51**, 1015 (1983)
3. Baron, M., Rauch, H.: AIP Conf. Proc. **1327**, 89 (2011)
4. Blanchard, P., Jadczyk, A.: Phys. Lett. A **175**, 157 (1993)
5. Clothier, R., Kaiser, H., Werner, S.A., Rauch, H., Wölwich, H.: Phys. Rev. A **44**, 5357–5368 (1991)
6. Cohen-Tannoudji, C., Diu, B., Laloe, F.: Quantum Mechanics, vol. I. Wiley, New York (1977)
7. Eder, G., Zeilinger, A.: Il Nuovo Cim. **34B**, 76 (1976)
8. Englert, B.G.: Phys. Rev. Lett. **77**, 2154 (1996)
9. Englert, B.G.: Eur. Phys. J. **67**, 238 (2013)
10. Giulini, D., Joos, E., Kiefer, C., Kupsch, J., Stamatescu, I.O., Zeh, H.D.: Decoherence and the Appearance of a Classical World in Quantum Theory. Springer, Berlin (1996)
11. Glauber, R.J.: Phys. Rev. **130**, 2529; **131**, 2766 (1963)

12. Glauber, R.J.: In: Cohen, E.G.D. (ed.) Fundamental Problems in Statistical Mechanics, p. 140. Noth-Holland, Amsterdam (1968)
13. Greenberger, D.M., Yasin, A.: Phys. Lett. **128**, 391 (1988)
14. Haag, R.: Nucl. Phys. **B18**, 135 (1990)
15. Jacobson, D.L., Werner, S.A., Rauch, H.: Phys. Rev. **A49**, 3196 (1994)
16. Mandel, L.: In: Wolf, E. (ed.) Progress in Optics, vol. 2, p.183. North-Holland, Amsterdam (1963)
17. Namiki, M., Pascazio, S., Nakazato, H.: Decoherence and Quantum Measurements. World Scientific, Singapore (1997)
18. Prigogine, I.: In: Proc. Ecol. Phys. Chem., p. 8. Elsevier, Amsterdam (1991)
19. Rauch, H.: Phys. Lett. **A173**, 240 (1993)
20. Rauch, H.: J. Phys. Conf. Ser. **36**, 164 (2006)
21. Rauch, H., Werner, S.A.: Neutron Interferometry, 2nd edn. Oxford University Press, Oxford (2015)
22. Rauch, H., Woelwitsch, H., Kaiser, H., Clothier, R., Werner, S.A.: Phys. Rev. **A53**, 902 (1996)
23. Stern, A., Aharonov, Y., Imry, Y.: Phys. Rev. **A41**, 3436 (1990)
24. Summhammer, J., Rauch, H., Tuppinger, D.: Phys. Rev. **A36**, 4447 (1987)
25. Summhammer, J., Hamacher, K.A., Kaiser, H., Weinfurter, H., Jacobson, D.L., Werner, S.A.: Phys. Rev. Lett. **75**, 3206 (1995)
26. Suyok, G., Hasegawa, Y., Klepp, J., Lemmel, H., Rauch, H.: Phys. Rev. **A81**, 053609 (2010)
27. Suyok, G., Lemmel, H., Rauch, H.: Phys. Rev. **A85**, 033624 (2012)
28. Zawisky, M., Rauch, H., Hasegawa, Y.: Phys. Rev. **A50**, 5000 (1994)
29. Zeh, H.D.: Found. Phys. **1**, 69 (1970)
30. Zurek, W.D.: Phys. Today (October), 36 (1991)

# Chapter 13
# Classical-Like Trajectories of a Quantum Particle in a Cloud Chamber

G. Dell'Antonio, R. Figari, and A. Teta

## 13.1 The Wilson Cloud Chamber

The cloud chamber, *"the most original and wonderful instrument in scientific history"* according to Ernest Rutherford, was devised and made available by Charles Thomson Rees Wilson at the end of the first decade of last century.

The main idea of Wilson was to make visible the "ionizing radiation" by condensing water on the ions produced by $\alpha$ and $\beta$ rays in air supersaturated with water-vapor. His aim was to give a conclusive experimental verification of the particle nature of the $\alpha$ and $\beta$-radiations.

On one hand, some experimental problems were overwhelmingly complicate to be confronted by. A perfect synchronization of the sudden cooling, of the illumination of the chamber and of the photographic capture was required.

On the other end, the overall intuitive picture of what was happening in the chamber was perfectly depicted: the $\alpha$-particle, emitted by a radioactive source (in an unpredictable although specific direction), ionizes molecules of the gas filling the chamber. In turn, ions become condensation nuclei for water-vapor giving rise to a sequence of small drops of water along the trajectory of the $\alpha$-particle.

G. Dell'Antonio (✉)
Dipartimento di Matematica, Sapienza Università di Roma, P.le A. Moro 5, 00185 Roma, Italy
e-mail: gianfa@sissa.it

Scuola Internazionale di Studi Superiori Avanzati, Via Bonomea 265, 34136 Trieste, Italy

R. Figari
Dipartimento di Scienze Fisiche, Università di Napoli, Napoli, Italy

Istituto Nazionale di Fisica Nucleare Sezione di Napoli, Complesso Universitario
di Monte S. Angelo, Via Cintia, Edificio 6, 80126 Napoli, Italy

A. Teta
Dipartimento di Matematica, Sapienza Università di Roma, P.le A. Moro 5, 00185 Roma, Italy

© Springer-Verlag Berlin Heidelberg 2015                 291
P. Blanchard, J. Fröhlich (eds.), *The Message of Quantum Science*, Lecture Notes
in Physics 899, DOI 10.1007/978-3-662-46422-9_13

In the early days of quantum mechanics, it was immediately noticed that the explanation of the phenomenon, as it was given by physicists at the time of Wilson, seemed at variance with some of the cornerstones of the new theory.

In fact, inside the framework of the orthodox theory, a classical trajectory of a quantum particle could only be the result of repeated "collapses" of its wave function due to repeated measurements. Pushing to the limit, each scattering event should be considered a measurement process. On the other end, in the orthodox approach, a measurement apparatus *must* be considered as a classical object whose evolution *has* to be described and predicted using classical kinematics and dynamics. Classical observables like position or trajectory can only be properties emerging as a consequence of the interaction of the microscopic system with the classical measurement apparatus. By no means, they are properties possessed by the system before the measurement.

It is clear that this point of view is hardly compatible with the interpretation of a simple scattering event as a measurement process.

The apparent contradiction shows that the issue of describing what "really" happens in a cloud chamber highlights the most subtle problems of the Copenhagen interpretation:

- Where should the frontier between the quantum and the classical realm be placed? In particular,
- how large should be an array of atoms to behave as a classical system?
- When and where does the collapse process take place?

It is well known that the final formalization of quantum mechanics, given by John von Neumann in 1932 [34], assumes two different evolution laws for a microscopic system: the continuous unitary evolution given by the Schrödinger equation, as long as the system remains isolated, and a stochastic and/or non-linear sudden change driving to the collapse of the wave function, during the measurement process.

As is clear, von Neumann's dynamical assumptions leave the issues listed above unanswered.

In 1929, Charles Galton Darwin proposed a completely new way to look at the formation of classical tracks in a cloud chamber. In retrospect, one can say that the following simple Darwin's reflection was extremely far-reaching: the Schrödinger equation describes the evolution in the configuration space of the whole system. Taking into account the electrons of the gas atoms together with the particle, no contradiction exists in principle with the existence of solutions which show the $\alpha$-particle moving in a small cone with apex in the radioactive source and only atoms in the cone ionized.

In the same year, Sir Nevill Francis Mott concretized Darwin's proposal analyzing the long time behavior of the Schrödinger equation solutions relative to the $\alpha$-particle and the electrons of two hydrogen atoms playing the role of the gas in the cloud chamber.

Mott's paper did not receive the attention that it would have deserved and it remained almost unknown also during the 1960s when the interest in probing the frontier between classical and quantum behavior had a new start.

By now, a significant amount of experimental and theoretical work in the field is available. More and more sophisticated experiments have pushed the border quantum-classical towards larger and larger system sizes. At the same time, new theoretical investigations analyzed the loss of quantum coherence due to the interaction of a microscopic system with a large environment (see, e.g., [1, 2, 8, 26–29, 36] and references therein).

Main purpose of this report is to present further attempts to investigate whether a purely quantum mechanical treatment can justify the (experimentally verified) statement that in a cloud chamber filled with supersaturated gas an $\alpha$-decay produces at most one sequence of liquid droplets (track) and that this track is compatible with the trajectory of a classical particle. In summary, we want to test the compatibility of the experimental outputs with the following rough statement: the $\alpha$-wave is turned, by the interaction with the environment, into an $\alpha$-particle of the same energy as the initial wave and with a momentum direction having a definite orientation.

## 13.2   The Earliest Theoretical Investigations

In 1928 Gamow [20] and Condon and Gurney [11] made the first attempt to approach the $\alpha$-decay phenomenon according to quantum mechanics. A crucial point of their analysis was that, at the time of the emission, the $\alpha$-particle state had to be a spherical wave centered in the radioactive nucleus, with a highly isotropic average momentum. As pointed out before, it was immediately manifest that the assumption of a spherical wave as initial state makes the explanation of the tracks observed in a cloud chamber rather problematic.

Such a difficulty had already been mentioned by Born who, during the general discussion at the Solvay Conference in 1927 [5], noted: *"Mr. Einstein has considered the following problem: A radioactive sample emits $\alpha$-particles in all directions;*

*these are made visible by the method of the Wilson cloud chamber. Now, if one associates a spherical wave with each emission process, how can one understand that the track of each α-particle appears as a (very nearly) straight line? In other words: how can the corpuscular character of the phenomenon be reconciled here with the representation by waves?"*

According to Born, the question could be answered using the notion of *"reduction of the probability packet"* induced by "observation" by means of light, discussed by Heisenberg in [22]. Indeed, Born claimed: *"As soon as such ionization is shown by the appearance of cloud droplets, in order to describe what happens afterwards one must reduce the wave packet in the immediate vicinity of the drops. One thus obtains a wave packet in the form of a ray, which corresponds to the corpuscular character of the phenomenon"*. It should be stressed that, according to this point of view, the evolution of an α-particle in a cloud chamber can be described as the result of the interaction of a quantum system (the α-particle) with a classical measurement apparatus (the atoms of the vapor). Such interaction is responsible for the "reduction" of the spherical wave to a wave packet with definite position and momentum.

As a possible alternative description of the process, Born also considered an approach where both the α-particle and the atoms of the vapor are considered part of a unique quantum system, described by a wave function depending on the coordinates of all the particles of the system. In particular, he proposed a simplified one-dimensional model consisting of a test particle (the α-particle) and two harmonic oscillators placed at fixed positions (the atoms of the vapor). At initial time, the test particle is described by a superposition state of two wave packets with opposite momentum and position close to the origin, while the harmonic oscillators are in their ground states. A qualitative discussion of such a model led Born to the claim that the test particle has a very low probability of exciting both oscillators unless they are located on the same half-line starting from the origin (for a quantitative analysis see e.g. [13, 19]).

The analysis of the model is then completed with a statement involving once again the reduction postulate: *"To the reduction of the wave packets corresponds the choice of one of the two directions of propagations"*, and the choice is done when the excitation of an oscillator is observed. Only after the observation, the evolution of the test particle can be assimilated to that of a classical particle propagating along a definite trajectory.

In conclusion, Born conceded that an analysis of the quantum evolution of the α-particle in interaction with the (quantum) environment made of the vapor atoms is possible, *"but this does not lead us further as regards the fundamental questions"*, in the sense that the reduction postulate is anyway required for a complete description.

A further fundamental contribution to the theoretical analysis of the cloud chamber problem was given by Heisenberg in his lectures at the University of Chicago in 1929, published in [23]. He pointed out that a quantum description of an experimental situation always requires to fix an arbitrary border between the quantum system under consideration and the (classical) measuring apparatus. For any fixed border one has different, but equivalent, descriptions of the phenomenon.

In the case of the cloud chamber Heisenberg considered the following two reasonable choices: (a) the quantum system consists of the $\alpha$-particle alone (and then the molecules of the vapor play the role of the measurement apparatus); (b) the quantum system consists of the $\alpha$-particle and of the molecules of the vapor.

In case (a) the analysis proceeds as follows. The $\alpha$-particle, evolving in the chamber as a spherical wave, collides with a molecule of the vapor which acts as a measuring device of the particle position. Therefore, immediately after the collision, the state of the $\alpha$-particle is reduced to a narrow wave packet concentrated around the position of the molecule. Furthermore, one knows that at time zero the $\alpha$-particle starts from the position of the radioactive nucleus and that the momentum is conserved. This implies that the wave packet has an average momentum along the direction $\gamma$ joining the radioactive nucleus and the position of the molecule (notice that in this way Heisenberg states that the measurement actualize a posteriori the momentum possessed by the particle before the measurement itself).

Such a wave packet emerging from the molecule propagates in the chamber according to the free Schrödinger dynamics, with an inevitable spreading in position. However, the $\alpha$-particle collides with the next molecule placed along $\gamma$ and a new position measurement takes place, determining a refocusing of the wave packet along $\gamma$. The process is repeated a large number of times and the result is that the wave packet remains concentrated along $\gamma$, which is identified as the observed "trajectory" of the $\alpha$-particle.

In case (b) the molecules of the vapor and the $\alpha$-particle form a many-particle quantum system, whose dynamics is governed by the Schrödinger equation. Heisenberg notes that in this case the physical description "*is more complicated than the preceding method, but has the advantage that the discontinuous change of the probability function recedes one step and seems less in conflict with intuitive ideas.*"

In order to give a qualitative idea of the behavior of the whole system, Heisenberg considers a three-particle model, made of the $\alpha$-particle and two molecules with centers of mass fixed in the positions $a_I$, $a_{II}$. At initial time the $\alpha$-particle is described by a plane wave with momentum $p$ and the molecules are in their ground states.

The problem is now reduced to find an approximate solution of the Schrödinger equation and to compute the probability that both molecules are excited. The procedure is only briefly outlined and many mathematical details are neglected, but the result, based on a deep physical intuition, is clearly stated: the probability that both molecules are excited is significantly different from zero only if the momentum $p$ is parallel to the line joining $a_I$ and $a_{II}$. Such a result can be considered a satisfactory explanation of the observed trajectory of an $\alpha$-particle in the chamber, since the trajectory can only be observed through the excitations of the molecules.

However Heisenberg stresses that also in case (b) the reduction postulate is used when one "observes" the excitation of the molecules. This means that in case (b) the border between system and apparatus has only been moved to include the molecules in the system.

The conclusion is that, according to Heisenberg, the approaches in cases (a) and (b) are conceptually equivalent since in both cases the reduction postulate must be invoked for a complete description of the physical situation.

In 1929, Darwin [12] presented a different interesting inspection of the cloud chamber problem. The approach proposed by Darwin is entirely based on the Schrödinger equation and it is surprisingly close to the one of decoherence theory developed in the last decades of the last century.

More precisely, he stresses that a satisfactory description of a quantum system $S$ (like the $\alpha$-particle) is achieved only if one takes into account its interaction with (part of) the environment $\mathcal{E}$. As a consequence, one has to compute the evolution of a wave function $\psi$ depending on the coordinates of $S$ and of $\mathcal{E}$. Given such $\psi$, the probabilistic predictions on the system $S$ can be obtained by taking an average over all possible final configurations of the environment $\mathcal{E}$. Such a strategy is surely *"discouragingly complicated"*, but it can provide an explanation of the particle behavior of $S$ without any reference to the reduction postulate.

In the case of the cloud chamber, the wave function $\psi$ is a function of the coordinates of the $\alpha$-particle and of the atoms in the chamber. At the initial time, such $\psi$ can be reasonably assumed to be a product of the spherical wave for the $\alpha$-particle and of the ground states for the atoms. *"But the first collision changes this product into a function in which the two types of coordinates are inextricably mixed, and every subsequent collision makes it worse."* A detailed computation of such $\psi$ is impossible but one can obtain for $\psi$ an integral representation containing a phase factor and *"without in the least seeing the details, it looks quite natural to expect that this phase factor will have some special character, such as vanishing, when the various co-ordinates satisfy a condition of collinearity."* It should be noted that in these words it is clearly outlined the stationary phase method as the correct technical tool to prove the emergence of the particle behavior.

Darwin's view can be summarized by saying that the wave function $\psi$ is the only crucial object of quantum theory and all the particle and the wave properties of a system can be derived from an accurate analysis of $\psi$.

## 13.3 Mott's Analysis

Darwin's program was concretely realized by Mott in his seminal paper [31]. Despite its importance, this work of Mott's does not seem to be sufficiently well known and therefore, in the following, we shall describe it in some details (we also refer the reader to [7, 10, 15–17, 30] for further critical considerations on the paper).

In the introduction, Mott acknowledges having been inspired by Darwin's paper. He admits that the perspective outlined by Darwin seems counterintuitive at first, since *"it is a little difficult to picture how it is that an outgoing spherical wave can produce a straight track; we think intuitively that it should ionise atoms at random throughout space"*. Like Heisenberg, Mott points out that the crucial point is to establish the border line between the system under consideration and the measuring device. In a first possible approach (corresponding to case (a)) in Heisenberg's approach), the $\alpha$-ray is the system and the gas of the chamber is the measurement device by which we observe the particle. Here, the $\alpha$-ray must be considered a particle immediately after the disintegration process, since at that moment the gas (i.e., the device) reduces the initial spherical wave to a narrow wave packet with a definite momentum. In the other approach (case (b) in Heisenberg), the $\alpha$-particle and the gas are considered together as the system under consideration. In this case, the ionized atoms are the entities to be observed and the wave function $\psi$ of the system should provide the ionization probability. Only after the ionization has been observed are we allowed to consider the $\alpha$-ray as a particle.

According to this point of view, the mentioned intuitive difficulty can be overcome, since it arises from our erroneous *"tendency to picture the wave as existing in ordinary three dimensional space, whereas we are really dealing with wave functions in multispace formed by the co-ordinates both of the $\alpha$-particle and of every atom in the Wilson chamber."*

In the rest of his paper, Mott discusses a simple model showing how this second approach actually works. The model is essentially the same as the one considered by Heisenberg and it consists of the $\alpha$-particle, initially described by a spherical wave centered at the origin, and only two hydrogen atoms. The nuclei of the atoms are supposed as at rest in the fixed positions $a_1$, $a_2$, with $|a_1| < |a_2|$. It is assumed that the $\alpha$-particle does not interact with the nuclei, and the interaction between the two electrons is also neglected. Moreover, the interaction between the $\alpha$-particle and the electrons is treated as a small perturbation. The main result of the paper can be summarized in the following statement:

> the two hydrogen atoms cannot both be excited (or ionized) unless $a_1$, $a_2$ and the origin lie on the same straight line.

We shall outline the way Mott derives the result under suitable assumptions, trying to follow his original notation and line of reasoning. We suggest the reader who is not interested in mathematical details to skip the next few pages and proceed directly to the final remarks at the end of this section.

The main object of the investigation are periodic solutions $F(R, r_1, r_2)e^{iEt/\hbar}$ of the Schrödinger equation for the three-particle system, where $R$, $r_1$, $r_2$ denote the coordinates of the $\alpha$-particle and of the two electrons of the hydrogen atoms respectively. Such $F$ is the solution of the stationary Schrödinger equation

$$-\frac{\hbar^2}{2M}\Delta_R F + \left(-\frac{\hbar^2}{2m}\Delta_{r_1} - \frac{e^2}{|r_1 - a_1|}\right)F + \left(-\frac{\hbar^2}{2m}\Delta_{r_2} - \frac{e^2}{|r_2 - a_2|}\right)F$$
$$-\left(\frac{2e^2}{|R - r_1|} + \frac{2e^2}{|R - r_2|}\right)F = E\,F \tag{13.1}$$

where $\Delta_x$ is the laplacian with respect to the coordinate $x$, $M$ is the mass of the $\alpha$-particle, $m$ is the mass of the electron, $-e$ is the charge of the electron and $2e$ is the charge of the $\alpha$-particle.

The solution of Eq. (13.1) can be conveniently expanded in series of the eigenfunctions of the two hydrogen atoms. More precisely, let $\psi_j$ be the $j$-th eigenfunction of a hydrogen atom centered in the origin, with $\psi_0$ denoting the ground state. Then the corresponding eigenfunctions of the atoms in $\mathbf{a}_1$, $\mathbf{a}_2$ are

$$\Psi_j^I(\mathbf{r}_1) = \psi_j(\mathbf{r}_1 - \mathbf{a}_1), \qquad \Psi_j^{II}(\mathbf{r}_2) = \psi_j(\mathbf{r}_2 - \mathbf{a}_2) \tag{13.2}$$

Note that here it seems tacitly assumed that the index $j$ can be an integer or a real positive number (and, correspondingly, $\psi_j$ is a proper eigenfunction or a generalized eigenfunction).

Taking advantage of completeness of the system of the eigenfunctions, we have the following representation for $F$

$$F(\mathbf{R}, \mathbf{r}_1, \mathbf{r}_2) = \sum_{j_1, j_2} f_{j_1 j_2}(\mathbf{R})\Psi_{j_1}^I(\mathbf{r}_1)\Psi_{j_2}^{II}(\mathbf{r}_2) \tag{13.3}$$

The Fourier coefficients $f_{j_1 j_2}(\mathbf{R})$ of the expansion have a direct physical interpretation. Indeed, using Born's rule, the probability for finding the first atom in the state labeled by $j_1$ and the second atom in the state labeled by $j_2$ is

$$\int d\mathbf{R} \, |f_{j_1 j_2}(\mathbf{R})|^2 \tag{13.4}$$

According to this interpretation, one might loosely say that the "wave function" of the $\alpha$-particle is $f_{00}(\mathbf{R})$ if both atoms remain in the ground state, $f_{j_1 0}(\mathbf{R})$, $j_1 \neq 0$, if the first atom is in the $j_1$-th excited (or ionized) state and the second in the ground state, $f_{j_1 j_2}(\mathbf{R})$, $j_1, j_2 \neq 0$, if both atoms are excited (or ionized).

The analysis shows that $f_{00}(\mathbf{R})$ is a (slightly deformed) spherical wave and $f_{j_1 0}(\mathbf{R})$, $j_1 \neq 0$, is a wave packet emerging from $\mathbf{a}_1$ with a momentum along the line $\overline{O\mathbf{a}_1}$. This means that the second atom can be excited by such wave packet only if $\mathbf{a}_2$ lies on the line $\overline{O\mathbf{a}_1}$. Thus the desired result will follow, i.e. $f_{j_1 j_2}(\mathbf{R})$, $j_1, j_2 \neq 0$, is approximately zero unless the condition of collinearity is satisfied.

The computation is carried out using second order perturbation theory and treating the interaction of the $\alpha$-particle with the two electrons as a small perturbation. Then one writes

$$F = F^{(0)} + F^{(1)} + F^{(2)} + \cdots \tag{13.5}$$

and each term of the series is computed by the method of successive approximation, in the form of a diverging spherical wave multiplied the ground state of the two atoms

$$F^{(0)}(\mathbf{R}, \mathbf{r}_1, \mathbf{r}_2) = f_{00}^{(0)}(\mathbf{R})\Psi_0^I(\mathbf{r}_1)\Psi_0^{II}(\mathbf{r}_2), \qquad f_{00}^{(0)}(\mathbf{R}) = \frac{e^{ik|\mathbf{R}|}}{|\mathbf{R}|}$$

$$k = \frac{\sqrt{2M(E - 2E_0)}}{\hbar} \tag{13.6}$$

where $E_j$ denotes the $j$-th eigenvalue of the hydrogen atom. We see that the context of the stationary Schrödinger equation forces Mott to choose a solution not in $L^2$, which, strictly speaking, is not legitimate. In particular the probabilistic interpretation (13.4) fails for $f_{00}^{(0)}$.

For the first order term

$$F^{(1)}(\boldsymbol{R}, \boldsymbol{r}_1, \boldsymbol{r}_2) = \sum_{i_1, i_2} f_{i_1 i_2}^{(1)}(\boldsymbol{R}) \Psi_{i_1}^I(\boldsymbol{r}_1) \Psi_{i_2}^{II}(\boldsymbol{r}_2) \tag{13.7}$$

one finds

$$f_{j_1 j_2}^{(1)}(\boldsymbol{R}) = \frac{M}{2\pi\hbar^2} \int d\boldsymbol{R}' \, K_{j_1 j_2}(\boldsymbol{R}') \frac{e^{\pm ik'|\boldsymbol{R}-\boldsymbol{R}'|}}{|\boldsymbol{R}-\boldsymbol{R}'|}, \qquad k' = \frac{\sqrt{2M(E - E_{j_1} - E_{j_2})}}{\hbar} \tag{13.8}$$

where

$$K_{j_1 j_2}(\boldsymbol{R}) = f_{00}^{(0)}(\boldsymbol{R}) \left( \delta_{0 j_2} V_{j_1 0}(\boldsymbol{R} - \boldsymbol{a}_1) + \delta_{j_1 0} V_{0 j_2}(\boldsymbol{R} - \boldsymbol{a}_2) \right) \tag{13.9}$$

$$V_{ij}(\boldsymbol{x}) = - \int d\boldsymbol{y} \, \frac{2e^2}{|\boldsymbol{x} - \boldsymbol{y}|} \psi_i(\boldsymbol{y}) \psi_j(\boldsymbol{y}) \tag{13.10}$$

Note that, from (13.9), one has $K_{j_1 j_2}(\boldsymbol{R}) = 0$ if both $j_1$ and $j_2$ are different from zero and therefore, by (13.8), one also has $f_{j_1 j_2}^{(1)} = 0$ if both $j_1$ and $j_2$ are different from zero.

From these preliminary considerations a first conclusion can be drawn:

At first order in perturbation theory the probability that both atoms are excited is always zero.

The result is not surprising since, as Mott remarks, in perturbation theory the probability that one atom is excited is a first order quantity and the probability that both atoms are excited is a second order quantity. This explains why the second order term $F^{(2)}$ is required in order to obtain an estimate of the double excitation occurrence.

The further point is to give an approximate expression for $f_{j_1 0}^{(1)}$ and $f_{0 j_2}^{(1)}$. From (13.8), and (13.9), for $f_{j_1 0}^{(1)}$ one has

$$f_{j_1 0}^{(1)}(\boldsymbol{R}) = \frac{M}{2\pi\hbar^2} \int d\boldsymbol{y} \, f_{00}^{(0)}(\boldsymbol{y} + \boldsymbol{a}_1) V_{j_1 0}(\boldsymbol{y}) \frac{e^{ik'|\boldsymbol{R} - \boldsymbol{a}_1 - \boldsymbol{y}|}}{|\boldsymbol{R} - \boldsymbol{a}_1 - \boldsymbol{y}|}, \qquad j_1 \neq 0 \tag{13.11}$$

and analogously for $f_{0j_2}^{(1)}$. In order to find the required approximate expression Mott introduces the following assumptions:

(a) the "observation point" $R$ is far away from the origin and the atom, i.e. $|a_1| \ll |R|$;
(b) the collision for the $\alpha$-particle is almost elastic, i.e. $k - k' \ll k$;
(c) the $\alpha$-particle has a high momentum $k$.

Using assumption (a) one obtains the asymptotic formula

$$f_{j_10}^{(1)}(R) \simeq \frac{e^{ik'|R-a_1|}}{|R-a_1|} \frac{M}{2\pi\hbar^2} \int dy \, f_{00}^{(0)}(y+a_1)V_{j_10}(y)\, e^{-ik'u_1(R)\cdot y} \qquad (13.12)$$

where

$$u_1(R) = \frac{R-a_1}{|R-a_1|} \qquad (13.13)$$

Using the explicit expression of $f_{00}^{(0)}$ (see (13.6)) and assumption (b) one can write

$$f_{j_10}^{(1)}(R) \simeq \frac{e^{ik'|R-a_1|}}{|R-a_1|} \mathcal{I}(u_1(R)) \qquad (13.14)$$

$$\mathcal{I}(u_1(R)) = \frac{M}{2\pi\hbar^2} \int dy \, \frac{V_{j_10}(y)}{|y+a_1|}\, e^{ik(|y+a_1|-u_1(R)\cdot y)} \qquad (13.15)$$

One sees that $f_{j_10}^{(1)}(R)$ has the form of a wave diverging from $a_1$, whose amplitude $\mathcal{I}$ is given by the integral in (13.15) and it is explicitly dependent on the direction $u_1(R)$.

The crucial point is now to evaluate such amplitude. By assumptions (c), the integral in (13.15) is a highly oscillatory integral and then stationary phase arguments can be used. The leading term of the asymptotic expansion for $k \to \infty$ is determined by the value of the integrand at the critical points of the phase, i.e. for points $y$ such that

$$\nabla_y \Big(|y+a_1| - u_1(R)\cdot y\Big) = \frac{y+a_1}{|y+a_1|} - u_1(R) = 0 \qquad (13.16)$$

On the other hand, the integrand in (13.15) is very small except in a neighborhood of $y = 0$. Therefore one obtains the condition

$$u_1(R) \simeq \frac{a_1}{|a_1|} \qquad (13.17)$$

Using condition (13.17) in (13.13) one can deduce that the amplitude $\mathcal{I}$ is significantly different from zero only for those $R$ such that $R - a_1$ is (almost) parallel to $a_1$, i.e. the observation point $R$ must be (almost) aligned with the first atom and the origin.

From the above argument one concludes that $f_{j_1 0}^{(1)}(R)$ is approximately given by a wave diverging from $a_1$ with an amplitude very small except for $R$ given by (13.17), i.e. except in a small cone with vertex in $a_1$ and pointing away from the origin. A completely analogous analysis is valid for $f_{0 j_2}^{(1)}(R)$.

We will not give here details of Mott's estimate of the second order term. The final result is a straightforward consequence of the "focusing" of the first order term in the direction connecting the origin to the first atom.

If one agrees that the (amplified) effect of the excitations of the atoms is the true observed phenomenon in a cloud chamber then the result can be rephrased to say that one can only observe straight tracks. In this sense, Mott provides a satisfactory explanation of the straight tracks observed in the chamber based entirely on the Schrödinger equation.

It is worth emphasizing that the analysis Mott developed is based on a deep physical intuition. Indeed, the three-body problem discussed in his paper is an extremely simple but non-trivial model and it is especially well suited for highlighting the emergence of the qualitatively behavior of the $\alpha$-particle without unnecessary complications.

Even though it is not particularly stressed in Mott's paper, another important aspect is the fact that the result is valid under specific physical assumptions (large value of $k$ and quasi-elastic interaction). In other words, the observed behavior of the $\alpha$-particle in a cloud chamber is far from being universal.

In this sense, Mott's analysis can be considered the original prototype of the modern approach to the theory of environment-induced decoherence. In fact, the classical behavior (the trajectory) of the system ($\alpha$-particle) emerges as an effect of the interaction with the environment (vapor atoms in the chamber) under suitable assumptions on the physical parameters of the model.

It should also be noted that there is surely a gap in the mathematical rigor of Mott's paper. For instance, the stationary phase theorem is used without an accurate control of the conditions of applicability. Another unsatisfactory aspect is the use of the stationary Schrödinger equation, which prevents a more transparent time-dependent description of the evolution of the whole system.

## 13.4   The Role of Semi-classical Analysis

In the study of the classical limit of quantum mechanics, the use of stationary phase techniques is suggested by analogy with geometrical optics since the initial wave has a very high frequency (momentum).

In fact, from the technical point of view the situation is similar to the case of optics, since we compare a theory based on point particles propagating along trajectories (classical mechanics corresponding to geometrical optics) with a theory based on wave solutions of the Schrödinger equation (quantum mechanics in analogy with "physical" optics). Instead, it at once becomes clear that the situation is quite different from the point of view of the physical interpretation. In quantum mechanics, the wave representing the quantum state does not describe a physical object distributed in ordinary space, as in optics. Rather, it is a probability amplitude in the (classical) configuration space associated to the (whole) quantum system under consideration. Its role is to provide the statistical distribution of the outputs of repeated experiments. Moreover, the superposition principle introduces a crucial difficulty since no definite meaning can be given to the configuration of a quantum system in a superposition state. In other words, the standard formulation of quantum mechanics does not provide a space-time description of the behavior of a quantum object easily comparable with the classical one. For these reasons, the classical limit of quantum mechanics is, both technically and conceptually, hard to ascertain.

The traditional approach is essentially based on the analysis of the solutions of the Schrödinger equation in the limit "$\hbar \to 0$" for a suitable choice of the initial state. We recall that in this context the limit "$\hbar \to 0$" simply means that the typical action of the system is large with respect to the Planck's constant.

Usually one considers two possible kinds of initial states, chosen by analogy with the case of optics: WKB states and coherent states.

The former are defined by an amplitude independent of $\hbar$ and a highly oscillating phase for $\hbar$ small. In this case one can show, for $\hbar$ small and for short times, that the corresponding solution of the Schrödinger equation has the same form, with amplitude and phase governed by the classical transport and Hamilton-Jacobi equations respectively. This means that in the limit the quantum state propagates like a classical fluid and in this sense the classical description is recovered.

Coherent states are wave packets well concentrated in position and momentum around a point $(x_0, p_0)$ in classical phase space for $\hbar$ small. One can prove that the time evolution, for $\hbar$ small and a time interval not too long, is again described by a wave packet well concentrated in position and momentum around the classical trajectory starting from $(x_0, p_0)$.

Making available precise statements and mathematical proofs to detail the above qualitative pictures has required a great deal of technical work. Many refined and detailed results and summaries of the theory are at one's disposal in the literature (see, e.g. [33] and [4] and references therein).

Despite their mathematical elegance, such kind of results cannot be considered conclusive for a complete understanding of the classical limit of quantum mechanics. The reason is that the approach is crucially based on the choice of specific, essentially classical, initial states. Given the characteristic of the quantum system and of its environment, we expect that a classical behavior should emerge also starting from a genuine quantum state, like a superposition state. In such a case the usual procedure $\hbar \to 0$ turns out to be insufficient.

The idea behind this theoretical analysis is that quantum coherence between the components of a superposition state is very fragile. Even a weak interaction of the system with the environment can significantly reduce coherence and a classical behavior of the system can emerge.

In the case of a cloud chamber, the initial state of the $\alpha$-particle is a spherical wave, i.e., a (continuous) superposition of semi-classical states of the type mentioned above.

To explain the classical trajectories that are observed one has to analyze in detail the decoherence effect induced by the environment. The intuitive picture is clear-cut: each semi-classical component of the $\alpha$-particle initial state evolves, according to semi-classical theory, along an almost straight line interacting in the meanwhile with a small section of the environment. The crucial point is that different semi-classical components interact with different parts of the environment. As a consequence, the state of entire system becomes an almost incoherent sum of states supported in distant regions of the classical configuration space.

Our attempt is to give a rigorous version of this picture. To achieve this goal we have to quantify the response of a model quantum environment to the particle passage.

It is worth emphasizing that a modification of the environment is the only experimental output one can observe. Contrary to what is often stated, one should not "trace out" the environment degrees of freedom, but rather those of the particle. That is exactly what Mott did when trying to estimate the multiple ionization probability of atoms in a cloud chamber.

The results we obtained during the last decade [9, 13, 14, 19, 32] suggest an effective strategy for dealing with the study of quantum mechanical microscopic systems in interaction with quantum environments. It consists in building up simple models of quantum environments and analyzing their evolution under specific hypotheses on the physical parameters of the models.

## 13.5   Mott's Analysis Revisited

In [14] we propose a rigorous, time-dependent version of the original Mott's result. We consider a three-particle quantum system consisting of the $\alpha$-particle, initially described by a spherical wave centered in the origin, and two model atoms placed at fixed positions $a_1$, $a_2 \in \mathbb{R}^3$, with $0 < |a_1| < |a_2|$. For the sake of simplicity, each model atom is described by a particle subject to an attractive point interaction placed in $a_i$, $i = 1, 2$. For a detailed study of an Hamiltonian with a point interaction we refer to [3]. Here we only recall that it is a solvable model, i.e., spectrum and eigenfunctions can be explicitly computed. In particular the absolutely continuous spectrum is the positive real semi-axis, the singular continuous spectrum is empty and, in the attractive case, there is one negative, non degenerate eigenvalue. We also recall that a Hamiltonian with a given smooth interaction potential $V$ can be reconstructed as the limit of Hamiltonians with many randomly distributed point interactions [18].

At initial time we assume that both atoms are in their ground state (the eigenvector corresponding to the negative eigenvalue). Furthermore, we assume that the $\alpha$-particle and the atoms interact via a smooth two-body potential.

A crucial step for the analysis is a precise specification of the assumptions on the physical parameters of the model. We assume that:

(i) the wavelength associated to the $\alpha$-particle at time zero is much smaller than the spatial localization of the spherical wave (semi-classical regime);

(ii) the spatial localization of the spherical wave, the "diameter" of the atoms and the effective range of the interaction between $\alpha$-particle and atoms are much smaller than the macroscopic distance $|a_1|$;

(iii) the ionization energy of the atoms and the strength of the interaction between $\alpha$-particle and atoms are much smaller than the initial kinetic energy of the $\alpha$-particle (quasi-elastic regime).

We also introduce the time $\tau_j$, $j = 1, 2$, as the time spent by a classical particle, starting from the origin with velocity equal to the mean isotropic velocity of the spherical wave, to reach the atom in $a_j$. We remark that, under the above assumptions, it is reasonable to consider $\tau_1$ and $\tau_2$ as the collision times of the spherical wave emerging from the origin with the first atom in $a_1$ and the second atom in $a_2$.

Our aim is the consider the time evolution of the three-particle system up to second order in perturbation theory and to compute the probability $\mathcal{P}_2(t)$ that both atoms are ionized for $t > \tau_2$.

The result we find is in agreement with the original Mott's analysis and it can be roughly summarized as follows:

$\mathcal{P}_2(t)$ is negligible unless the positions $a_1$, $a_2$ of the atoms are aligned with the origin.

For a more precise formulation of the result and for the proof we refer to [14, 17]. Here we only add some remarks.

– The assumptions (i)–(iii) are crucial for the validity of the result, in the sense that a different qualitative behavior of the system must be expected if the assumptions on the physical parameters of the model are modified.

– Since in a cloud chamber one observes the (amplified) effect of the ionization of the atoms, the result shows that one can only observe straight trajectories.

– The method of the proof is essentially based on a representation formula for $\mathcal{P}_2(t)$ in terms of highly oscillatory integrals and on the asymptotic analysis of such integrals using stationary and non-stationary phase methods.

In the rest of this section we shall give some technical details of a result obtained in [32] (see also [17]) which, in our opinion, clarifies the dynamical mechanism that underlies Mott's result. Once more, the material presented in the rest of this section can be passed over by the reader who wants to avoid mathematical details.

Let us consider a simpler model of a non relativistic quantum system made of only two spinless particles in dimension three of masses $M$ and $m$. The latter is bound by an harmonic potential of frequency $\omega$ around the equilibrium position $a$. The first particle plays the role of the $\alpha$-particle while the harmonically bounded particle plays the role of an electron in a very simplified version of model-atom with fixed nucleus. The interaction between the test particle and the harmonic oscillator is described by a smooth two-body potential $V$.

Denoting by $R$ the position coordinate of the $\alpha$-particle and by $r$ the position coordinate of the harmonic oscillator, the Hamiltonian of the system in $L^2(\mathbb{R}^6)$ is given by

$$H = H_0 + \lambda\, V(\delta^{-1}(R - r)), \qquad \lambda, \delta > 0 \qquad (13.18)$$

where $H_0$ is the free Hamiltonian of the system

$$H_0 = -\frac{\hbar^2}{2M}\Delta_R - \frac{\hbar^2}{2m}\Delta_r + \frac{1}{2}m\omega^2(r - a)^2 \qquad (13.19)$$

We recall that the eigenfunctions of the harmonic oscillator are

$$\varphi_{\underline{n}}(r) = \gamma^{-3/2}\phi_{\underline{n}}(\gamma^{-1}(r - a)), \qquad \gamma = \sqrt{\frac{\hbar}{m\omega}}, \qquad \phi_{\underline{n}}(x) \equiv \phi_{n_1}(x_1)\phi_{n_2}(x_2)\phi_{n_3}(x_3) \qquad (13.20)$$

where $\underline{n} = (n_1, n_2, n_3) \in \mathbb{N}^3$ and $\phi_{n_k}$ is the Hermite function of order $n_k$. In particular the ground state corresponds to $\underline{n} = \underline{0} = (0, 0, 0)$.

Under the same kind of assumptions (i)–(iii) made above, we analyze the evolution of this system when the initial state is a product state of a spherical wave for the $\alpha$-particle and the ground state for the oscillator. In order to satisfy the assumptions, it is convenient to introduce a small parameter $\varepsilon > 0$ and to fix

$$\hbar = \varepsilon^2 \qquad M = 1 \qquad \sigma = \varepsilon \qquad m = \varepsilon \qquad \omega = \varepsilon^{-1} \qquad \delta = \varepsilon \qquad \lambda = \varepsilon^2 \qquad (13.21)$$

where $\sigma$ is the spatial localization of the spherical wave. Under this scaling the Hamiltonian becomes

$$H^\varepsilon = H_0^\varepsilon + \varepsilon^2\, V\left(\varepsilon^{-1}(R - r)\right) \qquad (13.22)$$

where

$$H_0^\varepsilon = -\frac{\varepsilon^4}{2}\Delta_R + \frac{1}{\varepsilon}\left[-\frac{\varepsilon^4}{2}\Delta_r + \frac{1}{2}(r - a)^2\right] \qquad (13.23)$$

The rescaled initial state takes the form

$$\Psi_0^\varepsilon(R, r) = \psi^\varepsilon(R)\varphi_0^\varepsilon(r) \qquad (13.24)$$

$$\varphi_{\underline{n}}^\varepsilon(r) = \frac{1}{\varepsilon^{3/2}}\phi_{\underline{n}}\left(\varepsilon^{-1}(r - a)\right) \qquad \underline{n} \in \mathbb{N}^3 \qquad (13.25)$$

In (13.24) the spherical wave $\psi^\varepsilon$ is explicitly given by

$$\psi^\varepsilon(R) = \frac{\mathcal{N}_\varepsilon}{\varepsilon^{5/2}\pi^{3/4}} \, e^{-\frac{|R|^2}{2\varepsilon^2}} \int_{S^2} d\hat{u} \, e^{\frac{i}{\varepsilon^2} v_0 \hat{u} \cdot R} \tag{13.26}$$

where $v_0 > 0$ is the mean isotropic velocity of the spherical wave and $\mathcal{N}_\varepsilon$ is a normalization constant, with

$$\lim_{\varepsilon \to 0} \mathcal{N}_\varepsilon = \mathcal{N}_0 \equiv \frac{v_0}{4\pi} \tag{13.27}$$

Notice that the spherical wave is obtained considering a wave packet localized in momentum around $v_0 \hat{u}$, where $\hat{u}$ is a generic unit vector, and then taking an average over all possible unit vectors of the sphere $S^2$.

We are interested in asymptotic behavior for $\varepsilon \to 0$ of the solution of the Schrödinger equation of the system

$$\mathcal{U}^\varepsilon(t)\Psi_0^\varepsilon, \qquad \mathcal{U}^\varepsilon(t) = e^{-i\frac{1}{\varepsilon^2}H^\varepsilon} \tag{13.28}$$

for $t > \tau$, where

$$\tau = \frac{|a|}{v_0} \tag{13.29}$$

is the (classical) collision time of the $\alpha$-particle with the oscillator.

In order to formulate the result, we fix a reference frame such that

$$\hat{a} = (0, 0, 1), \qquad \hat{a} \equiv \frac{a}{|a|} \tag{13.30}$$

and we introduce the following definition.

Let $P^\varepsilon = P^\varepsilon(R, r)$ be the function

$$P^\varepsilon(R, r) = \sum_{\underline{n}} P_{\underline{n}}^\varepsilon(R) \, \varphi_{\underline{n}}^\varepsilon(r) \tag{13.31}$$

where $P_{\underline{n}}^\varepsilon$ is the wave packet for the $\alpha$-particle given by

$$P_{\underline{n}}^\varepsilon(R) \equiv P_{\underline{n}}^\varepsilon(R_1, R_2, R_3) = \frac{C_{\underline{n}}^\varepsilon}{\varepsilon^{3/2}} \, \mathcal{F}_{\underline{n}}\left(\frac{R_1}{\varepsilon}, \frac{R_2}{\varepsilon}, 0\right) e^{-\frac{1}{2\varepsilon^2}\left(R_3 - z_{\underline{n}}^\varepsilon\right)^2 + \frac{i}{\varepsilon^2} v_{\underline{n}}^\varepsilon R_3} \tag{13.32}$$

$$C_{\underline{n}}^\varepsilon = \frac{2\pi^{5/4}}{i|a|^2} \, e^{\frac{i}{\varepsilon}|n|\tau + i\frac{|n|^2\tau}{2v_0^2}} \tag{13.33}$$

$$\mathcal{F}_{\underline{n}}(y) \equiv \mathcal{F}_{\underline{n}}(y_1, y_2, y_3) = e^{-i\frac{|y|^2}{2\tau}}\left(\widetilde{\phi_{\underline{n}}\phi_0} \cdot \tilde{V}\right)\left(-\frac{y_1}{\tau}, -\frac{y_2}{\tau}, -\frac{y_3}{\tau} - \frac{|n|}{v_0}\right)$$

(13.34)

$$\mathcal{Z}_{\underline{n}}^{\varepsilon} = \frac{|n|\tau}{v_0}\varepsilon$$

(13.35)

$$v_{\underline{n}}^{\varepsilon} = v_0 - \frac{|n|}{v_0}\varepsilon$$

(13.36)

In (13.34) we have used the notation $\tilde{f}$ for the Fourier transform of a function $f$.

Let us briefly comment on the above definition. The function $P^{\varepsilon}$ is an infinite linear combination of product states, made of stationary states of the harmonic oscillator and wave packets $P_{\underline{n}}^{\varepsilon}$ of the $\alpha$-particle. Each $P_{\underline{n}}^{\varepsilon}$ is well concentrated, for $\varepsilon$ small, in position around

$$R(0) = (0, 0, \mathcal{Z}_{\underline{n}}^{\varepsilon}) \equiv \mathcal{Z}_{\underline{n}}^{\varepsilon}\,\hat{a}$$

(13.37)

and in momentum around

$$P(0) = (0, 0, v_{\underline{n}}^{\varepsilon}) \equiv v_{\underline{n}}^{\varepsilon}\,\hat{a}$$

(13.38)

The free evolution at time $t$ of $P_{\underline{n}}^{\varepsilon}$, for $\varepsilon$ small, is again a wave packet well concentrated in position around

$$R(t) = \mathcal{Z}_{\underline{n}}^{\varepsilon}\,\hat{a} + v_{\underline{n}}^{\varepsilon}\,\hat{a}\,t$$

(13.39)

and in momentum around $P(0)$ (the momentum is conserved). In particular, at time $t = \tau$ the wave packet is well localized in position around

$$R(\tau) = \mathcal{Z}_{\underline{n}}^{\varepsilon}\,\hat{a} + v_{\underline{n}}^{\varepsilon}\,\hat{a}\,\tau = a$$

(13.40)

i.e., around the position of the oscillator.

The wave packets $P_{\underline{n}}^{\varepsilon}$ play a crucial role in the asymptotic expression of the wave function of the system for $\varepsilon$ small. More precisely, the following result holds.

**Theorem** *Let us fix $t > \tau$. Then there exists $C(t) > 0$, independent of $\varepsilon$, such that*

$$\mathcal{U}^{\varepsilon}(t)\Psi_0^{\varepsilon} = \mathcal{U}_0^{\varepsilon}(t)\Psi_0^{\varepsilon} + \varepsilon^2\,\mathcal{U}_0^{\varepsilon}(t)P^{\varepsilon} + \mathcal{E}^{\varepsilon}(t)$$

(13.41)

*where*

$$\mathcal{U}_0^{\varepsilon}(t) = e^{-i\frac{t}{\varepsilon^2}H_0^{\varepsilon}}$$

(13.42)

*and*

$$\|\mathcal{E}^{\varepsilon}(t)\| \leq C(t)\,\varepsilon^{3} \tag{13.43}$$

For the proof (in the more general case of $N \geq 1$ harmonic oscillators) we refer to [32]. Here we only comment on the result.

Using the expressions for the free propagator $\mathcal{U}_0^{\varepsilon}(t)$, the initial state $\Psi_0^{\varepsilon}$ and the function $P^{\varepsilon}$, from (13.41) one has

$$\left(\mathcal{U}^{\varepsilon}(t)\Psi_0^{\varepsilon}\right)(\boldsymbol{R},\boldsymbol{r}) \simeq e^{-i\frac{t}{\varepsilon^2}E_{\underline{0}}^{\varepsilon}}\left[\left(e^{-i\frac{t}{\varepsilon^2}h_0^{\varepsilon}}\psi^{\varepsilon}\right)(\boldsymbol{R}) + \varepsilon^2\left(e^{-i\frac{t}{\varepsilon^2}h_0^{\varepsilon}}P_{\underline{0}}^{\varepsilon}\right)(\boldsymbol{R})\right]\varphi_{\underline{0}}^{\varepsilon}(\boldsymbol{r})$$

$$+\varepsilon^2\sum_{\underline{n}\neq\underline{0}}e^{-i\frac{t}{\varepsilon^2}E_{\underline{n}}^{\varepsilon}}\left(e^{-i\frac{t}{\varepsilon^2}h_0^{\varepsilon}}P_{\underline{n}}^{\varepsilon}\right)(\boldsymbol{R})\,\varphi_{\underline{n}}^{\varepsilon}(\boldsymbol{r}) \tag{13.44}$$

where $E_{\underline{n}}^{\varepsilon}$ denotes the energy level of the oscillator

$$E_{\underline{n}}^{\varepsilon} = \varepsilon\left(|n| + \frac{3}{2}\right), \qquad |n| = n_1 + n_2 + n_3 \tag{13.45}$$

and $h_0^{\varepsilon}$ is the free Hamiltonian for the $\alpha$-particle

$$h_0^{\varepsilon} = -\frac{\varepsilon^4}{2}\Delta_R \tag{13.46}$$

In (13.44) the approximate wave function for $t > \tau$ has been written as the sum of two terms, distinguished for the different behavior of the oscillator (unperturbed or excited). In the first one, the oscillator remains in its ground state and the $\alpha$-particle evolves according to

$$\left(e^{-i\frac{t}{\varepsilon^2}h_0^{\varepsilon}}\psi^{\varepsilon}\right)(\boldsymbol{R}) + \varepsilon^2\left(e^{-i\frac{t}{\varepsilon^2}h_0^{\varepsilon}}P_{\underline{0}}^{\varepsilon}\right)(\boldsymbol{R}) \tag{13.47}$$

i.e., the free evolution of the initial spherical wave slightly deformed by the free evolution of the small wave packet $P_{\underline{0}}^{\varepsilon}$, emerging from the oscillator. The second term is a sum over all possible excited states of the oscillator. In each term of the sum, the evolution of the $\alpha$-particle is given by $\varepsilon^2$ times

$$\left(e^{-i\frac{t}{\varepsilon^2}h_0^{\varepsilon}}P_{\underline{n}}^{\varepsilon}\right)(\boldsymbol{R}) \tag{13.48}$$

i.e., the free evolution of the wave packet $P_{\underline{n}}^{\varepsilon}$, $\underline{n} \neq \underline{0}$. As we already remarked, each wave packet emerges at $t = \tau$ from the excited oscillator with momentum $v_{\underline{n}}^{\varepsilon}\,\hat{\boldsymbol{a}}$ and, for $t > \tau$, it will be concentrated around the uniform classical motion (13.39), which can be more conveniently rewritten as

$$\boldsymbol{R}(t) = \boldsymbol{a} + \left(v_0 - \frac{|n|}{v_0}\varepsilon\right)(t - \tau)\,\hat{\boldsymbol{a}} \tag{13.49}$$

We emphasize that the result expressed in the theorem provides a physical explanation of Mott's result in the three-particle model ($\alpha$-particle plus two atoms in $a_1$, $a_2$). In fact, if the collision with the first atom in $a_1$ produces excitation of the atom then, according to the above result, the $\alpha$-particle is described by a localized wave packet emerging from $a_1$, with momentum along the direction $\hat{a}_1$. As a consequence, the requirement that also the second atom in $a_2$ is excited can be satisfied only if such atom is hit by the wave packet, and this happens only if $a_2$ lies on the direction $\hat{a}_1$.

This explains why excitation of both atoms can occur only if their positions are aligned with the origin.

## 13.6  Different Approaches and Open Problems

The simple models we analyzed should be considered as first steps in a branch of research that we consider relevant and promising. In our opinion, there are different strategies that might be exploited and open problems that would be well worth the study.

As we mentioned above the initial spherical wave packet can be seen as a continuous superposition of coherent states pointing toward all possible radial directions. In fact, (13.26) is the rigorous formulation of this claim.

The partition (slicing) of the incoming $\alpha$-wave in fuzzy coherent slices that move as semiclassical waves (keeping their coherence until one of them interacts with a real atom) can be deemed to be artificial. But the wave itself is a probability wave and therefore has no objective reality. We are slicing something that does not exist as a physical entity!

Indeed we are manipulating mathematical objects that enter into the mathematical framework by which quantum mechanics describes outcomes of experimental observations; only part of this mathematical framework is given a direct physical correspondence with experiments. The remaining part is there in order to give a meaningful dynamics (meaningful = as close as possible to the dynamics of material bodies). What counts is that the mathematical description we give be consistent and our results indicate that the family of coherent states we suggest as a "basis" in order to analyze the initial state of the $\alpha$-particle is in fact the right "pointer basis" for the problem under investigation.

In our opinion, this qualitative description can be turned into an effective technical tool to examine details of the asymptotic evolution of the $\alpha$-particle and of the environment in models of a cloud chamber.

We recall that semiclassical waves propagate, under the Schrödinger equation, keeping their barycenter on a classical path. Their shape changes slightly and their dispersion is of order $\sqrt{\hbar}$. As soon as we write the initial state as a superposition of such states, the problem is reduced to the interaction of a semiclassical wave with an atom, leading to ionization. It should be noted that the semiclassical wave packet remains such after the interaction with an atom only if the momentum transfer within the process is a little percentage of the initial momentum (almost zero-angle scattering). In the physical problem we are considering, this implies that after the interaction either the atom is in an excited state or the atom is ionized and low energy electrons are emitted. In the real Wilson chamber, this last condition is needed in order that the ionized atoms, acting as condensation seeds, may cause the formation of liquid droplets (once more we stress that all we see and measure is the formation of tracks of droplets in the cloud chamber). It is therefore reasonable to restrict attention to the case in which the semiclassical wave packet retains its identity during interaction, changing perhaps slightly its shape.

While the evolution of a semiclassical wave packet interacting with an external potential has been extensively investigated [21, 25, 33, 35], semiclassical inelastic scattering has not received, at the best of our knowledge, comparable attention and any result in this direction would be welcome. We are planning to examine this approach to the Mott's problem in further work. Here, we want to focus briefly on the connection of the strategy outlined above with the analysis of Michel Bauer and Denis Bernard on the wave function collapse in repeated quantum non demolition measurements [6]. In their paper these authors investigate the evolution of a microscopic system together with a probe. The latter is meant to perform a sequence of "nondemolition" measurements on the system.

The nondemolition character of the measurement is turned into the main assumption of authors' analysis: let $\mathcal{H} \equiv \mathcal{H}_S \otimes \mathcal{H}_P$ ($S$ standing for system and $P$ for probe) be the Hilbert space of the states of the whole system. The assumption reads: there exist an orthonormal basis $\phi_n$ in $\mathcal{H}_S$ and unitary operators $U_n$ in the Hilbert space of the probe $\mathcal{H}_P$ such that the evolution of the whole system, starting from an initial state $\phi_n \otimes \Psi$, is given by the unitary operator $\mathbb{I} \otimes U_n$. In short, each state in the chosen basis of the microscopic system evolves unaltered, whereas the state of the probe evolves according to a Schrödinger dynamics whose generator depends on the microscopic system state.

From the assumption stated above, the authors proceed to prove a list of interesting results making use only of classical probability tools. In summary, after a large number of repeated measurements processes, obeying at each time the main assumption,

- the state of the system tends to one of the states $\phi_n$ with probability $|(\phi_n, \phi)|^2$ if $\phi$ is the initial state of the system.
- A limit probability measure on the states of the probe is uniquely associated to each state $\phi_n$ of the system.

The authors give also results on the rate of convergence of the sequences to the final states.

Let us associate the chosen basis with a "suitable slicing" of the radial initial condition in a cloud chamber model and each $U_n$ with the response of a model atom in the chamber to each scattering event with a particle initially in the state $\phi_n$. It is reasonable to predict (and it should be feasible to prove) that, in the limit of quasi-elastic, small angle scattering, the system $\alpha$-particle plus model-atoms fulfill the assumption of Bauer and Bernard. In this scenario, the $\alpha$-particle is submitted to a sequence of non demolition measurements by the atoms of the gas in the chamber. The coherent states in which the initial state is analyzed are the "pointer basis" of the alpha particle, whereas the tracks are the pointer states of the probe.

For a given model of environment, it would be interesting to extend the results to more general initial conditions of the microscopic system and to cases in which external fields act on the microscopic system.

Another crucial step ahead would be to consider more realistic models of quantum environments. As was already pointed out by Hepp in [24], complete decoherence requires an infinite time in models where a quantum particle evolves in an environment made up of non-interacting quantum subsystems. An alternative way to reach complete decoherence would be to consider an ever larger number of environment constituents in a finite region.

This idea brought us to consider models of environment made up of multi-channel point interactions (one can think either of point atoms with a finite number of energetic levels or of localized spins). There are many advantages in working with such a kind of solvable models. To mention the most important, a non-perturbative theory is practicable (the environment is not an unmanageable multi-particle system) and it is possible to investigate the asymptotic limit of infinitely many scattering centers in a finite volume. A quite detailed introduction to these models is given in Chapter 3 of the book [17].

A more realistic choice would be to model the environment with self-interacting fields (e.g., spins ferromagnetically interacting among them), initially in a genuine meta-stable state. The non-linear self-interaction would enhance the response of the environment, which might show macroscopic modifications in finite time.

In conclusion a complete description of the mechanism of production of the tracks in a cloud chamber still escapes us, but we have outlined the role of quantum mechanics, of semi-classical analysis and of stationary phase techniques within the time-dependent Schrödinger equation in shedding some light on the investigation of the classical-quantum border.

**Acknowledgements** Part of the contents of this note were presented by one of us (G.F. Dell'Antonio) in a stimulating Conference in Bielefeld on the foundations of quantum mechanics. We are grateful to Ph. Blanchard and J. Fröhlich for organizing that influential and exceptional event and for several inspiring discussions with all of us.

# References

1. Adami, R., Figari, R., Finco, D., Teta, A.: On the asymptotic behavior of a quantum two-body system in the small mass ratio limit. J. Phys. A **37**, 7567–7580 (2004)
2. Adami, R., Figari, R., Finco, D., Teta, A.: On the asymptotic dynamics of a quantum system composed by heavy and light particles. Commun. Math. Phys. **268**(3), 819–852 (2006)
3. Albeverio, S., Gesztesy, F., Høegh-Krohn, R., Holden, H.: Solvable Models in Quantum Mechanics, 2nd edn. AMS Chelsea, Providence (2005)

4. Albeverio, S., Høegh-Krohn, R., Mazzucchi, S.: Mathematical Theory of Feynman Path Integrals. An Introduction. Lecture Notes in Mathematics, 2nd edn. Springer, New York (2008)
5. Bacciagaluppi, G., Valentini, A.: Quantum Theory at the Crossroads: Reconsidering the 1927 Solvay Conference. Cambridge University Press, Cambridge (2009)
6. Bauer, M., Bernard, D.: Convergence of repeated quantum nondemolition measurements and wave-function collapse. Phys. Rev. A **84**, 044103 (2011)
7. Bell, J.: Speakable and Unspeakable in Quantum Mechanics. Cambridge University Press, Cambridge (1987)
8. Blanchard, Ph., Giulini, D., Joos, E., Kiefer, C., Stamatescu, I.-O. (eds.) Decoherence: Theoretical, Experimental and Conceptual Problems. Lecture Notes in Physics, vol. 538. Springer, New York (2000)
9. Cacciapuoti, C., Carlone, R., Figari, R.: Spin-dependent point potentials in one and three dimensions. J. Phys. A **40**, 249–261 (2007)
10. Carazza, B., Kragh, H.: Classical behavior of macroscopic bodies from quantum principles: early discussions. Arch. Hist. Ex. Sci. **55**, 43–56 (2000)
11. Condon, E., Gurney, R.: Quantum mechanics and radioactive disintegration. Nature **122**, 439 (1928)
12. Darwin, C.G.: A collision problem in the wave mechanics. Proc. R. Soc. Lond. A **124**, 375–394 (1929)
13. Dell'Antonio, G., Figari, R., Teta, A.: Joint excitation probability for two harmonic oscillators in dimension one and the Mott problem. J. Math. Phys. **49**(4), 042105 (2008)
14. Dell'Antonio, G., Figari, R., Teta, A.: A time dependent perturbative analysis for a quantum particle in a cloud chamber. Ann. H. Poincaré **11**(3), 539–564 (2010)
15. Falkenburg, B.: The analysis of particle tracks: a case for trust in the unity of physics. Stud. Hist. Philos. Mod. Phys. **27**(3), 337–371 (1996)
16. Figari, R., Teta, A.: Emergence of classical trajectories in quantum systems: the cloud chamber problem in the analysis of Mott (1929). Arch. Hist. Ex. Sci. **67**(2), 215–234 (2013)
17. Figari, R., Teta, A.: Quantum Dynamics of a Particle in a Tracking Chamber. SpringerBriefs in Physics. Springer, Berlin/Heidelberg (2014)
18. Figari, R., Holden, H., Teta, A.: A law of large numbers and a central limit theorem for the Schrödinger operator with zero-range potentials. J. Stat. Phys. **51**, 205–214 (1988)
19. Finco, D., Teta, A.: Asymptotic expansion for the wave function in a one-dimensional model of inelastic interaction. J. Math. Phys. **52**(2), 022103 (2011)
20. Gamow, G.: Quantum theory of atomic nucleus. Z. Phys. **51**, 204 (1928)
21. Hagedorn, G.A., Joye, A.: Exponentially accurate semiclassical dynamics: propagation, localization, Ehrenfest times, scattering, and more general states. Ann. H. Poincaré **1**, 837–883 (2000)
22. Heisenberg, W.: Über den anschaulichen Inhalt der quantentheoretischen Kinematik und Mechanik. Z. Phys. **43**, 172–198 (1927); Eng. trans. reprinted in: Wheeler, J.A., Zurek, W.: Quantum Theory and Measurement. Princeton University Press, Princeton (1983)
23. Heisenberg, W.: The Physical Principles of the Quantum Theory. The University of Chicago Press, Chicago (1930)
24. Hepp, K.: Quantum theory of measurement and macroscopic observable. Helv. Phys. Acta **45**, 237–248 (1972)
25. Hepp, K.: The classical limit for quantum mechanical correlation functions. Commun. Math. Phys. **35**(4), 265–277 (1974)
26. Hornberger, K.: Introduction to decoherence theory. In: Buchleitner, A., Viviescas, C., Tiersch, M. (eds.) Entanglement and Decoherence. Foundations and Modern Trends. Lecture Notes in Physics, vol. 768, pp. 221–276. Springer, New York (2009)
27. Hornberger, K., Sipe, J.E.: Collisional decoherence reexamined. Phys. Rev. A **68**, 012105 (2003)
28. Joos, E., Zeh, H.D.: The emergence of classical properties through interaction with the environment. Z. Phys. B **59**, 223–243 (1985)

29. Joos, E., Zeh, H.D., Kiefer, C., Giulini, D., Kupsch, J., Stamatescu, I.O.: Decoherence and the Appearance of a Classical World in Quantum Theory, 2nd edn. Springer, New York (2003)
30. Leone, M., Robotti, N.: A note on the Wilson cloud chamber (1912). Eur. J. Phys. **25**, 781–791 (2004)
31. Mott, N.F.: The wave mechanics of $\alpha$-ray tracks. Proc. R. Soc. Lond. A **126**, 79–84 (1929)
32. Recchia, C., Teta, A.: Semiclassical wave-packets emerging from interaction with an environment (2013). arXiv:1305.0784 [math-ph]
33. Robert, D.: Semi-classical approximation in quantum mechanics. A survey of old and recent mathematical results. Helv. Phys. Acta **71**, 44–116 (1998)
34. von Neumann, J.: Mathematische Grundlagen der Quantenmechanik. Springer, New York (1932); Engl. trans. Mathematical Foundations of Quantum Mechanics, Princeton University Press, Princeton (1955)
35. Yajima, K.: The quasiclassical limit of quantum scattering theory. Commun. Math. Phys. **69**(2), 101–129 (1979)
36. Zurek, W.H.: Decoherence, einselection and quantum origins of the classical. Rev. Mod. Phys. **75**, 715 (2003)

# Chapter 14
# Quantum Mechanics of Time

**Andreas Ruschhaupt and Reinhard F. Werner**

We start by summarising some basic concepts in the research field of quantum mechanics of time. In the following, we will review some recent results namely a new version of an energy-time uncertainty relation and a no-go theorem about tunnelling times.

## 14.1 Introduction

The question "Where is a particle at a given time $t$?" is discussed theoretically in every quantum mechanics book. On the other hand, the similar question "When is a particle at a given position $x$?" is rarely discussed in these books. In fact, the theoretical study of time observables has often been in practice almost abandoned if not totally banned. The main reason for this banning has been Pauli's theorem which states that time operators cannot be self-adjoint. This has led to the phrase "time is only a parameter" in quantum mechanics which is still repeated by some physicists when discussing arrival times.

The situation is different in experiments: in a majority of experiments, there are detectors in the laboratories waiting for a particle to arrive, while measurements of positions at a fixed time are met less often. For a long time, the motion of particles in these experiments could be approximated with high accuracy by a classical trajectory and there has been no need to use a quantum mechanical description

A. Ruschhaupt (✉)
Department of Physics, University College Cork, Cork, Ireland
e-mail: aruschhaupt@ucc.ie

R.F. Werner
Institut für Theoretische Physik, Leibniz Universität Hannover, Appelstr. 2, 30167 Hannover, Germany

© Springer-Verlag Berlin Heidelberg 2015                                     315
P. Blanchard, J. Fröhlich (eds.), *The Message of Quantum Science*, Lecture Notes in Physics 899, DOI 10.1007/978-3-662-46422-9_14

of arrival times. Nevertheless, with increasing control possibilities in quantum experiments and the ability to reach lower and lower velocities, the question of a quantum mechanical description of time quantities will become more important and a detailed investigation of these might be essential.

Note, there is unfortunately a confusion with the expression "time-of-flight" measurements. These measurements are quite standard in the cold-atom community and they essentially monitor the position probability density and not a quantum-mechanical time-of-arrival density.

In the next section we give a brief overview about some general strategies and approaches to deal theoretically with arrival times and tunnelling times in quantum mechanics. For more extensive reviews on the arrival time and time in quantum mechanics see [1, 2]. In Sect. 14.3, we review some recent results about time-energy uncertainty relations. We present some recent results about tunnelling times in Sect. 14.4.

## 14.2  Overview

### 14.2.1  Arrival Times

Consider a quantum system with Hilbert space $\mathcal{H}$ and time-independent Hamiltonian $H$. Starting from some initial state $\psi_0 \in \mathcal{H}$ at time $t = 0$, we would like to determine the probability distribution of arrival times $\rho(t)$ at some detector or counter which is at some fixed position $x_D$. For simplicity, we will assume in the following a one-dimensional setting and let $x_D = 0$. There are different approaches to this problem, varying in the degree of detail with which the detector is described.

**Covariant Arrival Observables** The obvious approach might be to find an "idealised" observable $T$ describing arrival times in the same manner as there is, for example, an "idealised" position observable $X$, idealised in the sense that $X$ is independent of details of the real detector.

Pauli's theorem states that this arrival-time observable $T$ cannot be represented by a self-adjoint operator (as already mentioned, this fact is often misinterpreted by drawing the incorrect conclusion that "time is only a parameter" and "there is no arrival-time observable"). Pauli's theorem just means that the arrival-time observable has to be represented by a positive operator valued measure (POVM) [3–5], a generalisation of the "usual" way of describing observables by a self-adjoint operator.

A requirement for arrival-time observables is often its transformation behaviour: starting from the time-evolved state $\psi_{t_0} = \exp(-iHt_0/\hbar)\psi_0$ we should get the same arrival time distribution $\rho$, but shifted by $t_0$. Covariant arrival observables with this property have been studied extensively [6–9]. The simplest example for the free evolution $H = P^2/(2m)$ is Kijowski's distribution [6]

$$\rho(t) = \left| \int_0^\infty dp \sqrt{\frac{p}{2\pi m\hbar}} e^{-ip^2 t/(2m\hbar)} \psi_0(p) \right|^2.$$

However, transformation behaviour alone is not sufficient to single out a convincing model for a given experimental counter array. The most general expression of a covariant arrival-time observable has been derived in [7]. Moreover, this approach requires the Hamiltonian to have a purely continuous spectrum and is hence limited to infinite dimensional Hilbert spaces.

**Absorptive Arrival Times** While the previous approach is independent of the details of the detection process and the detector, the following approach is based on a phenomenological modelling of the detector and the detection process. It was first worked out in detail in [10]. The detector is described phenomenologically by a non-Hermitian term (also called complex potential) $-iD$ (where $D$ is some self-adjoint operator) added to the Hamiltonian $H$, which is thus replaced by $K = H - iD$. Thereby the unitary time evolution operator $U_t = \exp(-iHt/\hbar)$ is modified to a semigroup of contractions, i.e., operators $B_t = \exp(-iKt/\hbar)$ ($t \geq 0$) such that the norm of the wavefunction $\psi_t = B_t \psi_0$ is no longer preserved; it is decreasing, i.e. the wavefunction is "absorbed" by the detector. The loss of the norm of the wavefunction, $N_t = 1 - \|\psi_t\|^2$, is interpreted as the probability that the particle did arrive (or has been detected) before time $t$.

Thus, given an initial quantum state $\psi_0$, we get the arrival probability density $\rho(t)$ for $t > 0$ by

$$\rho(t) = -\frac{1}{p}\frac{d}{dt}N_t = \frac{1}{p}\langle \psi_t | D\psi_t \rangle \tag{14.1}$$

where $p = 1 - \lim_{t\to\infty} N_t$ is the probability that the particle is detected at all, i.e. the total absorption probability. More details can be also found in [11, 12]. An immediate advantage of the absorptive arrival time approach is that it also applies directly to certain finite dimensional models in contrast to covariant arrival-time observables.

**Detector Models** At the other extreme we can make a more detailed detector model. For example, in [13] and in following publications, a more realistic atom-laser model describing an arrival-time measurement by quantum-optical means has been proposed and examined, see [14] for a review. The basic idea is that a region of space is illuminated by a laser. We assume a two-level atom being initially in its ground state and that its excited state should decay very fast. Moreover, the laser should be on resonance with the transition ground-excited state. If the atom is now entering the illuminated region the atom will start emitting photons which are detected. The first photon emission can be taken as a measure of the arrival time of the atom in that region. It is possible to connect this model to the approaches described previously in certain limits, see the review [14] for details.

## 14.2.2  Tunnelling Times

One could ask also for other time quantities than the single arrival time. In the simple case of a massive particle moving in one dimension through a localised (tunnel) potential, the question of the "time spent in the tunnel" is especially interesting, and has given rise to extensive discussion some time ago (see e.g. [15–19] and the references therein).

An old observation related to tunnelling times is the so called Hartman effect [15], which states that the transmitted part of a wave function appears to move faster through the tunnel than the corresponding wave function in the free state: More precisely, after a long rectangular barrier and for a wave function of narrow momentum distribution, in leading order the transmitted pulse appears at the end of the tunnel instantaneously. However, all this is true only for the shape of the wave function disregarding normalisation. But obviously, especially for long tunnels, for which the gain in speed would only be noticeable, the transmission probability is exponentially small.

It has been suggested that the Hartman effect allows superluminal signal transport [20], nevertheless to utilise the Hartman effect for a faster signal transmission, we would have to analyse the trade-off between transmission probability and transmission speed, see Sect. 14.4 for more details.

## 14.2.3  Event-Enhanced Quantum Theory and Time Properties

Blanchard and Jadczyk proposed the phenomenological formalism of Event-Enhanced Quantum Theory (EEQT) [21–24]. The main idea of EEQT is that the total system is split in a quantum part and a classical part. The states of the quantum part are wave functions, which are not directly observable, they are the "hidden" variables of the theory. On the other hand, the states of the classical parts can be observed without disturbing them. "Events" are changes of the classical state (which in general also comes together with a change of the quantum state); they do not need an observer for their generation.

Let us summarise briefly the formalism of EEQT. We assume that the classical part has only $m$ possible discrete states. Statistical states of the total system can be represented by $m \times m$ diagonal matrices where the diagonal elements $\rho_j$ are positive operators on the Hilbert space $H_Q$ corresponding to the quantum part and fulfil $\sum_{j=1}^{m} \rho_j = 1$. The coupling between classical and the quantum system is described phenomenologically by linear operators $g_{\alpha\beta} : H_Q \to H_Q$ with $g_{\alpha\alpha} = 0$. Moreover there are Hamiltonians $\hat{H}_\alpha$. The time evolution of the total system is given by a set of master equations

$$\dot{\rho}_\alpha = -i[\hat{H}_\alpha, \rho_\alpha] + \sum_\beta g_{\alpha\beta}\, \rho_\beta g_{\alpha\beta}^+ - \frac{1}{2}\{\sum_\beta g_{\beta\alpha}^+ g_{\beta\alpha}, \rho_\alpha\} \tag{14.2}$$

where $\{,\}$ stands for the anticommutator. Another starting point is given by a Piecewise Deterministic Process algorithm [24]. Repeating the algorithm and averaging the outcome leads to a solution of the above set of master equations.

EEQT has also been used to discuss the question of arrival time [25] and tunnelling times [26, 27]. In the case of arrival time it has also been shown how the phenomenological couplings can be derived from a detection model [28]. While the original formalism is non-relativistic, there are also relativistic extensions [29, 30] which have been used to discuss relativistic arrival times [31].

## 14.3  Energy-Time Uncertainty Relation

From Heisenberg's seminal 1927 paper [32], uncertainty relations have been recognised as a fundamental feature of quantum mechanics. Heisenberg gives a semi-classical heuristic discussion and his "uncertainties" are conceptually different in different parts of his paper. Modern textbooks all agree on "the" uncertainty relation, namely a version stated and proved by Kennard [33] in the same year Heisenberg's paper appeared. Kennard achieved an important clarification both conceptually, by defining the uncertainties as the standard deviation of operationally well-defined probability distributions, and also quantitatively, so it becomes possible to say that a particular experiment realises an uncertainty product within 3 % of the absolute minimum.

This clarification was so successful that other aspects of Heisenberg's paper, like his discussion of the precision of a position measurement by microscope versus the momentum disturbance by the measurement, fell into disrepute. Nevertheless, these ideas are not only heuristically meaningful, but can be made operationally precise and proved as theorems in the quantum formalism [34, 35].

The situation becomes more obscure in the literature when we go to energy-time uncertainty relations. In most cases, energy-time uncertainty relations are invoked in a very handwaving fashion only. There are very few conceptually clear and quantitatively meaningful formulations (see [36] for a review). For covariant arrival-time observables a Kennard-like uncertainty relation can be proved strictly [37].

It was shown recently that this is also true for absorptive arrival times [38]; we will review this result in the following. The goal is to find a relation which is related as closely as possible to Kennard's interpretation of $\Delta P$ and $\Delta Q$. One the one hand, let $\Delta E$ be the standard deviation of the energy observable (the Hamiltonian) of the system at initial time, defined through $(\Delta E)^2 = \langle \psi_0 | H^2 \psi_0 \rangle - \langle \psi_0 | H \psi_0 \rangle^2$. On the other hand, there should be a detection process modelled phenomenologically by an absorbing potential leading to the arrival-time probability distribution $\rho(t)$, see Eq. (14.1). We denote by $\langle T \rangle$ and $\langle T^2 \rangle$ the first and second moment of this probability distribution $\rho(t)$ and set $(\Delta T)^2 = \langle T^2 \rangle - \langle T \rangle^2$.

We can hence look for a universal lower bound on the product $\Delta T \cdot \Delta E$. Without further conditions such a lower bound cannot hold. Indeed, $\Delta T$ can be computed knowing $H - iD$ and $\psi_0$, whereas $\Delta E$ depends on $H$ and $\psi_0$ only. For example, if we set $D = \alpha \mathbb{1}$, we get $p = 1$ and $\rho(t) = 2\alpha e^{-2\alpha t}$ independently of $\psi_0$. Clearly, this cannot imply any constraint on the energy distribution.

Nevertheless, surprisingly, only a mild and natural assumption on the initial state is required to achieve an uncertainty relation, namely (heuristically) that at initial time the particle is not already in the detector! More formally, if $D$ describes the detector as explained above, we want $D\psi = 0$, or $H\psi = K\psi$ where $K = H - iD$. Since these operators are usually unbounded, we also have to specify the domains. Writing dom $X$ for the domain of the operator $X$, we require that $\psi \in$ dom $K^2 \cap$ dom $H$ and $H\psi = K\psi$. Under these conditions it has been proved in [38] that

$$\Delta T \cdot \Delta E > \frac{\hbar}{2} \sqrt{p} \; . \tag{14.3}$$

where $p$ is the total detection probability. The dependence on $p$ might be unexpected having in mind heuristic derivations. Nevertheless, this dependence can be understood qualitatively by the fact that for a small detection operator $D$ only a few particles are ever detected ($p \approx 0$), so observing arrival times cannot imply a strong constraint on $\Delta E$. The exact power $\sqrt{p}$ can only be explained in the mathematical proof in [38].

The arrival time distribution $\rho(t)$ is always supported by the positive time axis $\mathbb{R}_+$. Therefore, the mean arrival time $\langle T \rangle$ is always positive, and can take the place of $\Delta T$ in the uncertainty relation. Under the same conditions it was proved in [38] that

$$\langle T \rangle \cdot \Delta E \geq C \hbar \sqrt{p} , \tag{14.4}$$

where $C = 2(-Z_1/3)^{(3/2)} \approx 1.376$ is a numerical constant involving the first negative zero $Z_1$ of the Airy function.

The basic idea of the proofs of Eqs. (14.3) and (14.4) is a so-called dilation construction; the system is mapped to a "larger" Hilbert space in such a way that energy and time become conjugate self-adjoint operators. In this "larger" system, standard theorems about conjugated self-adjoint operators (like position and momentum) can be used. For the details we refer to [38].

As an example, let us consider the simplest possible system to which the relations (14.3) and (14.4) apply. Unlike the covariant observable approach (in which the spectrum of $H$ has to be continuous) finite dimensional systems are now included. We consider a two-level system with

$$H = \frac{\hbar}{2} \begin{pmatrix} 0 & \Omega \\ \Omega & 0 \end{pmatrix} , \qquad D = \frac{\hbar}{2} \begin{pmatrix} 0 & 0 \\ 0 & \gamma \end{pmatrix} \tag{14.5}$$

where $\Omega, \gamma > 0$ are parameters. The relevant quantity is $\gamma/\Omega$, since we could make $\Omega/2 = \hbar = 1$ by a choice of units. As the initial state we take $\psi_0 = (1, 0)^T$, so our basic assumption $D\psi_0 = 0$ is satisfied. Obviously, $\langle H \rangle = 0$ and $\langle H^2 \rangle = \frac{\hbar^2}{4}\Omega^2$, so $\Delta E = \frac{\hbar}{2}\Omega$. One can explicitly calculate $\psi_t = \exp(-i(H - iD)t/\hbar)\psi_0$, and hence compute the probability density $\rho(t)$. The moments are also readily calculated. Both $\langle T \rangle$ and $\Delta T$ attain their minimum when $\gamma = \sqrt{2}\Omega$. Then the uncertainty inequalities are satisfied as $1/\sqrt{2} \approx 0.707 > 0.500$ for Eq. (14.3), and $\sqrt{2} \approx 1.414 > 1.376$ for Eq. (14.4), so $\langle T \rangle \Delta E$ reaches the minimum to within 3 %!

## 14.4  Tunnelling Time

Let us look again at tunnelling time and the Hartman effect. As we said before, in any attempt to utilise the Hartman effect for a faster signal transmission, we would have to analyse the trade-off between transmission probability and transmission speed. It has been proved recently in [39] that this trade-off is always trivial: when damping is taken into account, transmission through a tunnel will always slow down the signal. So even if the arrival time probability density for the transmitted particles peaks earlier than for the free particles, we must look at the integrated density (i.e., the probability for the particles to arrive before a given deadline $t$) and compare this to the corresponding quantity for the free particles.

**Setting** We want to compare two settings. In the first case, we have free evolution in one dimension described the Hamiltonian $H_0 = P^2$ (with $m = 1/2, \hbar = 1$). In the second setting we have also a tunnel potential $V$ such that the Hamiltonian is $H_V = H_0 + V$. $V$ should be compactly supported and bounded. In addition, it is essential that $H$ has no bound states. The particle should be prepared on the left-hand side of the potential and the detection should happen on the right-hand side.

**Rules of the Race** First we have to fix the rules of the race between the free and the tunnelling setting: we need to choose a precise notion of arrival detection, of the equality of initial wave functions, and of the equality of detectors for tunnelling and free dynamics. The equality of initial states is a non-trivial issue, because the two states are subject to different dynamical evolutions. So at least we need to fix a reference time. In the following we will choose $t = -\infty$, i.e., asymptotic equality of the incoming states in the two settings in the sense of scattering theory. We fix the direction from which the particles are coming by choosing input states with positive momentum.

On the detection side, we will choose a covariant arrival-time observable to describe the detectors, see Sect. 14.2. Again this raises the issue of how to compare the two cases, because the covariance condition explicitly depends on the time evolution, and an observable can be covariant with respect to only one of them. Again scattering theory helps, by defining a bijective correspondence between the respective sets of covariant observables: we identify observables, which give the same probability distributions on states coinciding for $t \to +\infty$.

For a fixed initial state, a fixed tunnelling potential and a fixed arrival-time observable, the arrival probability $\rho_0(t)$ resp. $\rho_V(t)$ can be calculated in the free resp. tunnelling setting. In addition, if a final time $t_f$ is fixed, the total probability that the particle has been arrived can be calculated by

$$P_{0/V} = \int_{-\infty}^{t_f} dt\, \rho_{0/V}(t) \tag{14.6}$$

The comparison is now easy, if $P_V > P_0$ (i.e. the probability that the particle has been detected with tunnel potential is greater than the detection probability in the free case) then the tunnelling setting has "won" and tunnelling might be useful for sending some information for example encoded in the internal state of the particle. On the other hand, if $P_V < P_0$ then the tunnelling setting is "useless" for increasing the detection probability.

**Result of the Race** The question which has been studied in [39] is now: is there any initial wavefunction, any tunnel potential, any covariant arrival-time observable and any final time such that $P_V > P_0$ (i.e. such that the total detection probability is increased by tunnelling and the tunnelling setting might be useful for sending information faster than in the free case)?

The answer is: NO! We always have $P_0 \geq P_V$. This result is perhaps not surprising; nevertheless, what is surprising is the generality in which the result can be proved (using general results from scattering theory and the theory of covariant observables): the result is true for any incoming state, any tunnel potential, any detector and any final time!

# References

1. Muga, J.G., Sala Mayato, R., Egusquiza, I.L. (eds.): Time in Quantum Mechanics. Lecture Notes in Physics, vol. 72. Springer, Berlin (2002)
2. Muga, J.G., Ruschhaupt, A., del Campo, A. (eds.): Time in Quantum Mechanics – Vol. 2. Lecture Notes in Physics, vol. 789. Springer, Berlin (2009)
3. Holevo, A.S.: Probabilistic and Statistical Aspects of Quantum Theory. North-Holland, Amsterdam (1982)
4. Ludwig, G.: Foundations of Quantum Mechanics, vol. I. Springer, Berlin (1983)
5. Busch, P., Grabowski, M., Lahti, P.: Operational Quantum Physics, 2nd edn. Springer, Berlin (1997)
6. Kijowski, J.: Rep. Math. Phys. **6**, 361 (1974)
7. Werner, R.F.: J. Math. Phys. **27**, 793 (1986)
8. Muga, J.G., Leavens, C.R.: Phys. Rep. **338**, 353 (2000)
9. Egusquiza, I.L., Muga, J.G., Baute, A.D.: 'Standard' quantum mechanical approach to times of arrival. In: Muga, J.G., Sala Mayato, R., Egusquiza, I.L. (eds.) Time in Quantum Mechanics. Lecture Notes in Physics, vol. 72. Springer, Berlin (2002)
10. Allcock, G.R.: Ann. Phys. **53**, 253 (1969)
11. Werner, R.F.: Ann. Inst. H. Poincaré Phys. Théor. **47**, 429 (1987)
12. Muga, J.G., Palao, J.P., Navarro, B., Egusquiza, I.L.: Phys. Rep. **395**, 357 (2004)

13. Damborenea, J.A., Egusquiza, I.L., Hegerfeldt, G.C., Muga, J.G.: Phys. Rev. A **66**, 052104 (2002)
14. Ruschhaupt, A., Muga, J.G., Hegerfeldt, G.C.: Detector models for the quantum time of arrival. In: Muga, J.G., Ruschhaupt, A., del Campo, A. (eds.) Time in Quantum Mechanics – Vol. 2. Lecture Notes in Physics, vol. 789. Springer, Berlin (2009)
15. Hartman, T.E.: J. Appl. Phys. **33**, 3427 (1962)
16. Büttiker, M., Landauer, R.: Phys. Rev. Lett. **49**, 1739 (1982)
17. Enders, A., Nimtz, G.: J. Phys. I France **2**, 1693 (1992)
18. Steinberg, A.M., Kwiat, P.G., Chiao, R.Y.: Phys. Rev. Lett. **71**, 708 (1993)
19. Eckle, P., et al.: Science **322**, 1525 (2008)
20. Nimtz, G., Heitmann, W.: Prog. Quantum Electron. **21**, 81 (1987)
21. Blanchard, Ph., Jadczyk, A.: Phys. Lett. A **203**, 260 (1995)
22. Blanchard, Ph., Jadczyk, A.: Rep. Math. Phys. **36**, 235 (1995)
23. Blanchard, Ph., Jadczyk, A.: Event-enhanced formalism of quantum theory or columbus solution to the quantum measurement problem. In: Belavkin, V.P., et al. (eds.) Quantum Communications and Measurement. Plenum Press, New York (1995)
24. Blanchard, Ph., Jadczyk, A.: Ann. Physik **4**, 583 (1995)
25. Blanchard, Ph., Jadczyk, A.: Helv. Phys. Acta **69**, 613 (1996)
26. Palao, J.P., Muga, J.G., Brouard, S., Jadczyk, A.: Phys. Lett. A **233**, 227 (1997)
27. Ruschhaupt, A.: Phys. Lett. A **250**, 249 (1998)
28. Ruschhaupt, A., Damborenea, J.A., Navarro, B., Muga, J.G., Hegerfeldt, G.C.: Europhys. Lett. **67**, 1 (2004)
29. Blanchard, Ph., Jadczyk, A.: Found. Phys. **26**, 1669 (1996)
30. Ruschhaupt, A.: J. Phys. A **35**, 9227 (2002)
31. Ruschhaupt, A.: J. Phys. A **35**, 10429 (2002)
32. Heisenberg, W.: Z. Phys. **43**, 172 (1927)
33. Kennard, E.: Z. Phys. **44**, 326 (1927)
34. Werner, R.F.: Quantum Inf. Comput. **4**, 546 (2004)
35. Busch, P., Heinonen, T., Lahti, P.: Phys. Rep. **452**, 155 (2007)
36. Busch, P.: The time-energy uncertainty relation. In: Muga, J.G., Sala Mayato, R., Egusquiza, I.L. (eds.) Time in Quantum Mechanics. Lecture Notes in Physics, vol. 72. Springer, Berlin (2002)
37. Holevo, A.: Rep. Math. Phys. **16**, 385 (1979)
38. Kiukas, J., Ruschhaupt, A., Schmidt, P.O., Werner, R.F.: J. Phys. A **45**, 185301 (2012)
39. Kiukas, J., Ruschhaupt, A., Werner, R.F.: Found. Phys. **39**, 829 (2009)

# Chapter 15
# Localization and Entanglement in Relativistic Quantum Physics

Jakob Yngvason

## 15.1 Introduction

These notes are a slightly expanded version of a lecture presented in February 2012 at the workshop "The Message of Quantum Science—Attempts Towards a Synthesis" held at the ZIF in Bielefeld. The participants were physicists with a wide range of different expertise and interests. The lecture was intended as a survey of a *small selection* of the insights into the structure of relativistic quantum physics that have accumulated through the efforts of many people over more than 50 years.[1] This contribution discusses some facts about relativistic quantum physics, most of which are quite familiar to practitioners of *Algebraic Quantum Field Theory* (AQFT)[2] but less well known outside this community. No claim of originality is made; the goal of this contribution is merely to present these facts in a simple and concise manner, focusing on the following issues:

- Explaining how quantum mechanics (QM) combined with (special) relativity, in particular an upper bound on the propagation velocity of effects, leads naturally to systems with an infinite number of degrees of freedom (relativistic quantum fields).

---

[1] Including, among many others, R. Haag, H. Araki, D. Kastler, H.-J. Borchers, A. Wightman, R. Streater, B. Schroer, H. Reeh, S. Schlieder, S. Doplicher, J. Roberts, R. Jost, K. Hepp, J. Fröhlich, J. Glimm, A. Jaffe, J. Bisognano, E. Wichmann, D. Buchholz, K. Fredenhagen, R. Longo, D. Guido, R. Brunetti, J. Mund, S. Summers, R. Werner, H. Narnhofer, R. Verch, G. Lechner, . . . .

[2] Also known as *Local Quantum Physics* [55].

J. Yngvason (✉)
Faculty of Physics, University of Vienna, Boltzmanngasse 5, 1090 Vienna, Austria
e-mail: jakob.yngvason@univie.ac.at

© Springer-Verlag Berlin Heidelberg 2015
P. Blanchard, J. Fröhlich (eds.), *The Message of Quantum Science*, Lecture Notes in Physics 899, DOI 10.1007/978-3-662-46422-9_15

- A brief summary of the differences in mathematical structure compared to the QM of finitely many particles that emerge form the synthesis with relativity, in particular different localization concepts, type III von Neumann algebras rather than type I, and "deeply entrenched" [36] entanglement,
- Comments on the question whether these mathematical differences have significant consequences for the physical interpretation of basic concepts of QM.

## 15.2   What is Relativistic Quantum Physics?

According to E. Wigner's groundbreaking analysis from 1939 of relativistic symmetries in the quantum context [92] any relativistic quantum theory should contain as *minimal ingredients*

- A Hilbert space $\mathcal{H}$ of state vectors.
- A unitary representation $U(a, \Lambda)$ of the inhomogeneous (proper, orthochronous) Lorentz group (Poincaré group) $\mathcal{P}_{+}^{\uparrow}$ on $\mathcal{H}$.[3] Here $a \in \mathbb{R}^4$ denotes a translation of Minkowski space and $\Lambda$ a Lorentz transformation.[4]

The representations were completely classified by Wigner in [92].

The representation $U(a, 1) =: U(a)$ of the translations leads directly to the observables *energy* $P^0$ and *momentum* $P^\mu$, $\mu = 1, 2, 3$ as the corresponding infinitesimal generators[5]:

$$U(a) = \exp\left( i \sum_{\mu=0}^{3} P^\mu a_\mu \right). \tag{15.1}$$

The *stability requirement* that the energy operator, $P^0$, should be bounded below implies that the joint spectrum of the commuting operators $P^\mu$ is contained in the forward light cone $V^+$. This is called the relativistic *spectrum condition*. The operator of the mass is $M = (\sum_{\mu=0}^{3} P^\mu P_\mu)^{1/2}$. In an irreducible representations

---

[3]More precisely, also representations "up to a phase" are allowed, which amounts to replacing $\mathcal{P}_{+}^{\uparrow}$ by its universal covering group $ISL(2, \mathbb{C})$.

[4]For simplicity of the exposition we refrain from discussing the possibility that the Lorentz transformations act only as automorphisms on the algebra of observables but are not unitarily implemented on the Hilbert space of states under consideration, as can be expected in charged superselection sectors of theories with massless particles [23, 50].

[5]Here, and in the following, units are chosen so that Planck's constant, $\hbar$, and the velocity of light, $c$, are equal to 1. The metric on Minkowski space is $g_{\mu\nu} = \mathrm{diag}(1, -1, -1, -1)$.

of $\mathcal{P}_+^\uparrow$ fulfilling the spectrum condition the mass has a sharp value $m \geq 0$. These representations fall into three classes:

1. The massive representations $[m, s]$ with the mass $m > 0$ and the spin $s = 0, \frac{1}{2}, 1, \ldots$ labeling the irreducible representations of the stabilizer group $SU(2)$ of the energy-momentum vector in the rest frame.
2. The massless representations $[0, h]$ of finite helicity $h = 0, \pm\frac{1}{2}, \pm 1, \ldots$ corresponding to one dimensional representations of the stabilizer group of a light-like energy-momentum vector (the two-dimensional Euclidean group $E_2$).
3. Massless representations of unbounded helicity ("infinite spin representations") corresponding to infinite dimensional representations of the stabilizer group $E_2$.

Besides energy and momentum, further observables, namely the angular momentum operators and the generators of Lorentz boosts, follow from the representation $U(0, \Lambda)$ of the homogeneous Lorentz group. In Sect. 15.3.5 below we shall see how an intrinsic localization concept ("modular localization") can be associated with the latter, but first we discuss the problems that arise if one tries to mimic the procedure in non-relativistic QM and define localization via position operators for particles.

### 15.2.1  Problems with Position Operators

In non-relativistic quantum mechanics, spatial *localization* of state vectors is determined through the spectral projectors of position operators. For instance, a single particle state with wave functions $\psi(\mathbf{x})$ is localized in a domain $\Delta \subset \mathbb{R}^3$ if and only if the support of $\psi$ lies in $\Delta$, which means that $E_\Delta \psi = \psi$ with $E_\Delta$ the multiplication operator by the characteristic function of $\Delta$. Time evolution generated by the non-relativistic Hamiltonian $H = \frac{1}{2m}\mathbf{P}^2 = -\frac{1}{2m}\nabla^2$ immediately spreads out the localization in the sense that for any pair $\Delta, \Delta'$ of disjoint domains so that $E_\Delta E_{\Delta'} = 0$ we have $\exp(itH)E_\Delta \exp(-itH)E_{\Delta'} \neq 0$ for all $t \neq 0$, no matter how far $\Delta$ and $\Delta'$ are from each other. Since there is no upper bound to the velocity of propagation of effects in non-relativistic QM this is not a surprise. In a relativistic theory, on the other hand, for instance with $H = (c^2\mathbf{P}^2 + m^2c^4)^{1/2}$, one might expect that $\exp(itH)E_\Delta \exp(-itH)E_{\Delta'}$ stays zero as long as $c|t|$ is smaller than the spatial distance between the two domains. This, however, is *not* the case, due to the analyticity implied by the relativistic spectrum condition:[6]

**Theorem 1 (Localization via Position Operators is in Conflict with Causality)**
*Suppose there is a mapping $\Delta \mapsto E_\Delta$ from subsets of space-like hyperplanes in*

---

[6]In this form the result was first published by J.F. Perez and I.F. Wilde in [75]. See also [65] for the same conclusion under slightly weaker premises.

*Minkowski space into projectors on $\mathcal{H}$ such that*

*(1)* $U(a)E_\Delta U(a)^{-1} = E_{\Delta+a}$
*(2)* $E_\Delta E_{\Delta'} = 0$ *if* $\Delta, \Delta'$ *space-like separated.*

*Then* $E_\Delta = 0$ *for all* $\Delta$.

*Proof* The spectrum condition implies that the function $a \mapsto U(a)\Psi$ has, for every $\Psi \in \mathcal{H}$, an *analytic continuation* into $\mathbb{R}^4 + iV^+ \subset \mathbb{C}^4$. The second condition (2) means that $\langle E_\Delta\Psi, U(a)E_\Delta\Psi \rangle = \langle \Psi, E_\Delta E_{\Delta+a}U(a)\Psi \rangle = 0$ on an open set in Minkowski space. But an analytic function that is continuous on the real boundary of its analyticity domain and vanishing on an open subset of this boundary vanishes identically.[7]                                                                                           □

The conclusion that can be drawn from this result is that localization in terms of position operators is *incompatible with causality* in relativistic quantum physics.[8]

This dilemma is resolved by realizing that the relevant concept in relativistic quantum physics is localization of *quantum fields* rather than localization of wave functions of particles in position space. The space-time points $x$ appear as variables of the quantum field operators $\Phi_\alpha(x)$ (with $\alpha$ a tensor or spinor index). Causality manifests itself in *commutativity* (or anticommutativity) of the operators at space-like separation of the variables.

Taken together, covariance w.r.t. space-time translations, the spectrum condition and local (anti)commutativity imply that the dependence of field operators on the coordinates is by necessity singular, and well defined operators are only obtained after smearing with test functions. This means that quantum fields are operator-valued distributions rather than functions [83, 91], i.e., only "smeared" operators $\Phi_\alpha(f)$ with $f$ a test function on space-time are well defined. Localization of field operators *at a point* is thus a somewhat problematic concept,[9] while localization in a domain of space-time (the support of the test function $f$) has a clear meaning.

These ideas are incorporated in the general conceptual framework of *Algebraic Quantum Field Theory* (AQFT), also called *Local Quantum Physics* (LQP) [6, 25, 55, 56]. Here the emphasis is on the collection ("net") of operator algebras generated by quantum field operators localized in different domains of space-time. The quantum fields themselves appear as auxiliary objects since many different quantum fields can generate the same net of algebras. The choice of some definite field to describe a given net is somewhat analogous to the choice of a coordinate system in differential geometry. In some cases the net is even defined without any reference to quantum fields in the traditional sense, and important general results of the theory do not, in fact, rely on a description of the net in terms of operator-valued distributions.

---

[7]This follows from the "edge of the wedge" theorem, that is a generalization of the Schwarz reflection principle to several complex variables, see, e.g., [83].

[8]This objection does not exclude *approximate* localization in the sense of Newton and Wigner [73].

[9]Field operators at a point can, however, be defined as *quadratic forms* on vectors with sufficiently nice high energy behavior.

## 15.3  Local Quantum Physics

### 15.3.1  The General Assumptions

The basic ingredients of a model in LQP are:

- A separable Hilbert space $\mathcal{H}$ of state vectors.
- A Unitary representation $U(a, \Lambda)$ of the Poincaré group $\mathcal{P}_+^\uparrow$ on $\mathcal{H}$.[10]
- An invariant, normalized state vector $\Omega \in \mathcal{H}$ (vacuum), unique up to a phase factor.
- A family of *-algebras $\mathcal{F}(\mathcal{O})$ of operators[11] on $\mathcal{H}$ (a "field net"), indexed by regions $\mathcal{O} \subset \mathbb{R}^4$ with $\mathcal{F}(\mathcal{O}_1) \subset \mathcal{F}(\mathcal{O}_2)$ if $\mathcal{O}_1 \subset \mathcal{O}_2$ (*isotony*).

The requirements are:

R1.  *Local (anti-)commutativity*: $\mathcal{F}(\mathcal{O}_1)$ commutes with $\mathcal{F}(\mathcal{O}_2)$ if $\mathcal{O}_1$ and $\mathcal{O}_2$ space-like separated. (Or, in the case of Fermi statistics, commutes after a "twist" of the fermionic parts of the algebras, cf. [11], Eq. 33b.)

R2.  *Covariance*: $U(a, \Lambda)\mathcal{F}(\mathcal{O})U(a, \Lambda)^{-1} = \mathcal{F}(\Lambda\mathcal{O} + a)$.

R3.  *Spectrum condition*: The energy momentum spectrum, i.e., the joint spectrum of the generators of the translations $U(a)$ lies in $V^+$.

R4.  *Cyclicity of the vacuum*: $\cup_\mathcal{O}\mathcal{F}(\mathcal{O})\Omega$ is dense in $\mathcal{H}$.

*Remarks*

- The operators in $\mathcal{F}(\mathcal{O})$ can intuitively be thought of as generating physical operations that can be carried out in the space-time region $\mathcal{O}$.
- Usually (but not always!) $\mathcal{F}(\mathcal{O})$ is nontrivial for all open regions $\mathcal{O}$.
- Associated with the field net $\{\mathcal{F}(\mathcal{O})\}_{\mathcal{O}\subset\mathbb{R}^4}$ there is usually another net of operator algebras, $\{\mathcal{A}(\mathcal{O})\}_{\mathcal{O}\subset\mathbb{R}^4}$, representing local *observables* and commuting with the field net and itself at space-like separations. Usually this is a *subnet* of the field net, selected by invariance under some (global) gauge group.[12]

---

[10]More generally, a representation of the covering group $ISL(2, \mathbb{C})$.

[11]For mathematical convenience we assume that the operators are bounded and that the algebras are closed in the weak operator topology, i.e., that they are *von Neumann algebras*. The generation of such algebras from unbounded quantum field operators $\Phi_\alpha(f)$ is in general a nontrivial issue that is dealt with, e.g., in [18, 45]. In cases when the real and imaginary parts of the field operators are essentially self-adjoint, one may think of the $\mathcal{F}(\mathcal{O})$ as generated by bounded functions (e.g., spectral projectors, resolvents, or exponentials) of these operators smeared with test functions having support in $\mathcal{O}$. More generally, the polar decomposition of the unbounded operators can be taken as a starting point for generating the local net of von Neumann algebras.

[12]In the theory of superselection sectors, initiated by Borchers in [14] and further developed in particular by Doplicher et al. in [39–42], the starting point is the net of observables while the field net and the gauge group are derived objects. For a very recent development, applicable to theories with long range forces, see [26].

Thus, the operators in $\mathcal{F}(\mathcal{O})$ can have two roles:

1. They implement local transformations of states[13] $\omega$ in the sense of Kraus [62], i.e.,

$$\omega \mapsto \sum_i \omega(K_i^* \cdot K_i) \tag{15.2}$$

   with $K_i \in \mathcal{F}(\mathcal{O})$.
2. The self-adjoint elements of $\mathcal{A}(\mathcal{O})$ correspond to physical properties of the system that can, at least in principle, be measured in $\mathcal{O}$.

Already in the mid 1950s Rudolf Haag [53, 54] had the fundamental insight that information about interactions between *particles*, that emerge asymptotically for large positive or negative instances of time but are usually not unambiguously defined at finite times, is already *encoded in the field net*. In order to determine the particle spectrum of a given theory and compute scattering amplitudes it is not necessary to attach specific interpretations to specific operators in $\mathcal{F}(\mathcal{O})$ besides their localization.

### 15.3.2   Construction Methods

Traditionally, the main methods to construct models in quantum field theory have been:

- Lagrangian field theory plus canonical quantization. This leads rigorously to free fields and variants like generalized free fields, Wick-powers of such fields etc. that satisfy the Wightman-Gårding axioms [13, 60, 83, 91]. Perturbation theory plus renormalization leads (also rigorously!) to theories with *interactions* defined in terms of *formal power series* in a coupling constant. (See [21] for a modern, rigorous version of perturbation theory for quantum fields.)
- Constructive QFT (J. Glimm, A. Jaffe and others) [51]. Here the renormalization of certain lagrangian field theories is carried out rigorously, without recourse to perturbation theory. In this way, models of interacting fields in space-time dimensions $1+1$ and $1+2$ have been obtained, *but so far not in $1+3$ dimensions*.
- Conformal QFT in $1+1$ space-time dimensions based on Virasoro algebras and other algebraic structures, see, e.g., [61] and references cited therein.

---

[13]Here and in the sequel, a *state* means a positive, normalized linear functional on the algebra in question, i.e., a linear functional such that $\omega(A^*A) \geq 0$ for all $A$ and $\omega(\mathbf{1}) = 1$. We shall also restrict the attention to *normal* states, i.e., $\omega(A) = \text{trace}\,(\rho A)$ with a nonnegative trace class operator $\rho$ on $\mathcal{H}$ with trace 1.

It is a big challenge in QFT to develop new methods of construction and classification. Recently, progress has been achieved through *deformations* of known models [35, 52]. In particular, G. Lechner has shown that a large class of integrable models in $1 + 1$ dimensions, and many more, can be obtained from deformations of free fields [63]. See also [1]. There is also a very recent approach due to Barata et al. [9] based on Tomita–Takesaki modular theory [19, 87]. The latter concerns operator algebras with a cyclic and separating vector, and applies to the local algebras of relativistic quantum field theory because of the Reeh–Schlieder theorem discussed next.

### 15.3.3 The Reeh–Schlieder Theorem

The Reeh–Schlieder Theorem was originally derived in the context of Wightman quantum field theory in [80]. For general nets of local algebras as in Sect. 15.3.1, one additional assumption is needed, *weak additivity*: For every fixed open set $O_0$ the algebra generated by the union of all translates, $\mathcal{F}(O_0 + x)$, is dense in the union of all $\mathcal{F}(O)$ in the weak operator topology. If the net is generated by Wightman fields this condition is automatically fulfilled.

**Theorem 2 (Reeh–Schlieder)** *Under the assumption of weak additivity, $\mathcal{F}(O)\Omega$ is dense in $\mathcal{H}$ for all open sets $O$, i.e., the vacuum is cyclic for every single local algebra and not just for their union.*

*Proof* Write $U(a)$ for $U(a, 1)$. Pick $O_0 \subset O$ such that $O_0 + x \subset O$ for all $x$ with $|x| < \varepsilon$, for some $\varepsilon > 0$. If $\Psi \perp \mathcal{F}(O)\Omega$, then $\langle \Psi, U(x_1)A_1 U(x_2 - x_1) \cdots U(x_n - x_{n-1})A_n\Omega \rangle = 0$ for all $A_i \in O_0$ and $|x_i| < \varepsilon$. Then use the analyticity of $U(a)$ to conclude that this must hold for *all* $x_i$. The theorem now follows by appealing to weak additivity. □

**Corollary** *The vacuum is a separating vector of $\mathcal{F}(O)$ for every $O$ such that its causal complement $O'$ has interior points, i.e., $A\Omega = 0$ for $A \in \mathcal{F}(O)$ implies $A = 0$. Moreover, if $A$ is a positive operator in $\mathcal{F}(O)$ and $\langle \Omega, A\Omega \rangle = 0$, then $A = 0$.*

*Proof* If $O_0 \subset O'$, then $AB\Omega - BA\Omega - 0$ for all $B \in \mathcal{F}(O_0)$ [14] But $\mathcal{F}(O_0)\Omega$ is dense if $O_0$ is open, so $A = 0$. The last statement follows because the square root of a positive $A \in \mathcal{F}(O)$ belongs also to $\mathcal{F}(O)$. □

---

[14]For simplicity we have assumed local commutativity. In the case of Fermi fields the same conclusion is drawn by splitting the operators into their bosonic and fermionic parts.

*Remarks*

- The Reeh–Schlieder Theorem and its Corollary hold, in fact, not only for the vacuum vector $\Omega$ but for any state vector $\psi$ that is an *analytic vector for the energy*, i.e. such that $e^{ix_0 P^0} \psi$ is an analytic function of $x_0$ in a whole complex neighborhood of 0. This holds in particular if $\psi$ has bounded energy spectrum.
- *No violation of causality* is implied by the Reeh–Schlieder Theorem. The theorem is "just" a manifestation of unavoidable *correlations* in the vacuum state (or any other state given by an analytic vector for the energy) in relativistic quantum field theory. (See the discussion of entanglement in Sect. 15.5.)
- Due to the *cluster property*, that is a consequence of the uniqueness of the vacuum, the correlation function

$$F(x) = \langle \Omega, AU(x)B\Omega \rangle - \langle \Omega, A\Omega \rangle \langle \Omega, B\Omega \rangle \tag{15.3}$$

for two local operators $A$ and $B$ tends to zero if $x$ tends to space-like infinity. If there is a mass gap in the energy momentum spectrum the convergence is exponentially fast [48]. Thus, although the vacuum cannot be a product state for space-like separated local algebras due to the Reeh–Schlieder Theorem, the correlations become very small as soon as the space-like distance exceeds the Compton wavelength associated with the mass gap [86, 94].

### 15.3.4 Modular Structures and the Bisognano–Wichmann Theorem

A remarkable development in the theory of operator algebras was initiated 1970 when M. Takesaki published his account [88] of M. Tomita's theory of modular Hilbert algebras developed 1957–1967. In 1967 similar structures had independently been found by R. Haag, N. Hugenholz and M. Winnink in their study of thermodynamic equilibrium states of infinite systems [57], and in the 1970s the theory found its way into LQP. Various aspects of these developments are discussed in the review article [17], see also [85] for a concise account. On the mathematical side Tomita–Takesaki theory is the basis of A. Connes' groundbreaking work on the classification von Neumann algebras [37].

The Tomita–Takesaki modular theory concerns a von Neumann algebra $\mathcal{A}$ together with a cyclic and separating vector $\Omega$. To every such pair it associates a one parameter group of unitaries (the *modular group*) whose adjoint action leaves the algebra invariant, as well as an anti-unitary involution (the *modular conjugation*) that maps the algebra into its commutant $\mathcal{A}'$. The precise definition of these objects is as follows.

First, one defines an antilinear operator $S_0 \colon \mathcal{A}\Omega \to \mathcal{A}\Omega$ by

$$S_0 A\Omega = A^*\Omega. \tag{15.4}$$

This operator (in general unbounded) is well defined on a dense set in $\mathcal{H}$ because $\Omega$ is separating and cyclic. We denote by $S$ its closure, $S = S_0^{**}$. It has a polar decomposition

$$S = J\Delta^{1/2} = \Delta^{-1/2}J \tag{15.5}$$

with the *modular operator* $\Delta = S^*S > 0$ and the anti-unitary *modular conjugation* $J$ with $J^2 = 1$.

The basic facts about these operators are stated in the following Theorem. See, e.g., [19] or [88] for a proof.

**Theorem 3 (Modular Group and Conjugation; KMS Condition)**

$$\Delta^{it} \mathcal{A} \Delta^{-it} = \mathcal{A} \quad \text{for all } t \in \mathbb{R}, \quad J\mathcal{A}J = \mathcal{A}'. \tag{15.6}$$

*Moreover, for $A, B \in \mathcal{A}$,*[15]

$$\langle \Omega, AB\Omega \rangle = \langle \Omega, B\Delta^{-1}A\Omega \rangle. \tag{15.7}$$

Equation (15.7) is equivalent to the Kubo-Martin Schwinger (KMS) condition that characterizes thermal equilibrium states with respect to the "time" evolution $A \mapsto \alpha_t(A) := \Delta^{it} A \Delta^{-it}$ on $\mathcal{A}$ [57].[16]

Most of the applications of modular theory to LQP rely on the fact that the modular group and conjugation for an algebra corresponding to a space-like wedge in Minkowski space and the vacuum have a geometric interpretation. A *space-like wedge* $\mathcal{W}$ is, by definition, a Poincaré transform of the *standard wedge*

$$\mathcal{W}_1 = \{x \in \mathbb{R}^4 : |x_0| < x_1\}. \tag{15.8}$$

With $\mathcal{W}$ is associated a one-parameter family $\Lambda_{\mathcal{W}}(s)$ of Lorentz boosts that leave $\mathcal{W}$ invariant. The boosts for the standard wedge are in the $x_0$-$x_1$ plane given by the matrices

$$\Lambda_{\mathcal{W}_1}(s) = \begin{pmatrix} \cosh s & \sinh s \\ \sinh s & \cosh s \end{pmatrix}. \tag{15.9}$$

There is also a reflection, $j_{\mathcal{W}}$, about the edge of the wedge that maps $\mathcal{W}$ into the opposite wedge (causal complement) $\mathcal{W}'$. For the standard wedge $\mathcal{W}_1$ the reflection is the product of the space-time inversion $\theta$ and a rotation $R(\pi)$ by $\pi$ around the 1-axis. For a general wedge the transformations $\Lambda_{\mathcal{W}}(s)$ and $j_{\mathcal{W}}$ are obtained from

---

[15]Equation (15.7) is, strictly speaking, only claimed for $A, B$ in a the dense subalgebra of "smooth" elements of $\mathcal{A}$ obtained by integrating $\Delta^{it} A \Delta^{-it}$ with a test functions in $t$.

[16]Due to a sign convention in modular theory the temperature is formally $-1$, but by a scaling of the parameter $t$, including an inversion of the sign, can produce any value of the temperature.

those for $W_1$ by combining the latter with a Poincaré transformation that takes $W_1$ to $W$.

Consider now the algebras $\mathcal{F}(W)$ of a field net generated by a Wightman quantum field with the vacuum $\Omega$ as cyclic and separating vector. For simplicity we consider the case of Bose fields.[17] The modular objects $\Delta$ and $J$ associated with $(\mathcal{F}(W), \Omega)$ depend on $W$ but it is sufficient to consider $W_1$. As discovered by J. Bisognano and E. Wichmann in 1975 [10, 11] $\Delta$ and $J$ are related to the representation $U$ of the Lorentz group and the PCT operator $\Theta$ in the following way:

**Theorem 4 (Bisognano–Wichmann)**

$$J = \Theta U(R(\pi)) \quad and \quad \Delta^{it} = U(\Lambda_{W_1}(2\pi t)). \tag{15.10}$$

### 15.3.5 Modular Localization

*Modular localization* [20] is based on a certain converse of the Bisognano–Wichmann theorem. This concept associates a localization structure with any (anti-)unitary representation of the proper Poincaré group $\mathcal{P}_+$ (i.e., $\mathcal{P}_+^\uparrow$ augmented by space-time reflection) satisfying the spectrum condition, in particular the one-particle representations. Weyl quantization then generates naturally a local net satisfying all the requirements (1–4) in Sect. 15.3.1, including commutativity (or anti-commutativity) at space-like separation of the localization domains. A sketch of this constructions is given in this subsection, focusing for simplicity on the case of local commutativity rather than anti-commutativity.

Let $U$ be an (anti-)unitary representation of $\mathcal{P}_+$ satisfying the spectrum condition on a Hilbert space $\mathcal{H}_1$. For a given space-like wedge $W$, let $\Delta_W$ be the unique positive operator satisfying

$$\Delta_W^{it} = U(\Lambda_W(2\pi t)), \quad t \in \mathbb{R}, \tag{15.11}$$

and let $J_W$ to be the anti-unitary involution representing $j_W$. We define

$$S_W := J_W \Delta_W^{1/2}. \tag{15.12}$$

The space

$$\mathcal{K}(W) := \{\phi \in \text{domain } \Delta_W^{1/2} : S_W \phi = \phi\} \subset \mathcal{H}_1 \tag{15.13}$$

---

[17]Fermi fields can be included by means of a "twist" that turns anticommutators into commutators as in [11].

satisfies:

- $\mathcal{K}(\mathcal{W})$ is a closed *real* subspace of $\mathcal{H}_1$ in the real scalar product Re $\langle \cdot, \cdot \rangle$.
- $\mathcal{K}(\mathcal{W}) \cap i\mathcal{K}(\mathcal{W}) = \{0\}$.
- $\mathcal{K}(\mathcal{W}) + i\mathcal{K}(\mathcal{W})$ is dense in $\mathcal{H}_1$.[18]
- $\mathcal{K}(\mathcal{W})^{\perp} := \{\psi \in \mathcal{H}_1 : \text{Im} \langle \psi, \phi \rangle = 0 \text{ for all } \phi \in \mathcal{K}(\mathcal{W})\} = \mathcal{K}(\mathcal{W}')$.

The functorial procedure of Weyl quantization (see, e.g., [79]) now leads for any $\psi \in \bigcup_{\mathcal{W}} \mathcal{K}(\mathcal{W})$ to an (unbounded) field operator $\Phi(\psi)$ on the Fock space

$$\mathcal{H} = \bigoplus_{n=0}^{\infty} \mathcal{H}_1^{\otimes \text{symm}} \tag{15.14}$$

such that

$$[\Phi(\psi), \Phi(\phi)] = i \, \text{Im} \langle \psi, \phi \rangle \mathbf{1}. \tag{15.15}$$

In particular,

$$[\Phi(\psi_1), \Phi(\psi_2)] = 0 \tag{15.16}$$

if $\psi_1 \in \mathcal{K}(\mathcal{W})$, $\psi_2 \in \mathcal{K}(\mathcal{W}')$.

Finally, a net of algebras $\mathcal{F}(\mathcal{O})$ satisfying requirements R1-R4 is defined by

$$\mathcal{F}(\mathcal{O}) := \{\exp(i\Phi(\psi)) : \psi \in \cap_{\mathcal{W} \supset \mathcal{O}} \mathcal{K}(\mathcal{W})\}''. \tag{15.17}$$

These algebras are in [20] proved to have $\Omega$ as a cyclic vector if $\mathcal{O}$ is a space-like cone, i.e, a set of the form $x + \bigcup_{\lambda > 0} \lambda D$ where $D$ is a set with interior points that is space-like separated from the origin.

Although this construction produces only interaction free fields it is remarkable for at least two reasons:

- It uses as sole input a representation of the Poincaré group, i.e., it is *intrinsically quantum mechanical* and not based on any "quantization" of a classical theory. For the massive representations as well as those of zero mass and finite helicity the localization can be sharpened by using Wightman fields [82, 83], leading to nontrivial algebras $\mathcal{F}(\mathcal{O})$ also for *bounded*, open sets $\mathcal{O}$.
- The construction works also for the *zero mass, infinite spin representations*, that can *not* be generated by point localized fields, i.e., operator valued distributions satisfying the Wightman axioms [93]. Further analysis of this situation reveals that these representations can be generated by *string-localized* fields $\Phi(x, e)$ [67, 68] with $e$ being a space like vector of length 1, and $[\Phi(x, e), \Phi(x, e)] = 0$ if the "strings" (rays) $x + \lambda e$ and $x' + \lambda' e'$ with $\lambda, \lambda' > 0$ are space-like separated. After

---

[18]Such real subspaces of a complex Hilbert space are called *standard* in the spatial version of Tomita–Takesaki theory [81].

smearing with test functions in $x$ and $e$ these fields are localized in space-like cones.

*Remarks*

• Free string localized fields can be constructed for all irreducible representations of the Poincaré group and their most general form is understood [68]. Their correlation functions have a better high-energy behavior than for Wightman fields. One can also define string localized vector potentials for the electromagnetic field and a massless spin 2 field [66], and there are generalizations to massless fields of arbitrary helicities [77].
• Localization in cones rather than bounded regions occurs also in other contexts: (1) Fields generating massive particle states can always be localized in space-like cones and in massive gauge theories no better localization may be possible [24]. (2) In Quantum Electrodynamics localization in *light-like* cones is to be expected [26].

## 15.4   The Structure of Local Algebras

I recall first some standard mathematical terminology and notations concerning operator algebras. The algebra of all bounded, linear operators on a Hilbert space $\mathcal{H}$ is denoted by $\mathcal{B}(\mathcal{H})$. If $\mathcal{A} \subset \mathcal{B}(\mathcal{H})$ is a subalgebra (more generally, a subset), then its *commutant* is, by definition,

$$\mathcal{A}' = \{B \in \mathcal{B}(\mathcal{H}) : [A, B] = 0 \text{ for all } A \in \mathcal{A}\}. \tag{15.18}$$

A *von Neumann algebra* is an algebra $\mathcal{A}$ such that

$$\mathcal{A} = \mathcal{A}'', \tag{15.19}$$

i.e., the algebra is equal to its double commutant. Equivalently, the algebra is closed in the weak operator topology, provided the algebra contains $\mathbf{1}$ that will always be assumed. A (normal) *state* on a von Neumann algebra $\mathcal{A}$ is a *positive linear functional* of the form $\omega(A) = \text{trace}(\rho A)$, $\rho \geq 0$, trace $\rho = 1$. The state is a *pure state* if $\omega = \frac{1}{2}\omega_1 + \frac{1}{2}\omega_2$ implies $\omega_1 = \omega_2 = \omega$. Note that if $\mathcal{A} \neq \mathcal{B}(\mathcal{H})$ then $\rho$ is not unique and the concept of a pure state is *not* the same as that of a *vector state*, i.e. $\omega(A) = \langle \psi, A\psi \rangle$ with $\psi \in \mathcal{H}$, $\|\psi\| = 1$.

A vector $\psi \in \mathcal{H}$ is called *cyclic* for $\mathcal{A}$ if $\mathcal{A}\psi$ is dense in $\mathcal{H}$ and *separating* if $A\psi = 0$ for $A \in \mathcal{A}$ implies $A = 0$ (equivalently, if $\psi$ is cyclic for $\mathcal{A}'$).

A *factor* is a v.N. algebra $\mathcal{A}$ such that

$$\mathcal{A} \vee \mathcal{A}' := (\mathcal{A} \cup \mathcal{A}')'' = \mathcal{B}(\mathcal{H}) \tag{15.20}$$

which is equivalent to

$$\mathcal{A} \cap \mathcal{A}' = \mathbb{C}\mathbf{1}. \tag{15.21}$$

The original motivation of von Neumann for introducing and studying this concept together with Murray [69–71, 89] came from quantum mechanics: A "factorization" of $\mathcal{B}(\mathcal{H})$ corresponds to a splitting of a system into two subsystems. The simplest case is

$$\mathcal{H} = \mathcal{H}_1 \otimes \mathcal{H}_2, \tag{15.22}$$

$$\mathcal{A} = \mathcal{B}(\mathcal{H}_1) \otimes \mathbf{1}, \quad \mathcal{A}' = \mathbf{1} \otimes \mathcal{B}(\mathcal{H}_2), \tag{15.23}$$

$$\mathcal{B}(\mathcal{H}) = \mathcal{B}(\mathcal{H}_1) \otimes \mathcal{B}(\mathcal{H}_2). \tag{15.24}$$

This is the **Type I** case, familiar from non-relativistic quantum mechanics of systems with a finite number of particles and also Quantum Information Theory [74] where the Hilbert spaces considered are usually finite dimensional. This factorization is characterized by the existence of *minimal projectors* in $\mathcal{A}$: If $\psi \in \mathcal{H}_1$ and $E_\psi = |\psi\rangle\langle\psi|$, then

$$E = E_\psi \otimes \mathbf{1} \in \mathcal{A} \tag{15.25}$$

is a minimal projector, i.e., it has no proper subprojectors in $\mathcal{A}$.

At the other extreme is the **Type III** case which is defined as follows:

For every projector $E \in \mathcal{A}$ there exists an isometry $W \in \mathcal{A}$ with

$$W^*W = \mathbf{1}, \quad WW^* = E. \tag{15.26}$$

It is clear that for a type III factor, $\mathcal{A} \vee \mathcal{A}'$ is *not* a tensor product factorization, because a minimal projector cannot satisfy (15.26).

It is natural to ask whether we need to bother about other cases than type I in quantum physics. The answer is that is simply a *fact* that in LQP the algebras $\mathcal{F}(\mathcal{O})$ for $\mathcal{O}$ a double cone (intersection of a forward and a backward light cone) or a space like wedge, are in all known cases of type III. More precisely, under some reasonable assumptions, they are isomorphic to the *unique*, hyperfinite type $III_1$ factor in a finer classification due to Connes [32, 37, 58]. This classification is in terms of the intersection of the spectra of the modular operators for the cyclic and separating state vectors for the algebra (*Connes spectrum*). The characteristic of type $III_1$ is that the Connes spectrum is equal to $\mathbb{R}_+$.

Concrete example of type III factors can be obtained by considering infinite tensor products of $2 \times 2$ or $3 \times 3$ matrix algebras [7, 8, 78]. Thus, a type $III_\lambda$ factor with $0 < \lambda < 1$, which has the integral powers of $\lambda$ as Connes spectrum, is generated by the infinite tensorial power of the algebra $M_2(\mathbb{C})$ of complex $2 \times 2$ matrices in the Gelfand–Naimark–Segal representation [19] defined by the state

$$\omega(A_1 \otimes A_2 \otimes \cdots \otimes A_N \otimes \mathbf{1} \cdots) = \prod_n \mathrm{tr}(A_n \rho_\lambda) \tag{15.27}$$

with $A_n \in M_2(\mathbb{C}), 0 < \lambda < 1$ and

$$\rho_\lambda = \frac{1}{1+\lambda} \begin{pmatrix} 1 & 0 \\ 0 & \lambda \end{pmatrix}. \tag{15.28}$$

A type $\mathbf{III}_1$ factor is obtained from an analogous formula for the infinite product of complex $3 \times 3$ matrices in the representation defined by tracing with the matrix

$$\rho_{\lambda,\mu} = \frac{1}{1+\lambda+\mu} \begin{pmatrix} 1 & 0 & 0 \\ 0 & \lambda & 0 \\ 0 & 0 & \mu \end{pmatrix} \tag{15.29}$$

where $\lambda, \mu > 0$ are such that $\frac{\log \lambda}{\log \mu} \notin \mathbb{Q}$.

The earliest proof of the occurrence of type III factors in LQP was given by Araki in [2–4] for the case of a free, scalar field. Type III factors appear also in non-relativistic equilibrium quantum statistical mechanics in the thermodynamic limit at nonzero temperature [7].

General proofs that the local algebras of a relativistic quantum field in the vacuum representation are of type $\mathbf{III}_1$ rely on the following ingredients[19]:

- The Reeh–Schlieder Theorem.
- The Bisognano Wichmann Theorem for the wedge algebras $\mathcal{F}(\mathcal{W})$, that identifies their Tomita–Takesaki modular groups w.r.t. the vacuum with geometric transformations (Lorentz-boosts). The corresponding modular operators have $\mathbb{R}_+$ as spectrum. Moreover, by locality and the invariance of the wedge under dilations the spectrum is the same for other cyclic and separating vectors [5, 47]. Hence the wedge algebra is of type $\mathbf{III}_1$.
- Assumptions about non-triviality of scaling limits that allows to carry the arguments for wedge algebras over to double cone algebras [27, 47].

See also [43, 44, 64] and [16] for other aspects of the type question.

---

[19]The hyperfiniteness, i.e., the approximability by finite dimensional matrix algebras, follows from the split property considered in Sect. 15.5.1.

## 15.4.1 Some Consequences of the Type III Property

We now collect some important facts about local algebras that follow from their type III character. Since the focus is on the observables, we state the results in terms of the algebras $\mathcal{A}(\mathcal{O})$ rather than $\mathcal{F}(\mathcal{O})$.

### 15.4.1.1 Local Preparability of States

For every projector $E \in \mathcal{A}(\mathcal{O})$ there is an isometry $W \in \mathcal{A}(\mathcal{O})$ such that if $\omega$ is any state and $\omega_W(\cdot) := \omega(W^* \cdot W)$, then

$$\omega_W(E) = 1, \tag{15.30}$$

but

$$\omega_W(B) = \omega(B) \quad \text{for } B \in \mathcal{A}(\mathcal{O}'). \tag{15.31}$$

In words: Every state can be changed into an eigenstate of a local projector, by a local operation that is independent of the state and does not affect the state in the causal complement of the localization region of the projector.

This result is a direct consequence of the type III property. It is worth noting that in a slightly weaker form it can be derived from the general assumptions of LQP, without recourse to the Bisognano–Wichmann theorem and the scaling assumptions mentioned above: In [15] H.J. Borchers proved that the isometry $W$ can in any case be found in an algebra $\mathcal{A}(\mathcal{O}_\varepsilon)$ with

$$\mathcal{O}_\varepsilon := \mathcal{O} + \{x : |x| < \varepsilon\}, \quad \varepsilon > 0. \tag{15.32}$$

Equation (15.31) is then only claimed for $B \in \mathcal{A}(\mathcal{O}'_\varepsilon)$, of course.

In Sect. 15.5.1 we shall consider a strengthened version of the local preparability, but again with $W \in \mathcal{A}(\mathcal{O}_\varepsilon)$, under a further assumption on the local algebras.

### 15.4.1.2 Absence of Pure States

A type III factor $\mathcal{A}$ has *no pure states*,[20] i.e., for every $\omega$ there are $\omega_1$ and $\omega_2$, *different* from $\omega$, such that

$$\omega(A) = \frac{1}{2}\omega_1(A) + \frac{1}{2}\omega_2(A) \tag{15.33}$$

---

[20]Recall that "state" means here always *normal* state, i.e. a positive linear functional given by a density matrix in the Hilbert space where $\mathcal{A}$ operates. As a $C^*$ algebra $\mathcal{A}$ has pure states, but these correspond to disjoint representations on different (non separable) Hilbert spaces.

for all $A \in \mathcal{A}$. This means that for local algebras it is not meaningful to interpret statistical mixtures as "classical" probability distributions superimposed on pure states having a different "quantum mechanical" probability interpretation, as sometimes done in textbooks on non-relativistic quantum mechanics.

On the other hand *every state* on $\mathcal{A}$ *is a vector state*, i.e., for every $\omega$ there is a (non-unique!)[21] $\psi_\omega \in \mathcal{H}$ such that

$$\omega(A) = \langle \psi_\omega, A\psi_\omega \rangle \tag{15.34}$$

for all $A \in \mathcal{A}$ [88, Cor. 3.2, p. 336].

A type III factors has these mathematical features in common with the abelian von Neumann algebra $L^\infty(\mathbb{R})$ that also has no pure states whereas every state is a vector state in the natural representation on $L^2(\mathbb{R})$. But while $L^\infty(\mathbb{R})$ is decomposable into a direct integral of trivial, one-dimensional algebras, type III factors are noncommutative, indecomposable and of infinite dimension.

### 15.4.1.3 Local Comparison of States Cannot be Achieved by Means of Positive Operators

For $\mathcal{O}$ a subset of Minkowski space and two states $\varphi$ and $\omega$ we define their *local difference* by

$$D_\mathcal{O}(\varphi, \omega) := \sup\{|\varphi(A) - \omega(A)| : A \in \mathcal{A}(\mathcal{O}), \|A\| \le 1\}. \tag{15.35}$$

If $\mathcal{A}(\mathcal{O})$ were a type I algebra local differences could, for a dense set of states, be tested by means of positive operators in the following sense:

For a dense set of states $\varphi$ there is a positive operator $P_{\varphi,\mathcal{O}}$ such that

$$D_\mathcal{O}(\varphi, \omega) = 0 \quad \text{if and only if} \quad \omega(P_{\varphi,\mathcal{O}}) = 0. \tag{15.36}$$

For a type III algebra, on the other hand, such operators *do not exist for any state*.

Failure to recognize this has in the past led to spurious "causality problems" [59], inferred from the fact that for a positive operator $P$ an expectation value $\omega(e^{iHt}Pe^{-iHt})$ with $H \ge 0$ cannot vanish in an interval of $t$'s without vanishing identically.[22] In a semi-relativistic model for a gedankenexperiment due to Fermi [46] such a positive operator is used to measure the excitation of an atom due to a photon emitted from another atom some distance away. Relativistic causality requires that the excitation takes place only after a time span $t \ge r/c$ where $r$

---

[21] If the algebra is represented in a "standard form" in the sense of modular theory the vector can be uniquely fixed by taking it from the corresponding "positive cone" [19].

[22] This holds because $t \mapsto P^{1/2}e^{-iHt}\psi$ is analytic in the complex lower half plane for all vectors $\psi \in \mathcal{H}$ if $H \ge 0$.

is the distance between the atoms and $c$ the velocity of light. However, due to the mathematical fact mentioned, an alleged excitation measured by means of a positive operator is in conflict with this requirement. This is not a problem for relativistic quantum field theory, however, where the local difference of states defined by (15.35) shows perfectly causal behavior, while a positive operator measuring the excitation simply does not exist [30].[23]

#### 15.4.1.4   Remarks on the Use of Approximate Theories

The above discussion of "causality problems" prompts the following remarks:

Constructions of fully relativistic models for various phenomena where interactions play a decisive role are usually very hard to carry out in practice. Hence one must as a rule be content with some approximations (e.g., divergent perturbation series without estimates of error terms), or semi-relativistic models with various cut-offs (usually at high energies). Such models usually violate one or more of the general assumptions underlying LQP. Computations based on such models may well lead to results that are in conflict with basic principles of relativistic quantum physics such as an upper limit for the propagation speed of causal influence, but this is quite natural and should not be a cause of worry (or of unfounded claims) once the reason is understood.

### 15.5   Entanglement in LQP

If $\mathcal{A}_1$ and $\mathcal{A}_2$ commute, a state $\omega$ on $\mathcal{A}_1 \vee \mathcal{A}_2$ is by definition *entangled*, if it can *not* be approximated by convex combinations of product states.

Entangled states are ubiquitous in LQP due to the following general mathematical fact [36]: If $\mathcal{A}_1$ and $\mathcal{A}_2$ commute, are nonabelian, possess each a cyclic vector and $\mathcal{A}_1 \vee \mathcal{A}_2$ has a separating vector, then the entangled states form a *dense, open* subset of the set of all states.

This applies directly to the local algebras of LQP because of the Reeh–Schlieder Theorem. Thus the entangled states on $\mathcal{A}(\mathcal{O}_1) \vee \mathcal{A}(\mathcal{O}_2)$ are generic for space-like separated, bounded open sets $\mathcal{O}_1$ and $\mathcal{O}_2$.

The type III property implies even stronger entanglement:

If $\mathcal{A}$ is a type III factor, then $\mathcal{A} \vee \mathcal{A}'$ does not have *any* (normal) product states, i.e., *all* states are entangled for the pair $\mathcal{A}$, $\mathcal{A}'$.

---

[23]Already the Corollary to the Reeh–Schlieder Theorem in Sec. 15.3.3. implies that excitation cannot be measured by a local positive operator since the expectation value of such an operator cannot be zero in a state with bounded energy spectrum. The nonexistence of *any* positive operator satisfying (15.36) is a stronger statement.

*Haag duality* means by definition that $\mathcal{A}(\mathcal{O})' = \mathcal{A}(\mathcal{O}')$. Thus, if Haag duality holds, a quantum field in a bounded space-time region can never be disentangled from the field in the causal complement.

By allowing a small positive distance between the regions, however, disentanglement *is possible*, provided the theory has a certain property that mitigates to some extent the rigid coupling between a local region and its causal complement implied by the type III character of the local algebras. This will be discussed next.

### 15.5.1  Causal Independence and Split Property

A pair of commuting von Neumann algebras, $\mathcal{A}_1$ and $\mathcal{A}_2$, in a common $\mathcal{B}(\mathcal{H})$ is *causally (statistically) independent* if for every pair of states, $\omega_1$ on $\mathcal{A}_1$ and $\omega_2$ on $\mathcal{A}_2$, there is a state $\omega$ on $\mathcal{A}_1 \vee \mathcal{A}_2$ such that

$$\omega(AB) = \omega_1(A)\omega_2(B) \tag{15.37}$$

for $A \in \mathcal{A}_1$, $B \in \mathcal{A}_2$.

In other words: States can be *independently* prescribed on $\mathcal{A}_1$ and $\mathcal{A}_2$ and extended to a common, *uncorrelated* state on the joint algebra. This is really the von Neumann concept of independent systems.

The *split property* [38] for commuting algebras $\mathcal{A}_1, \mathcal{A}_2$ means that there is a type I factor $\mathcal{N}$ such that

$$\mathcal{A}_1 \subset \mathcal{N} \subset \mathcal{A}_2' \tag{15.38}$$

which again means: There is a tensor product decomposition $\mathcal{H} = \mathcal{H}_1 \otimes \mathcal{H}_2$ such that

$$\mathcal{A}_1 \subset \mathcal{B}(\mathcal{H}_1) \otimes \mathbf{1}, \quad \mathcal{A}_2 \subset \mathbf{1} \otimes \mathcal{B}(\mathcal{H}_2). \tag{15.39}$$

In the field theoretic context causal independence and split property are equivalent [22, 84, 90].

The split property for local algebras separated by a finite distance can be derived from a condition (*nuclearity*) that expresses the idea that the *local energy level density* (measured in a suitable sense) does not increase too fast with the energy [28, 29, 33, 34]. Nuclearity is not fulfilled in all models (some generalized free fields provide counterexamples), but it is still a reasonable requirement.

The split property together with the type III property of the strictly local algebras leads to a *strong version of the local preparability of states*. The following result is essentially contained in [31]. See also [84, 90].

**Theorem 5 (Strong Local Preparability)** *For every state $\varphi$ ("target state") and every bounded open region $\mathcal{O}$ there is an isometry $W \in \mathcal{A}(\mathcal{O}_\varepsilon)$ (with $\mathcal{O}_\varepsilon$ slightly larger than $\mathcal{O}$, cf. (15.32)) such that for an arbitrary state $\omega$ ("input state")*

$$\omega_W(AB) = \varphi(A)\omega(B) \tag{15.40}$$

*for $A \in \mathcal{A}(\mathcal{O})$, $B \in \mathcal{A}(\mathcal{O}_\varepsilon)'$, where $\omega_W(\cdot) \equiv \omega(W^* \cdot W)$.*

In particular, $\omega_W$ is uncorrelated and its restriction to $\mathcal{A}(\mathcal{O})$ is the target state $\varphi$, while in the causal complement of $\mathcal{O}_\varepsilon$ the preparation has no effect on the input state $\omega$. Moreover, $W$ depends *only on the target state* and not on the input state.

*Proof* The split property implies that we can write $\mathcal{A}(\mathcal{O}) \subset \mathcal{B}(\mathcal{H}_1) \otimes \mathbf{1}$, $\mathcal{A}(\mathcal{O}_\varepsilon)' \subset \mathbf{1} \otimes \mathcal{B}(\mathcal{H}_2)$.

By the type III property of $\mathcal{A}(\mathcal{O})$ we have $\varphi(A) = \langle \xi, A\xi \rangle$ for $A \in \mathcal{A}(\mathcal{O})$, with $\xi = \xi_1 \otimes \xi_2$. (The latter because every state on a type III factor is a vector state, and we may regard $\mathcal{A}(\mathcal{O})$ as a type III factor contained in $\mathcal{B}(\mathcal{H}_1)$.) Then $E := E_{\xi_1} \otimes \mathbf{1} \in \mathcal{A}(\mathcal{O}_\varepsilon)'' = \mathcal{A}(\mathcal{O}_\varepsilon)$.

By the type III property of $\mathcal{A}(\mathcal{O}_\varepsilon)$ there is a $W \in \mathcal{A}(\mathcal{O}_\varepsilon)$ with $WW^* = E$, $W^*W = \mathbf{1}$. The second equality implies $\omega(W^*BW) = \omega(B)$ for $B \in \mathcal{A}(\mathcal{O}_\varepsilon)'$.

On the other hand, $EAE = \varphi(A)E$ for $A \in \mathcal{A}(\mathcal{O})$ and multiplying this equation from left with $W^*$ and right with $W$, one obtains

$$W^*AW = \varphi(A)\mathbf{1} \tag{15.41}$$

by employing $E = WW^*$ and $W^*W = \mathbf{1}$. Hence

$$\omega(W^*ABW) = \omega(W^*AWB) = \varphi(A)\omega(B). \tag{15.42}$$

$\square$

The theorem implies also that *any state on $\mathcal{A}(\mathcal{O}) \vee \mathcal{A}(\mathcal{O}_\varepsilon)'$ can be disentangled by a local operation* in $\mathcal{A}(\mathcal{O}_\varepsilon)$:

Given a state $\omega$ on $\mathcal{A}(\mathcal{O}) \vee \mathcal{A}(\mathcal{O}_\varepsilon)'$ there is, by the preceding Theorem, an isometry $W \in \mathcal{A}(\mathcal{O}_\varepsilon)$ such that

$$\omega_W(AB) = \omega(A)\omega(B). \tag{15.43}$$

for all $A \in \mathcal{A}(\mathcal{O})$, $B \in \mathcal{A}(\mathcal{O}'_\varepsilon)$.

In particular: Leaving a security margin between a bounded domain and its causal complement, the global vacuum state $\omega(\cdot) = \langle \Omega, \cdot \, \Omega \rangle$, which, being cyclic and separating for the local algebras is entangled between $\mathcal{A}(\mathcal{O})$ and any $\mathcal{A}(\tilde{\mathcal{O}}) \subset \mathcal{A}(\mathcal{O}_\varepsilon)'$ [36, 72], can be disentangled by a local operation producing an uncorrelated state on $\mathcal{A}(\mathcal{O}) \vee \mathcal{A}(\mathcal{O}_\varepsilon)'$ that is identical to the vacuum state on each of the factors.

## 15.5.2   Conclusions

Here are some lessons that can be drawn from this brief survey of relativistic quantum physics:

* LQP provides a framework which resolves the apparent paradoxes resulting from combining the particle picture of quantum mechanics with special relativity. This resolution is achieved by regarding the system as composed of *quantum fields* in space-time, represented by a net of local algebras. A subsystem is represented by one of the local algebras, i.e., the fields in a specified part of space-time. "Particle" is a derived concept that emerges asymptotically at large times but (for theories with interactions) is usually not strictly defined at finite times. Irrespective of interactions particle states can never be created by operators strictly localized in bounded regions of space-time.
* The fact that local algebras have no pure states is relevant for interpretations of the state concept. It lends support to the interpretation that a state refers to a (real or imagined) ensemble of identically prepared copies of the system but makes it hard to maintain that it is an inherent attribute of an individual copy.
* The type III property is relevant for causality issues and local preparability of states, and responsible for "deeply entrenched" entanglement of states between bounded regions and their causal complements, that is, however, mitigated by the split property.

On the other hand, the framework of LQP does not per se resolve all "riddles" of quantum physics. Those who are puzzled by the violation of Bell's inequalities in EPR type experiments will not necessarily by enlightened by learning that local algebras are type III. Moreover, the terminology has still an anthropocentric ring ("observables", "operations") as usual in Quantum Mechanics. This is disturbing since physics is concerned with more than designed experiments in laboratories. We use quantum (field) theories to understand processes in the interior of stars, in remote galaxies billions of years ago, or even the "quantum fluctuations" that are allegedly responsible for fine irregularities in the 3K background radiation. In none of these cases "observers" were/are around to "prepare states" or "reduce wave packets"! A fuller understanding of the emergence of macroscopic "effects" from the microscopic realm,[24] without invoking "operations" or "observations", and possibly a corresponding revision of the vocabulary of quantum physics is still called for.[25]

---

[24] See [12] for important steps in this direction and [49] for a thorough analysis of foundational issues of QM.

[25] Already Max Planck in his Leiden lecture of 1908 speaks of the "Emanzipierung von den antrophomorphen Elementen" as a goal, see [76], p. 49.

**Acknowledgements** I thank the organizers of the Bielefeld workshop, Jürg Fröhlich and Philippe Blanchard, for the invitation that lead to these notes, Detlev Buchholz for critical comments on the text, Wolfgang L. Reiter for drawing my attention to [76], and the Austrian Science Fund (FWF) for support under Project P 22929-N16.

# References

1. Alazzawi, S.: Deformations of fermionic quantum field theories and integrable models. Lett. Math. Phys. **103**, 37–58 (2013)
2. Araki, H.: A lattice of von Neumann algebras associated with the quantum theory of a free Bose field. J. Math. Phys. **4**, 1343–1362 (1963)
3. Araki, H.: Von Neumann algebras of local observables for free scalar fields. J. Math. Phys. **5**, 1–13 (1964)
4. Araki, H.: Type of von Neumann algebra associated with free field. Prog. Theor. Phys. **32**, 956–965 (1964)
5. Araki, H.: Remarks on spectra of modular operators of von Neumann algebras. Commun. Math. Phys. **28**, 267–277 (1972)
6. Araki, H.: Mathematical Theory of Quantum Fields. Oxford University Press, Oxford (1999)
7. Araki, H., Woods, E.J.: Representations of the canonical commutation relations describing a non-relativistic infinite free Bose gas. J. Math. Phys. **4**, 637–662 (1963)
8. Araki, H., Woods, E.J.: A classification of factors. Publ. R.I.M.S., Kyoto Univ. **4**, 51–130 (1968)
9. Barata, J.C.A., Jäkel, C.D., Mund, J.: The $\mathcal{P}(\varphi)_2$ model on the de sitter space. arXiv:1311.2905
10. Bisognano, J.J., Wichmann, E.H.: On the duality condition for a Hermitian scalar field. J. Math. Phys. **16**, 985–1007 (1975)
11. Bisognano, J.J., Wichmann, E.H.: On the duality condition for quantum fields. J. Math. Phys. **17**, 303–321 (1976)
12. Blanchard, P., Olkiewicz, R.: Decoherence induced transition from quantum to classical dynamics. Rev. Math. Phys. **15**, 217–244 (2003)
13. Bogoliubov, N.N., Lugonov, A.A., Oksak, A.I., Todorov, I.T.: General Principles of Quantum Field Theory. Kluwer, Dordrecht (1990)
14. Borchers, H.J.: Local rings and the connection of spin with statistics. Commun. Math. Phys. **1**, 281–307 (1965)
15. Borchers, H.J.: A remark on a theorem of B. Misra. Commun. Math. Phys. **4**, 315–323 (1967)
16. Borchers, H.J.: Half-sided translations and the type of von Neumann algebras. Lett. Math. Phys. **44**, 283–290 (1998)
17. Borchers, H.J.: On revolutionizing quantum field theory with Tomita's modular theory. J. Math. Phys. **41**, 3604–3673 (2000)
18. Borchers, H.J., Yngvason, J.: From quantum fields to local von Neumann algebras. Rev. Math. Phys. Special Issue, 15–47 (1992)
19. Bratteli, O., Robinson, D.W.: Operator Algebras and Quantum Statistical Mechanics I. Springer, New York (1979)
20. Brunetti, R., Guido, D., Longo, R.: Modular localization and wigner particles. Rev. Math. Phys. **14**, 759–786 (2002)
21. Brunetti, R., Duetsch, M., Fredenhagen, K.: Perturbative algebraic quantum field theory and the renormalization groups. Adv. Theor. Math. Phys. **13**, 1541–1599 (2009)
22. Buchholz, D.: Product states for local algebras. Commun. Math. Phys. **36**, 287–304 (1974)
23. Buchholz, D.: Gauss' law and the infraparticle problem. Phys. Lett. B **174**, 331–334 (1986)
24. Buchholz, D., Fredenhagen, K.: Locality and the structure of particle states. Commun. Math. Phys. **84**, 1–54 (1982)

25. Buchholz, D., Haag, R.: The quest for understanding in relativistic quantum physics. J. Math. Phys. **41**, 3674–3697 (2000)
26. Buchholz, D., Roberts, J.E.: New light on infrared problems: sectors, statistics, symmetries and spectrum. Commun. Math. Phys. **330**, 935–972 (2014)
27. Buchholz, D., Verch, R.: Scaling algebras and renormalization group in algebraic quantum field theory. Rev. Math. Phys. **7**, 1195–1239 (1996)
28. Buchholz, D., Wichmann, E.: Causal independence and the energy level density of states in local quantum field theory. Commun. Math. Phys. **106**, 321–344 (1986)
29. Buchholz, D., Yngvason, J.: Generalized nuclearity conditions and the split property in quantum field theory. Lett. Math. Phys. **23**, 159–167 (1991)
30. Buchholz, D., Yngvason, J.: There are no causality problems in Fermi's two atom system. Phys. Rev. Lett. **73**, 613–616 (1994)
31. Buchholz, D., Doplicher, S., Longo, R.: On noether's theorem in quantum field theory. Ann. Phys. **170**, 1–17 (1986)
32. Buchholz, D., D'Antoni, C., Fredenhagen, K.: The universal structure of local algebras. Commun. Math. Phys. **111**, 123–135 (1987)
33. Buchholz, D., D'Antoni, C., Longo, R.: Nuclear maps and modular structures I. J. Funct. Anal. **88**, 233–250 (1990)
34. Buchholz, D., D'Antoni, C., Longo, R.: Nuclear maps and modular structures II: applications to quantum field theory. Commun. Math. Phys. **129**, 115–138 (1990)
35. Buchholz, D., Lechner, G., Summers, S.J.: Warped convolutions, rieffel deformations and the construction of quantum field theories. Commun. Math. Phys. **304**, 95–123 (2011)
36. Clifton, R., Halvorson, H.: Entanglement and open systems in algebraic quantum field theory. Stud. Hist. Philos. Mod. Phys. **32**, 1–31 (2001)
37. Connes, A.: Une classification des facteurs de type III. Ann. Sci. Ecole Norm. Sup. **6**, 133–252 (1973)
38. Doplicher, S., Longo, R.: Standard and split inclusions of von Neumann algebras. Invent. Math. **75**, 493–536 (1984)
39. Doplicher, S., Roberts, J.E.: Why there is a field algebra with a compact gauge group describing the superselection structure in particle physics. Commun. Math. Phys. **131**, 51–107 (1990)
40. Doplicher, S., Haag, R., Roberts, J.E.: Fields, observables and gauge transformations II. Commun. Math. Phys. **15**, 173–200 (1969)
41. Doplicher, S., Haag, R., Roberts, J.E.: Local observables and particle statistics I. Commun. Math. Phys. **23**, 199–230 (1971)
42. Doplicher, S., Haag, R., Roberts, J.E.: Local observables and particle statistics II. Commun. Math. Phys. **35**, 49–85 (1974)
43. Driessler, W.: Comments on lightlike translations and applications in relativistic quantum field theory. Commun. Math. Phys. **44**, 133–141 (1975)
44. Driessler, W.: On the type of local algebras in quantum field theory. Commun. Math. Phys. **53**, 295–297 (1977)
45. Driessler, W., Summers, S.J., Wichmann, E.H.: On the connection between quantum fields and von Neumann algebras of local operators. Commun. Math. Phys. **105**, 49–84 (1986)
46. Fermi, E.: Quantum theory of radiation. Rev. Mod. Phys. **4**, 87–132 (1932)
47. Fredenhagen, K.: On the modular structure of local algebras of observables. Commun. Math. Phys. **84**, 79–89 (1985)
48. Fredenhagen, K.: A remark on the cluster theorem. Commun. Math. Phys. **97**, 461–463 (1985)
49. Fröhlich, J., Schubnel, B.: Quantum probability theory and the foundations of quantum mechanics. arXiv:1310.1484v1 [quant-ph]
50. Fröhlich, J., Morchio, G., Strocchi, F.: Infrared problem and spontaneous breaking of the Lorentz Group in QED. Phys. Lett. B **89**, 61–64 (1979)
51. Glimm, J., Jaffe, A.: Quantum Physics: A Functional Integral Point of View. Springer, New York (1987)
52. Grosse, H., Lechner, G.: Wedge-Local Quantum Fields and Noncommutative Minkowski Space. J. High Energy Phys. **0711**, 012 (2007)

53. Haag, R.: On quantum field theories. Mat.-fys. Medd. Kong. Danske Videns. Selskab **29**, Nr.12 (1955)
54. Haag, R.: Quantum field theory with composite particles and asymptotic conditions. Phys. Rev. **112**, 669–673 (1958)
55. Haag, R.: Local Quantum Physics. Springer, Berlin (1992)
56. Haag, R., Kastler, D.: An algebraic approach to quantum field theory. J. Math. Phys. **5**, 848–861 (1964)
57. Haag, R., Hugenholtz, N.M., Winnink, M.: On the equilibrium states in quantum statistical mechanics. Commun. Math. Phys. **5**, 215–236 (1967)
58. Haagerup, U.: Connes' bicentralizer problem and uniqueness of injective factors of type III$_1$. Acta Math. **158**, 95–148 (1987)
59. Hegerfeldt, G.C.: Causality problems in Fermi's two atom system. Phys. Rev. Lett. **72**, 596–599 (1994)
60. Jost, R.: The General Theory of Quantized Fields. American Mathematical Society, Providence, RI (1965)
61. Kawahigashi, Y., Longo, R.: Classification of two-dimensional local conformal nets with $c < 1$ and 2-cohomology vanishing for tensor categories. Commun. Math. Phys. **244**, 63–97 (2004)
62. Kraus, K.: States, Effects and Operations. Springer, Berlin (1983)
63. Lechner, G.: Deformations of quantum field theories and integrable models. Commun. Math. Phys. **312**, 265–302 (2012)
64. Longo, R.: Notes on algebraic invariants for non-commutative dynamical systems. Commun. Math. Phys. **69**, 195–207 (1979)
65. Malament, D.B.: In defence of a dogma: why there cannot be a relativistic quantum mechanics of (localizable) particles. In: Clifton, R. (ed.) Perspectives on Quantum Reality, pp. 1–10. Kluwer, Dordrecht (1996); Montreal 2001
66. Mund, J.: String-localized quantum fields, modular localization and gauge theories. In: Sidoravicius, V. (ed.) New Trends in Mathematical Physics, pp. 495–508. Springer, New York (2009)
67. Mund, J., Schroer, B., Yngvason, J.: String-localized quantum fields from Wigner representations. Phys. Lett. B **596**, 156–162 (2004)
68. Mund, J., Schroer, B., Yngvason, J.: String-localized quantum fields and modular localization. Commun. Math. Phys. **268**, 621–672 (2006)
69. Murray, F.J., von Neumann, J.: On rings of operators. Ann. Math. **37**, 116–229 (1936)
70. Murray, F.J., von Neumann, J.: On rings of operators II. Trans. Am. Math. Soc. **41**, 208–248 (1937)
71. Murray, F.J., von Neumann, J.: On rings of operators IV. Ann. Math. **44**, 716–808 (1943)
72. Narnhofer, H.: The role of transposition and CPT operation for entanglement. Phys. Lett. A **310**, 423–433 (2003)
73. Newton, T.D., Wigner, E.P.: Localized states for elementary systems. Rev. Mod. Phys. **21**, 400–406 (1949)
74. Peres, A., Terno, D.R.: Quantum information and relativity theory. Rev. Mod. Phys. **76**, 93–123 (2004)
75. Perez, J.F., Wilde, I.F.: Localization and causality in relativistic quantum mechanics. Phys. Rev D. **16**, 315–317 (1977)
76. Planck, M.: Vorträge und Erinnerungen. S. Hirzel Verlag, Stuttgart (1949)
77. Plaschke, M., Yngvason, J.: Massless string fields for any helicity. J. Math. Phys. **53**, 042301 (2012)
78. Powers, R.T.: Representations of uniformly hyperfinite algebras and their associated von Neumann rings. Ann. Math. **86**, 138–171 (1968)
79. Reed, M., Simon, B.: Methods of Modern Mathematical Physics II. Academic, New York (1975)
80. Reeh, H., Schlieder, S.: Eine Bemerkung zur Unitärequivalenz von Lorentzinvarianten Feldern. Nuovo Cimento **22**, 1051–1068 (1961)

81. Rieffel, M.A., Van Daele, A.: A bounded operator approach to Tomita-Takesaki theory. Pac. J. Math. **69**, 187–221 (1977)
82. Schwartz, J.: Free quantized Lorentzian fields. J. Math. Phys. **2**, 271–290 (1960)
83. Streater, R.F., Wightman, A.S.: PCT, spin and statistics, and all that. W.A. Benjamin Inc., New York (1964)
84. Summers, S.J.: On the independence of local algebras in quantum field theory. Rev. Math. Phys. **2**, 201–247 (1990)
85. Summers, S.J.: Tomita-Takesaki Modular Theory. In: Francois, J.P., Naber, G.L., Tsun, T.S. (eds.) Encyclopedia of Mathematical Physics, pp. 251–257. Academic Press (2006)
86. Summers, S.J., Werner, R.F.: On Bell's inequalities and algebraic invariants. Lett. Math. Phys. **33**, 321–334 (1995)
87. Takesaki, M.: Tomita's Theory of Modular Hilbert Algebras and Its Applications. Lecture Notes in Mathematics, vol. 128. Springer, Berlin (1970)
88. Takesaki, M.: Theory of Operator Algebras II. Springer, New York (2003)
89. von Neumann, J.: On rings of operators III. Ann. Math. **41**, 94–161 (1940)
90. Werner, R.F.: Local preparability of states and the split property in quantum field theory. Lett. Math. Phys. **13**, 325–329 (1987)
91. Wightman, A.S., Gårding, L.: Fields as operator-valued distributions in relativistic quantum theory. Arkiv før Fysik **28**, 129–189 (1965)
92. Wigner, E.: On unitary representations of the inhomogeneous Lorentz Group. Ann. Math. Sec. Ser. **40**, 149–204 (1939)
93. Yngvason, J.: Zero-mass infinite spin representations of the Poincaré group and quantum field theory. Commun. Math. Phys. **18**, 195–203 (1970)
94. Zych, M., Costa, F., Kofler, J., Brukner, C.: Entanglement between smeared field operators in the Klein-Gordon vacuum. Phys. Rev. D **81**, 125019 (2010)